Theoretical Foundations of Travel Choice Modeling

Related Pergamon books

Daganzo
Fundamentals of Transportation and Traffic Operations

Ettema & Timmermans
Activity Based Approaches to Travel Analysis

Stopher & Lee-Gosselin
Understanding Travel Behaviour in an Era of Change

Related Pergamon journals

Transportation Research Part A:
Policy and Practice
Editor: Frank A. Haight

Transportation Research Part B:
Methodological
Editor: Frank A. Haight

Free specimen copies of journals available on request

Theoretical Foundations of Travel Choice Modeling

Edited by:

Tommy Gärling
Göteborg University, Sweden

Thomas Laitila
Umeå University, Sweden

and

Kerstin Westin
Umeå University, Sweden

1998
ELSEVIER
Amsterdam - Lausanne - New York - Oxford -Shannon - Singapore - Tokyo

ELSEVIER SCIENCE Ltd
The Boulevard, Langford Lane
Kidlington, Oxford OX5 1GB, UK

Library of Congress Cataloging-in-Publication Data

Theoretical foundations of travel choice modeling / editors, Tommy
Gärling, Thomas Laitila, Kerstin Westin.
p. cm.
Includes indexes.
ISBN 0-08-043062-7
1. Choice of transportation--Mathematical models. I. Gärling,
Tommy. II. Laitila, Thomas. III. Westin, Kerstin.
HE336.C5T468 1998
388--dc21 98-20224
CIP

Pergamon

British Library Cataloguing in Publication Data

Theoretical foundations of travel choice modeling
1.Choice of transportation
I.Gärling, Tommy II.Laitila, Thomas III.Westin, Kerstin
338.4'7'338

ISBN 0080430627

First edition 1998

ISBN: 0-08-043062-7

∞ The paper used in this publication meets the requirements of ANSI/NISO Z39.48-1992 (Permanence of Paper).

Printed in The Netherlands.

CONTRIBUTORS[1]

Staffan Algers
Department of Infrastructure and Planning, Royal Institute of Technology, SE-10044 Stockholm, Sweden

Kay W. Axhausen
Institute for Highway Engineering and Transport Planning, Leopold-Franzens University, A-6020 Innsbruck, Austria

Chandra R. Bhat
Department of Civil Engineeering, University of Texas at Austin, TX 78712-1076, USA

David Brownstone
Department of Economics, University of California, Irvine, CA 92697-5100, USA

Andrew Daly
Hague Consulting Group, 2585 GJ Den Haag, The Netherlands

André de Palma
Department of Economics, University of Cergy-Pontoise, Cergy-Pontoise, F-95011 France.

Tommy Gärling
Department of Psychology, Göteborg University, P.O. Box 500, SE-40530 Göteborg, Sweden

Reginald G. Golledge
Department of Geography, University of California, Santa Barbara, CA 93106-4060, USA

Phil Goodwin[2]
ESRC Transport Studies Unit, University College, London WC1E 6BT, UK

Thomas F. Golob
Institute of Transportation Studies, University of California, Irvine, CA 92717, USA

David A. Hensher
Institute of Transport Studies, University of Sydney, NSW 2006, Sydney, Australia

---

[1] Senior chapter authors who also participated in the conference.
[2] Unable to attend the conference.

Sergio R. Jara-Díaz
Department of Civil Engineering, University of Chile, Santiago, Chile

Ryuichi Kitamura
Department of Civil Engineering, Kyoto University, Kyoto, 606-01 Japan

Thomas Laitila
Department of Statistics and Transportation Research Unit, Umeå University, SE 90187 Umeå, Sweden

Hani S. Mahmassani
Department of Civil Engineering, University of Texas at Austin, TX 78712-1076, USA,

Daniel McFadden
Department of Economics, University of California, Berkeley, CA 94618, USA

Eric I. Pas[3]
Department of Civil and Environmental Engineering, Duke University, Durham, NC 27708-0287, USA

John Polak
Centre for Transport Studies, Imperial College, University of London, London SW7 2BU, UK

Eliahu Stern
Department of Geography and Environmental Development, Ben-Gurion University of the Negev, 84105 Beer-Sheva, Israel

Peter R. Stopher
Department of Civil and Environmental Engineering, Louisiana State University, Baton-Rouge, LA 70803-6405, USA

Ola Svenson
Department of Psychology, Stockholm University, SE 106 91 Stockholm, Sweden

Kerstin Westin
Department of Geography and Transportation Research Unit, Umeå University, SE 90187 Umeå, Sweden

[3] Died in 1997.

The contributors photographed at the conference site in Stockholm, August 1996

PREFACE

During the world conference on transport research (WCTR) in Sydney 1995, the idea came up to organize a small research conference in the summer of 1996. The original plan was to choose a conference site in the northern wilderness area of Sweden. However, transportation would pose an almost insurmountable problem! Therefore, we chose Stockholm which at least those living there believe is in the center of the world. This choice was rational although not really what we wanted. Yet, things turned out better than we were afraid they would. The weather was exceptionally good. We were fortunate to select a conference site in the Stockholm metropolitan area which made possible (during happy hours) observations of the incoming and outcoming boat traffic which, of course, was interesting to transport researchers. In addition the town council had decided to organize the Stockholm water festival at the same time as our conference. The "grand finale" of the conference therefore took care of itself!

After choosing a conference site, the next problem we faced was: What should the topic be of the conference? Of course, the topic of travel choice was given. After the very successful conference on activity-based approaches to travel choice modeling organized by Dick Ettema and Harry Timmermans at Eindhoven University of Technology in 1995, we realized that a different emphasize would be valuable. Our impression was that travel choice modeling has been dominated by consumer theory developed in economics. Coming from the disciplines of geography, psychology, and statistics, for us it was therefore natural to ask: What theory can these disciplines offer? In fact, what *are* the theoretical foundations of travel choice modeling, and what should they be? These very basic questions came to our minds. They are also the general issues which were addressed at the Stockholm conference on *Theoretical Foundations of Travel Choice Modeling* held in August 7-11, 1996. We are very pleased that so many of the outstanding researchers in the field believed it was worthwhile to come (although not all were able to). David Hensher, Kay Axhausen, and Chandra Bhat were also helpful in making the International Association of Travel Behavior Research (IATBR) sponsor the conference. We feel that this edited volume resulting from the conference provides interesting answers to our questions, and we hope and believe that the readers will think the same.

A number of persons contributed to making the conference successful. First of all we thank the paper presenters for allocating their valuable time to prepare chapter drafts before the conference, to international travel, to present papers and to actively take part in heated discussions during the conference, and (together with their co-authors) to produce the final chapter manuscripts after the conference just in time for us to be able to almost comply with the deadline. We would also like to thank all other participants who not only listened to the paper presentations and enjoyed the conference meals but were actively engaged in discussions with the chapter authors. Practical matters during the conference were efficiently taken care of by Lennart Westin. After the conference, Lotta Brännlund skillfully assisted us in converting the manuscripts to camera-ready book chapters, and Fredrik Gärling in compiling the indexes. We thank them all.

Financial support for organizing the conference and editing the book was obtained from the Swedish Communications and Transport Research Board (KFB), the Transportation Research Unit of Umeå University (TRUM), Sweden, and the Swedish Institute of Transport and Communications Analysis (SIKA). We are grateful to Jörgen Garvill, Director of TRUM, Per Norman, Officer at KFB, and Staffan Widlert, Director of SIKA, for their support and help.

All the contacts we have had with Chris Pringle at Elsevier have been through email. They have always been very friendly. Perhaps not surprisingly, telecommunication thus also conveys such impressions! Still, we hope to one day meet Chris in person.

Editing of the book was completed while the first editor (Tommy Gärling) in the Winter of 1998 visited University of California, Santa Barbara. The visit made salient to the other editors one of the few (although very important) shortcomings of telecommuting - the impossibility to transfer the weather.

Tommy Gärling
Thomas Laitila
Kerstin Westin
Santa Barbara, California, USA
and Umeå, Sweden
February 1998

CONTENTS

1

THEORETICAL FOUNDATIONS OF TRAVEL CHOICE MODELING: AN INTRODUCTION

Tommy Gärling, Göteborg University, Sweden
Thomas Laitila, Umeå University, Sweden
Kerstin Westin, Umeå University, Sweden

INTRODUCTION

In modern societies travel is vital to human welfare. Considerable investments are therefore made in transportation infrastructure. To be effective such investments require among other things knowledge of future demand of travel. In forecasting travel demand, use is frequently made of disaggregate models of current users′ revealed or stated travel choices (Ben-Akiva and Lerman, 1985; Ortúzar and Willumsen, 1994). Within the broader area of research on travel behavior, an interest in travel choice modeling has grown out of this basically practical need (Pas, 1990; Pipkin, 1995).

Current research on travel choice is the main focus of this book. Since choices made by people are targeted, modeling these choices relies on behavioral assumptions. Two fundamental issues the book aims at addressing are: Which are these behavioral assumptions (since they frequently are implicit), and are they founded on current systematic knowledge of behavior? There are several possible disciplines where such systematic knowledge is available. The book highlights microeconomic theory and behavioral decision theory. Microeconomic theory has clear connections to other research in the wider field of transportation (e.g., see de Palma, Chapter 2, and Goodwin, Chapter 5, in this volume). Furthermore, the principle of utility maximization in microeconomic theory is a cornerstone of the developments of models in the area (Ben-Akiva & Lerman, 1985; McFadden, 1973). However, much current modeling (Axhausen & Gärling, 1992; Ettema & Timmermans, 1997) is targeted at complex travel choices which may be governed by principles other than those proposed in microeconomic theory. Therefore, it is important to find out whether an alternative such as behavioral decision theory has the potential of making a contribution.

Theoretically founded behavioral assumptions are essential for the development of models of travel choice which can be used to forecast travel demand. These assumptions should furthermore be coordinated with the development of mathematical-statistical modeling tools. Is there a theoretical foundation of the tools needed to model complex travel choices which do not rely on the utility-maximization principle? Are methods available for collecting the type of data which are needed? These are the remaining general questions raised in the book.

In this introductory chapter we first attempt to characterize present-day daily household travel. This sets the stage for a discussion of different facets of travel choice which theory needs to deal with. An overview then follows of the different theoretical foundations of travel choice modeling which each defines a part of the book. A summary is finally made of the contents, the main points, and the conclusions of the following chapters.

TRAVEL BEHAVIOR

Today travel has become a necessary ingredient of most daily activities (Hanson and Schwab, 1995). As compared to 100 years ago, more people travel today, they use faster travel modes, and they travel longer distances. In addition daily household travel behavior has become much more complex due to several significant changes (Hanson, 1995; Levinson and Kumar, 1995). Work trips are no longer dominant. Other frequent trip purposes are shopping, various leisure and entertainment activities, social visits, and chauffeuring. Automobile is the dominant travel mode. In addition, the total number of trips has increased, trip chaining is more frequent, and traffic peaks are smoother. Although most easily observed in North American urban areas, these changes are present also in cities all over the world as well as in the rural parts of many countries. A number of driving forces of the observed changes are possible to identify. They include changes in sociodemographic factors, changes in residential patterns, changes in activity and employment patterns, and new planning policies.

In most Western economies the number of households has increased in the post-war period. The increase is in part due to population growth, in part a consequence of an increasing number of single-person households. These changes partly account for the increase in the total number of trips which in many parts of the world impose capacity limits on travel. Another tendency is the growing number of elderly persons whose daily activity patterns differ from those of the younger population. Due to age-related disabilities, the elderly also often have special transportation needs. Although not contributing to a growth of travel per se, it leads to different requirements on the transportation system.

New dispersed residential patterns are observed in urban areas. Residential areas are geographically separated from industrial areas and service facilities. Work places are being decentralized from central business districts to peripheral areas, still separated from residential areas. This implies new requirements on transportation services. In fact, improved and easily accessible modes of transportation (such as the automobile) have perhaps been prerequisites for these changes. Furthermore, localization of shopping malls close to main arteries and frequently not accessible by public transport also excludes other travel modes than the car. Another trend in settlement patterns is that former rural areas outside cities are transformed to middle-class and upper-class residential areas. These areas are smaller than the traditional suburbs and more individually developed. Public transport is frequently not available.

In modern life leisure time is both longer and clearly separated from work time. Because of more engagement in out-of-home leisure activities, nonwork-related travel increases. Changes at work have also affected travel. Work hours have become more flexible. An increasing number of people starts to telecommute. There is an increased unemployment in Western economies which changes daily activity and travel patterns. Women in the labor market increases in most industrialized countries, as well as in many developing countries. With both husbands and wives working outside the home, the activity patterns become more complex and require more travel. Furthermore, with a growing number of two-persons working households, the need for child daycare has increased and, consequently, also makes the activity/travel pattern more complex.

New goals of policy and planning has furthermore affected travel. For instance, a growing environmental awareness and concern have involved new laws and regulations to control negative effects of pollution, congestion, and safety. Agenda 21 codifies this concern for "a sustainable development" with clear implications for transport planning and management.

Is increased complexity of travel behavior paralleled by an increase in the complexity of travel choices faced by households? It may be recalled that Hägerstrand (1970) explicitly ascribed an explanatory role to choices people make. At the same time he focused attention on capacity, coupling, and authority constraints limiting the freedom of such choices. From this it may, on the one hand, be argued that since the number of options has objectively increased, households have more freedom of choice. On the other hand, engagement in an increasing number of activities at different locations involving many other people may decrease freedom of choice by imposing more constraints. However, this does not relieve households from making decisions. In fact, a case can easily be made for that travel choices are becoming more complex as they involve, for instance, more extensive search of information when the number of alternatives increases and the alternatives become more complex, an increasing frequency of joint and dynamic decisions being made when trip chaining increases, an increasing amount of contingency planning in the face of increased uncertainty (e.g., due to congestions), an increasing frequency of coordinating decisions because of increased social interdependence, and an increasing frequency of activity scheduling when the number of activities to be performed increases. This has consequences for the behavioral assumptions on which travel-choice models need to be founded. For instance, when choices are complex, utility maximization does no longer appear to be a tenable assumption. Furthermore, other information-processing activities such as search, problem solving, and planning or metadecisions become important components of the choice process which cannot be ignored.

TRAVEL CHOICE

Travel behavior is not invariably the outcome of a choice made by a household or an individual. Like many other behaviors, travel is sometimes executed almost without conscious awareness or automatically (e.g., driving the routine route to work). In other cases the choice set is empty (e.g., no alternative mode is available for the trip from work to the grocery store). In any analysis of travel choice, a first task is therefore to infer whether or not the traveler makes a choice. A next, closely related task is to infer the properties of the choice the traveler makes.

Inferring a travel choice may seem simple and straightforward. Yet, on second thought some complexity arises. Is the presence of a non-empty choice set a sufficient criterion or is the existence of deliberation a necessary ingredient? Does a choice between one alternative (e.g., to travel to work from home) and the current alternative (to stay home) qualify as a choice set, or is it necessary that the traveler can choose between several alternatives (e.g., using a bike, bus, or car) excluding or including the current alternative (e.g., walking)? Do choices of, for instance, residential location, starting extra work, or having a child qualify as travel (or mobility) decisions? Drawing on these and similar observations, we believe that in particular the following questions which can be asked about travel choices are important for a theory to address: (1) When, how and why are choice alternatives formed? (2) When, how, and why do degree and type of dependence on other choices vary? (3) When, how, and why do degree and type of deliberation vary? In addition to these specific issues concerning how travel choices are made, what has become known as the activity-based approach to travel choice modeling raises broader issues concerning people's motivation to travel. A theory also needs to address these broader issues.

The Formation of Choice Sets

An assumption commonly entertained (e.g., Thill, 1992) is that a set of alternatives or choice set is formed before the choice among the alternatives are made. Several specific questions which are relevant for the modeling of travel choices may be raised concerning the stage of forming a choice set. A basic question a theory must address is when and why choice sets are actually formed rather than being given? If they are formed, a subsequent question concerns whether the same or another principle is drawn on in their formation as in making the choice among the alternatives? Another set of questions concern the role of status quo (the current alternative). Is status quo invariably included in the choice set? If not, when is it and when is it not? Does status quo influence a choice even though it is not included in the choice set? If status quo is included in the choice set, is it evaluated in the same way as the alternatives to which it is compared?

Degree and Type of Dependencies

Recent developments of mathematical-statistical modeling tools (Daly, 1987; McFadden, 1978) have made it possible to model dependencies between different travel choices, for instance, of mode, destination, and departure time (e.g., Algers and Widlert, 1997; Arentze et al., 1993). Rather than predicting degree and type of dependencies between choices, the dependencies are inferred from the goodness of the model's fit to data. From the lack of theory follows that no principle is evoked to explain when, how, and why some choices are related whereas others are not.

Degree and Type of Deliberation

A majority of travel choices is made with some regularity (Huff and Hanson, 1990; Pas, 1988; Pas and Koppelman, 1987). It may therefore seem reasonable to conclude that such choices are predominantly habitual (Burnett and Hanson, 1982). A theory is however needed which specify how type and degree of deliberation vary with different antecedent factors (e.g., importance of the choice or degree of involvement, perceived degree of control) as well as have different consequences (e.g., information search, type of dependence, choice of decision rule or strategy).

Motivation to Travel

An important distinction illustrated in Figure 1.1 is between, on the one hand, choices

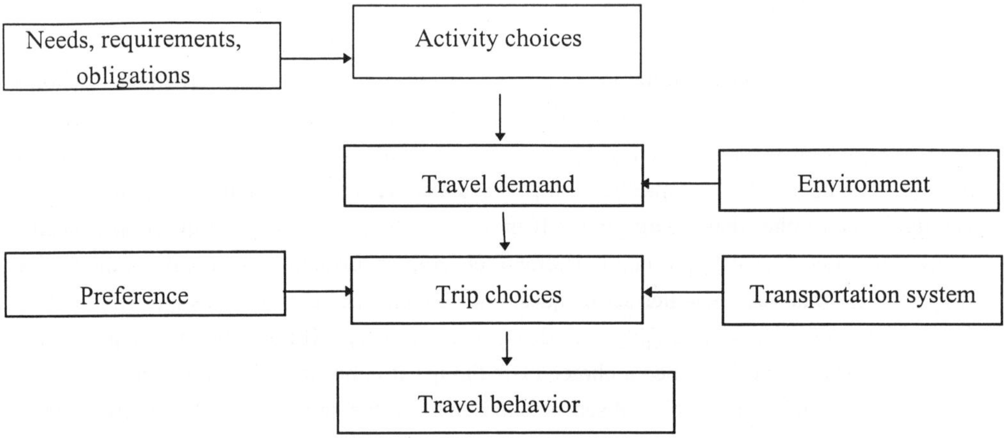

Figure 1.1. Determinants of travel behavior.

motivated by needs, obligations, and requirements to engage in out-of-home activities and, on the other hand, choices of trip characteristics such as destination, mode, departure time, and route. Choices of activities in a given environment determine the travel demand, trip choices are made with the goal of satisfying this demand for a given transportation system. Together they are determinants of the observed travel behavior. The distinction is maintained in most of the chapters of the book. It is the basis for identifying travel as a "derived demand" or means

which is instrumental for the attainment of activity goals (in the widest sense). It brings in sharp focus the necessity of analyzing why and how households allocate time to different activities. Such an analysis is germane to the activity-based approach to travel choice modeling (Axhausen & Gärling, 1992; Ettema & Timmermans, 1997; Jones et al., 1990; Kitamura, 1988). Which are the behavioral assumptions made by this approach? Are they founded in theory?

THEORETICAL FOUNDATIONS

Microeconomic Theory

What is Microeconomic Theory? Microeconomic theory (MET) focuses on the behavior of economic agents at the individual level, such as individual consumers, investors, managers, and others. Although there are many different types of economic agents operating in markets, the theory is limited to producers and consumers. The principal approach of MET to choice is to determine all available decision alternatives facing agents and evaluating the economic consequences for them. Based on the assumption that individual agents behave economically rational, making the choices which lead to the consequences that will maximize their utility, the choices they make can be predicted given knowledge of risk attitudes and budget restrictions. For instance, a production unit may have a fixed production technology and seeks to maximize profit with respect to production quantity. If the marginal cost of producing an extra unit is known, comparison of the marginal cost function to market demand determines the optimal quantity to produce. Similarly, in a given time period, a consumer decides to consume a quantity of a good given that he or she can afford the cost. The problem faced by the consumer is to select the quantity of goods that maximizes his or her utility for a given budget restriction. In addition, discrete choices are made between types of goods which vary in different characteristics.

The assumption of utility maximization is made in MET. It implies that consumers have perfect knowledge of all alternatives and their future consequences (e.g., Hargreaves Heap et al., 1992). These assumptions are not always tenable when the theory is used to predict consumer choice. Although humans adapt to the circumstances, because they have limited cognitive capacity (as will be elaborated upon below) they frequently use approximate solutions to complex choice problems. Some developments of MET have attempted to modify the utility-maximization assumption when it is untenable. As an example, de Palma et al. (1994) introduced the principle of bounded rationality (Simon, 1955, 1982) in a model of sequential decisions.

MET exemplifies a structural approach to decision making and choice (see Svenson, Chapter 7 in this volume). The theory has nothing or very little to say about the process of making choices but may nevertheless accurately predict the choices consumers actually make. However, in many cases it does not (e.g., Kahneman and Tversky, 1979; Simon, 1955, 1982, 1990). Facing that consumers make different choices even when theory predicts that they should not, random utility theory (Manski, 1977) was introduced. It is assumed that the *unsystematic* individual differences can be accounted for by taste variation among individuals who therefore place different weights on the attributes of the choice alternatives. Yet, the many *systematic* discrepancies from utility maximization is not accounted for, although there are attempts to account for these in several revised utility theories (Camerer, 1989).

How Has Micreconomic Theory Been Applied to Travel Choice Modeling? In the past MET has provided the major theoretical foundation for modeling travel choice or demand (Ben-Akiva and Lerman, 1985). However, as noted by de Palma (Chapter 2 in this volume), the part of the theory focusing on consumer choice is more frequently applied than the supply part of the theory. This imbalance is due to several factors, one being the complexity of modeling supply in the transportation market. Another one is the progress made in economic and econometric theory on discrete-choice modeling (McFadden, 1973). For transportation research this progress is particularly important since travelers′ often face discrete choices of travel mode, destination, and route.

Many of the discrete-choice models applied in transportation research is based on the principle of random utility maximization (see McFadden, Chapter 15 in this volume). In random utility models, the utility of choice alternatives consist of two components, one component determined by the attributes of the alternative (and possibly individual characteristics) and one random component. Differences among individuals are captured by the random utility component which must be uncorrelated with the attributes of the alternatives. One frequently applied random-utility model is the multinomial logit (MNL) model of discrete choices.

How Can Microeconomic Theory Be Applied to Travel Choice Modeling? Traditional MET may be extended by defining more broadly the bundle of goods among which choices are made. For instance, multidimensional choice sets may be delineated by including several choice dimensions such as activity, mode, and destination. On the other hand, it is sometimes more appropriate to assume that choices are sequential (Heckman, 1981). Although MET accomodates both possibilities, it does not predict when one or the other occurs. A possibility would of course be to assume that utility maximization predicts the occurence (Timmermans, 1988). As noted by Gärling et al. (1997a), simple enumeration of all the alternatives which then must be compared suggest however that this task sometimes overtaxes the capacity of humans making unaided decisions. When does the task overtax capacity and what does the choice maker then do? Are there ways of aiding households to make utility-maximizing choices and when and how can such decision aids be applied? These latter issues are alluded to in some chapters in this volume (Stern, Chapter 8, and Kwan et al., Chapter 12).

A universal choice set is assumed to consist of all the possible alternatives in a discrete choice problem. However, the actual choice set of a consumer is limited to a subset due to, for instance, budget restrictions. Awareness may also be an important reason. Although the theory does not predict this, there are several model developments which attempts to take this into account (Thill, 1992). For instance, Manski (1977) discusses the possibility of defining choice sets within the random utility model by assigning probabilities of inclusion to the available alternatives.

There is one stream of MET focusing on the time dimension of consumption (see Jara-Díaz, Chapter 3 in this volume). Time is a resource which is limited. If people allocate time to work, it can be converted to money, a resource necessary for goods consumption. Satisfaction is obtained by consumption which requires time. Thus, there is a tradeoff between allocating time to work and time to consumption. As elaborated by Jara-Díaz, this stream of MET is especially appropriate to apply to travel choice. Travel consumes time which limits time for other consumption.

Behavioral Decision Theory

What is Behavioral Decision Theory? Behavioral decision theory (BDT) refers to an empirical approach to the study of human decision making with the goal of describing and understanding how people actually make decisions. In a vein similar to MET (frequently referred to as normative decision theory or utility theory), BDT proposes that a good decision is a choice of an action which meets the decision makers′ objectives. As an example, it may be assumed that a good consumer decision should result in an outcome which makes the consumer satisfied. However, it is an empirical question whether or not consumers make good decisions. In fact, it has been argued (Oliver, 1997) that consumer choices are not influenced by the same attribute characteristics as is satisfaction. Observations such as these have led to research asking questions concerning the reasons for the discrepancy. Even more frequent has been research that ask why human decisions are frequently inconsistent with predictions from (multiattribute) utility theory (or expected utility theory in studies of decisions under risk and uncertainty[1]).

Although its strength is the empirical or descriptive orientation, BDT would make a smaller contribution unless it is capable of systematizing the empirical findings. Over the years several different theories have to this end been proposed. An overview of these theories is given by Svenson (Chapter 7 in this volume). A criticism sometimes voiced is that a proliferation of theories is no better than no theory. However, such a criticism also applies to utility theory (see, e.g., Camerer, 1989, for a description of several different utility theories). In fact an economist (Loomes, 1991) recently saw reasons to question the very existence of a single model able to describe human decision making. Only if it is proposed as a normative theory (thus rendering empirical falsification irrelevant), a single utility theory may be possible to entertain.

---

[1] In expected utility theory the utility of the outcomes are multiplied by their probability of occurrence. If the probability is subjective, the theory is referred to as subjective expected utility.

An important distinction can be made in BDT between structural and process approaches. In general the former are revisions of expected or subjective utility theory in the light of disproving empirical observations (sometimes referred to as "anomalies"). An example is Prospect Theory (Kahneman & Tversky, 1979; Tversky & Kahneman, 1992; Tversky & Fox, 1995). In this theory is introduced a nonlinear function relating decision weights to outcome probabilities. The form of the value or utility function furthermore depends on whether the outcome is an increase or decrease relative to status quo. Another example is Venture Theory (Hogarth & Einhorn, 1990) which goes farther in also assuming that the decision weights assigned to outcomes depend on the values of the outcomes. For instance, a positive risk attitude implies that better outcomes are weighed more than worse outcomes even though objectively they are equally likely (if this is possible to determine). From an extensive comparative test, the economist Camerer (1989) concludes that the changes made in Prospect Theory relative to other utility theories are sufficient to make it consistent with a number of frequently demonstrated "anomalies."

Another issue concerns whether the theories also adequately describe the process of making a choice (whether or not it maximizes utility). Prospect Theory makes a few assumptions concerning the process. Particularly important is the assumption that people react to changes in outcomes rather than their absolute values, thus they subjectively encode outcomes relative to status quo (or sometimes expectations) as losses or gains in an editing phase. The editing phase precedes the evaluation phase where values and decision weights are assigned to outcomes. Because the utility functions differ for gains and losses, reversals of choices of alternatives are observed depending on how the outcomes are encoded. The empirical evidence for .this are extensive (see a recent meta-analysis in Kühberger, 1998). An important issue in need of further research is what factors affect the encoding of outcomes as losses or gains.

Svenson (1990) (see also Chapter 7 in this volume) contributes to the understanding of the editing phase of Prospect Theory in introducing the notion of different levels of decisions. At the first level the decision maker has no conscious representation of the attractiveness[2] of the decision alternatives. Such decisions are based on recognizing the situation as similar to a previous situation leading to the same choice over again without reflecting over its consequences. At the second level the connection of an alternative to its attractiveness is automatically encoded implying that the decision maker knows without further reflection what are good or bad choice alternatives. It is first at the third level that deliberation starts to play an important role. Here, tradeoffs may if necessary be consciously made between

[2] The term attractiveness is used with the understanding that it does not presuppose a common utility or value scale.

attributes relying on a value or utility representation of the alternatives. At the fourth level decision making is more similar to problem solving (Newell & Simon, 1972) consisting of deliberate changes of the decision alternatives such as, for instance, forming new combinations by collapsing old ones or generating other alternatives. At this latter level a decision maker may have additional goals than choosing the most attractive alternative. Thus, a utility representation of decision alternatives may sometimes be insufficient to capture decision makers´ objectives.

Some theories (e.g., Montgomery, 1983; Payne et al., 1993) provide more detailed accounts of the different stages of the decision process. In this vein a decision strategy is defined by Payne et al. (p. 9) "as a sequence of mental and effector (actions on the environment) operations used to transform the initial state of knowledge into a final goal state of knowledge ..." Operations in decision strategies may be modeled by means of sequences of productions of the form "IF (condition 1 ... condition *n*), THEN (action 1 ... action *m*)." Such production-system models were introduced by Newell and Simon (1972) as a formalism for the modeling of higher cognitive processes. A basic tenet of the approach is that human thinking is a symbol-processing system. The conditions of a production is represented in a working memory with a limited capacity whereas the productions are stored in and retrieved from a permanent long-term memory with a high capacity (Anderson, 1993). The limited-capacity working memory is the main reason why decision makers are sometimes overtaxed. However, extensive empirical evidence (Ford et al., 1989) indicate that rather than being overwhelmed they are in general capable of adapting to such overtaxing circumstances by relatively accurate tradeoffs of accuracy against mental effort. In the theory proposed by Payne et al. (1993) which extends the principle of bounded rationality (Simon, 1955, 1982), such tradeoffs are assumed to depend on several factors. A primary factor is characteristics of the decision task, in particular the number of alternatives and the number of attributes of each alternative. Other task characteristics include how the information is presented and how much time is available for processing it. Also important are person characteristics such as the degree of prior knowledge and expertise. The social context imposes constraints in the form of requirements to account for or justify a decision which also affect accuracy-effort tradeoffs. As a consequence, the *same* individual is likely to use different decision strategies under different circumstances. There are also differences between individuals which however may be smaller than the differences within individuals (Camerer & Johnson, 1991).

Another basic tenet which has received empirical support is that there are several decision strategies or rules available to a human decision maker (Ford et al., 1989; Svenson, 1979). A frequent distinction is made between compensatory decision strategies in which all information is processed so that tradeoffs are made between different attributes (or between the utility of different possible outcomes) and noncompensatory decision strategies in which only partial information is processed precluding such tradeoffs. An example of the former is the additive-utility (or expected-utility) rule which is considered to be normative. Examples of the latter include the elimination-by-aspects (Tversky, 1972) and lexicographic decision strategies. In a computer simulation of a production-system model validated against the responses of a sample of human subjects, Payne et al. (1988) showed that different decision strategies differed in accuracy and mental effort (primarily reflected in the number of mental operations required). As one would expect, noncompensatory decision strategies were less effortful. However, less expected was that noncompensatory decision strategies under many circumstances were not much less accurate than compensatory decision strategies. In fact, under time pressure they were even more accurate. A crucial research question is how choices among decision strategies are made. As already mentioned, in the theory of Payne et al. (1993) it is assumed that the choice reflects an accuracy-effort tradeoff. This tradeoff is sometimes based on top-down processing drawing on prior knowledge of the decision problem. For instance, such knowledge may suggest that accuracy is less important. Another possibility is that the tradeoff is based on bottom-up or opportunistic processing where characteristics of the decision problem are important information (e.g., the number of alternatives suggesting the required effort).

A special although important class of process theories are those which make assumptions about distortions of the representation of the decision problem (e.g., Janis & Mann, 1977; Montgomery, 1983; Svenson, Chapter 7 in this volume). Drawing on the seminal work of Festinger (1957, 1964) on cognitive dissonance, these theories make the additional assumption that the mental representation of the decision problem is changed, primarily due to motivational factors. For instance, Montgomery (1983) makes the assumption that decision makers are motivated to avoid making tradeoffs since something desired must be given up. As an illustrative example, by downplaying the attractiveness of some attribute characteristics and bolstering others, a decision maker may create an illusion that he or she makes no tradeoffs (as when one alternative dominates the other alternatives). A similar idea which have received considerable empirical support is entailed by Differentiation and Consolidation Theory proposed by Svenson and described in Chapter 7 of this volume. The theory applies when the decision makers´ goal is not only to find the best alternative but also to avoid post-decisional regret.

How Has Behavioral Decision Theory Been Applied to Travel Choice Modeling? BDT has almost never been applied in a formal analysis or model of travel choice. A single exception is the attempt by Stern in Chapter 8 of this volume to apply Decision Field Theory (Busemeyer & Townsend, 1989) to choices made by road users facing congestions. Other exceptions include production-system modeling of activity scheduling (Ettema et al., 1993; Gärling et al., 1989a, 1998b). Still other exceptions are that models based on the conectionism theoretical framework for modeling low-level cognition (Rumelhart & McClelland, 1986) have been applied (Kitamura and Fujii, Chapter 11 in this volume). Such approaches are very promising since they address the fundamental theoretical question of how decision strategies (utility maximizing or not) are learnt. In addition there are some examples of the development of travel choice models based on noncompensatory decision rules (e.g., Mahmassani and Jou, Chapter 9 in this volume; Timmermans, 1983). The paucity of applications of BDT to travel choice modeling is in sharp contrast to other fields such as consumer research, marketing science, and management science where, for instance, Prospect Theory (Kahneman & Tversky, 1979) has found an abundance of applications (see, e.g., Kleindorffer et al., 1993). Also in economics, BDT has found applications, for instance, in experimental economics (Kagel & Roth, 1995) and in game theory (Camerer, 1990).

That BDT has not been applied in travel choice modeling does not mean that behavioral principles have not. Common-sense notions appear to have had much more impact than theory. In most studies of travel choice (see, e.g., Golob, Chapter 16 in this volume), the selection of endogenous and exogenous variables are not guided by theory. This is not to say that this should not be acceptable. However, the strength of such studies would increase if they were founded in theory.

How Can Behavioral Decision Theory be Applied to Travel Choice Modeling? It is unlikely that an empirical approach to human decision making will contribute such an elegant quantitative theory as normative decision theory (or MET) (von Neumann & Morgenstern, 1947). However, it must be asked in what sense a theory is quantitative if it fails to make accurate quantitative predictions? We feel that the contention of Simon (1990) should be taken seriously:

> ... some of the most important invariants in science are not quantitative at all, but ... ′laws of qualitative structure.′ For example, the germ theory of disease ... says only something like ... ′If you observe pathology, look for a microorganism - it might be causing the symptoms.′ ...
>
> ... in biology (including human) realms, systems change adaptively over time. Simple change is not the problem, for Newton showed how we can write invariant laws as differential equations ... But with adaptive change, which is as much governed by a system′s environment as by its internal constitution, it becomes more difficult to identify true invariants. As a result, evolutionary biology has a rather different flavor from physics, chemistry, or even molecular biology.
>
> ... Psychology does not much resemble classical mechanics, nor should it aim to do so. Its laws are, and will be, limited in range and generality and will be mainly qualitative. Its invariants are and will be of the kinds that are appropriate to

> adaptive systems. Its success must be measured not by how closely it resembles physics but by how well it describes and explains human behavior. (p. 2)
> ... (Hence,) to describe, predict and explain the behavior of a system of bounded rationality, we must both construct a theory of the system's processes and describe the environments to which it is adapting. (p. 6-7)

Thus, much of the basis for quantitative modeling and forecasting is undermined. Of course, this is not really news. The predictive ability of existing travel choice models are in fact not impressive. The realization of the problem we believe is evident in the increasing number of empirical approaches (see Pas, Chapter 10 in this volume) which do not presuppose the assumptions of MET (e.g., utility-maximization) but in an empirical spirit, similar to BDT, attempt to disentangle factors that affect aspects of travel behavior. What choices travelers make and for what reasons must however then be inferred. These approaches have also freed themselves from the statistical tools (i.e., the estimation of logit models) which relies on behavioral assumptions which are not always tenable. Can BDT provide a theoretical foundation for these approaches[3]? In the following some suggestions for and examples of potential contributions of BDT are given. Frequently they require extensions of existing theories, and none is a particularly rigid application.

Sometimes decision makers use phased decision strategies, for instance, the combination of the elimination-by-aspects and additive utility decision strategies, implying that they form a new choice set before making the final choice. From the accuracy-effort theoretical framework (Payne et al., 1993), qualitative predictions of when and how choice sets are formed are possible to make from combinations of task characteristics, person characteristics, and social context factors. The endowment effect (Kahneman et al., 1990) refers to that status quo is overvalued. Thus, if status quo is included in the choice set, deviations from utility-maximization will be observed for this reason alone. However, it is conceivable that the decision process sometimes has two phases where the first is a decision to replace status quo, the second a choice of alternative replacements. Of course, this has consequences for how to model the choices (see Daly and Rohr, Chapter 19 in this volume). Again the situation may largely determine whether the choice process is two-phased or not (e.g., "forced" replacements of something broken vs. desired replacements). Status quo will still play a role as a comparison standard or reference point (Kahneman & Tversky, 1979). Since alternatives are compared to the reference point, their relative values on an interval scale of attractiveness will be altered when the reference point changes.

---

[3] It should be acknowledged that there also exist other behavior theories which may apply.

As noted by Svenson (Chapter 7 in this volume), BDT has most to say about isolated choices which are deliberately made in a way so that values or utilities are assigned to outcomes. Although it is true that empirical findings seem to suggest that people frequently make isolated choices even if they are facing a sequence of decisions, there are important exceptions. A review of relevant research is found in Gärling et al. (1997b). If cognitive capacity is overtaxed, motivational factors may still drive decision makers to make simultaneous choices although in simplified ways. Even though temporal contiguity may play some role, it appears to be more plausible that the degree to which the choice outcomes are related to a common goal or objective is decisive. For instance, the choice of a destination is related to choice of an activity because it is a means of reaching the goal of performing the chosen activity. The *superordinate* principle suggested by Gärling et al. (1997a) states that the goal is chosen without considering the means to reach it whereas the means are chosen considering the goal. Gärling et al. also proposed the *loss-sensitivity* principle based on that people are strongly motivated to avoid negative outcomes. The principle implies that when facing a sequence of decisions, the worst case consisting of the combinations of negative outcomes is the only one considered.

Since many travel choices appear to be made with some regularity (Huff and Hanson, 1990; Pas, 1988; Pas and Koppelman, 1987), it is easy to jump to the conclusion that such choices are predominantly habitual. However, a behavior may be routine without being habitual in the sense that it is performed automatically. It has been proposed that a critical criterion is that decisions precede the observed choice or behavior (Ronis et al., 1989), although not necessarily in immediate temporal contiguity. A sequence of trip choices is perhaps made in the morning as part of an activity-scheduling process. These choices may then be carried out during the day without much further deliberation. In contrast, habitual choices are not preceded by decisions. Yet, the distinction becomes blurred since frequently habitual choices were once deliberate. An additional potential criterion is therefore that the goal (not necessarily conscious) of habitual choices is to reduce mental load. Of course, if they were once deliberate, habitual choices may also achieve other goals such as utility maximization. Not infrequently it may nevertheless be the case that they do not since the circumstances have changed. If the behavior is automatic or "mindless" such changes tend to go unnoticed. In a similar vein Svenson (Chapter 7 in this volume) notes that single decisions may differ in the degree of deliberation depending on whether or not the decision outcomes are explicitly mapped on utilities.

Another example of applying BDT to habitual choices is the proposal by Gärling and Garvill (1993) of a continuum from less to more habitual. In the case of dependent choices, deliberation was assumed to increase with the number of decisions that are made. For instance, choice of trip purpose may on some occasions entail a "script" which specifies destination and travel mode (Lindh, 1998; Verplanken et al., 1994, 1997). On other occasions the deliberation may be more extensive, for instance, if the automobile is unavailable because of maintenance service so that a travel-mode choice needs to be made. Gärling and Garvill (1993) noted that in general obstacles to or a low degree of perceived control over performing an activity affects the degree to which a choice is deliberated. Operationalizing the degree and type of deliberation is difficult, in particular since the number of previous choices of the same alternative may not necessarily be a good indicator (Verplanken et al., 1994, 1997). Innovative methods of collecting and analyzing data are called for (see Axhausen, Chapter 13 in this volume; Gärling et al., 1998a; Mahmassani and Jou, Chapter 9 in this volume).

As already noted, the degree of deliberation is also related to decision strategy, noncompensatory decision strategies requiring less deliberation than compensatory decision strategies. The frequency of the former increases with familiarity with a decision problem (Camerer & Johnson, 1991) but may also be chosen for other reasons as was discussed above.

The degree of deliberation preceding an observed choice may finally also be low if it is dominated by an affective reaction. Affect is characterized along the two primary, orthogonal dimensions arousal and pleasantness. As reviewed by Isen (1987), research shows that both dimensions affect deliberation. In a positive affect people deliberate more and are more concerned about potentially negative outcomes. The explanation is that in a positive affect people are not only evaluating the outcomes of the decision alternatives but also whether the outcomes are threatening their current positive mood state. Increased arousal leads to less deliberation because it limits attentional capacity. In a similar vein, Loewenstein (1996) recently proposed a theory of the role of visceral factors in decision making. One proposition is that a strong emotion will dominate a choice when being in conflict with utility. Another proposition states that an individual who makes plans underestimate the degree to which he or she is affected by emotions when actually carrying out a decision at some later time. As discussed by Gärling et al. (1998a) and Mahmassani and Jou (Chapter 9 in this volume), it is important that a measurement procedure both identify habitual and impulse travel choices since they will differ in several respects from each other and from other choices.

BDT would not treat activity choice differently from any other choice. However, this does not really address the important question of its motivational basis. A distinction made in recent research on attitude theory in social psychology (Gärling et al., 1998a; Gollwitzer, 1996) is between the formation of a goal intention and an implementation intention. Isolated or dependent trip choices would be means of implementing (activity) goals. Of course, these choices are to some extent predictable from characteristics of the transportation system and possible to understand from a disutility or cost-minimization perspective. Individual preferences or tastes (e.g., that an automobile is a more comfortable mode of transportation than public transit) are also likely to be important determining factors. Yet, as noted in several chapters in this volume (Chapter 3 by Jara-Díaz, Chapter 10 by Pas, and, in particular, in Chapter 13 by Axhausen), a more thorough understanding of travel choice is hardly achieved unless the motives for human activity choice on which they ultimately depend are understood. This is partly the realm of attitude and motivational theory in psychology (Ajzen, 1987; Eagly and Chaiken, 1993). There are also several examples of the development of conceptual frameworks with this point of departure (e.g., Bhat and Koppelman, 1993; Gärling et al., 1998b) as well as operational models (see Ettema et al., 1993, and, in this volume, Chapter 11 by Kitamura and Fujii and Chapter 16 by Golob). A potentially fruitful approach is also sketched in Chapter 13 by Axhausen.

In summary, an important contribution of BDT to the analysis of travel choice is by first focussing attention on that decisions preceding travel choices are made in many different ways, and, second, by identifying many of those factors (situational as well as personal) on which decisions are contingent. The identification of needs of and the design of decision aids is one important application. Another one is the development of travel choice models based on valid behavioral assumptions. In addition to forecasts of the magnitude of travel, the planning, design, and management of transportation systems also require as input knowledge of factors (e.g., properties of the system) which make consumers satisfied. Until now such information has largely been obtained from empirical investigations involving the estimation of travel-demand models. MET and BDT only provide an incomplete theoretical foundation for these studies. Therefore, this is an area where the theoretical foundation needs to be broadened, to include psychological theories of human motivation and value priorities (see, e.g., Gärling et al., 1989b).

Statistical Theory

Travel choice modeling entails analyses of mathematically formalized models. In order to make the models operational, numerical values must be assigned to unknown parameters. A simple example is that preference is a linear function of the attributes of a choice alternative. The weights (parameters or coefficients) which an individual places on the different attributes are unknown and estimated by fitting the model to data. After having estimated the parameters of the model, the researcher may want to test the effects of some attributes and/or estimate preferences for different combinations of attribute values. These problems are typically statistical problems. In more general terms, the goal of statistical theory is to provide a theoretical foundation for the generation, description, and analyses of data.

A basic assumption of statistical theory is that observations constitute realizations of random variables. If so, the distribution of statistics can, at least in principle, be deduced from assumptions about the distributions of the random variables. One familiar result is that the mean of a sample of independent normally distributed random variables is also normal, while the mean of a sample of independent non-normally distributed random variables is asymptotically normally distributed. These kinds of derivations from results of analyses of the sampling distribution of statistics are central and helps in evaluating the properties of statistics like estimators, tests, and predictors.

With regard to modeling and estimation of models, desirable properties of estimators are unbiasedness, consistency, and efficiency. Unbiasedness means that the expectation of the estimator equals the true unknown coefficients. Consistency means that as the sample size increases, the estimator converges to the true unknown coefficients. Efficiency means that there is no other estimator with a smaller expected mean squared error, with respect to the difference between estimated and true unknown coefficients. Also, knowledge of the sampling distribution of the estimator opens the possibility to make inferences about the model and its coefficients, that is, tests of coefficients and prediction intervals. Sometimes, the sampling distribution can be derived but in case of non-linear models it must frequently be approximated with the asymptotic distribution of the estimator.

Survey design and statistical estimation are strongly interconnected. For instance, the variance of the sampling distribution of the mean is the ratio of the variance to the number of observations. Thus, the precision of the estimator (the sample mean) increases with the sample size. In a travel diary, suppose respondents are asked to record all the distances of all automobile trips during a fixed period of time. The logarithm of travel distances recorded may in an analysis be related to, for instance, respondent characteristics in a linear regression model whose coefficients can be estimated with ordinary least squares (OLS). If the respondents were instructed not to record trips shorter than, say, 1 kilometer, a restriction motivated by reasons like respondent fatigue or the object of the study, a truncation is induced and OLS estimation may no longer be appropriate. Thus, the estimator needs not only account for the model under study but also for the components of the statistical model attributed to the survey design. If the design is badly chosen, the model may be difficult or even impossible to estimate. It is therefore necessary that sample designs are chosen such that they comply with assumptions of known estimation and other inference techniques.

There is a great variety of statistical models used in research on travel choice, but models for ordered and non-ordered categorical response variables are among the most frequently encountered. This is because choices are frequently conceptualized as discrete. Among these models, the binomial and multinomial logit models are the most popular. The multinomial logit (MNL) model (e.g., Maddala, 1983) has several appealing features which certainly account for its popularity. The model yields closed form solutions to choice probabilities making estimation and calculations easy. Not the least important feature is that the MNL model is consistent with random utility maximization (McFadden, 1981), thus giving the model a theoretical justification. One restrictive assumption of the MNL model is the independence of irrelevant alternatives (IIA) (Luce, 1959). The implausible IIA property of the MNL model has motivated the use of alternative models. One alternative is the multinomial probit (MNP) model, considered by, for instance, Hausman and Wise (1978). McFadden (1978, Chapter 15 in this volume) proposes a model based on the generalized extreme value distribution in which the MNL model is a special case.

In addition to discrete-choice models, studies of travel behavior frequently introduce models with latent attitudinal variables. Latent structure models are applied to this kind of data. In addition, attitudes have been measured in applications of stated preference (SP) research methods. For instance, Kroes and Sheldon (1986) compared attitude and SP models. SP methods are experimental where respondents give preference judgments of objects designed by the researcher. Although the methods mimic experiments in the sciences, the response variable should usually not be assumed to posses the properties of interval-scaled measurements. Alternative estimation and analysis techniques than analysis of variance has therefore partly been considered. Here again, the MNL model has played an important role. However, SP data present complex estimation problems due to dependence and heterogeneity between responses (e.g., Bradley and Daly, 1994; McFadden, 1986).

Construction of simulation models is often used in research on travel choice. Such models may have different purposes but their aim is to simulate outcomes of dependent variables conditional on values of a set of exogenous or independent variables. The outcome may then be used as a prediction of a future event or it may be used for evaluating effects of certain actions. There is an essential difference between simulators based on RP and SP data, respectively. Since RP data are obtained from actual choices, their external validity is less frequently questioned. Simulators based on SP data are therefore thought of being in need of "adjustments" to RP data in order to have valid simulated outcomes.

The increasing complexity of individual travel behavior has had an effect on the topics addressed in travel choice research. Consequently, models formulated have also become more complex and the need of more detailed data has increased. This development has had an effect on the design of household travel surveys. For instance, analysis of trip chaining choices may require a travel diary filled in by the respondent during a specific period of time. Furthermore, dynamic effects like adaptation to new information might require measurements during several consecutive periods (see Mahamassani and Jou, Chapter 9 in this volume, for an example). The change of focus to complex travel behavior increases therefore the demand for more complex designs of household travel surveys. Data resulting from these kinds of surveys may have much to offer but may at the same time present new problems. The problem of panel attrition (e.g., Hensher, 1985; Brownstone, Chapter 18 in this volume) is one example.

The trend towards complex surveys (Stopher, Chapter 17 in this volume) and models challenges transport researchers with new statistical problems. The fast development of computer technology provides means of solutions to some of these problems. For instance, data collection can be made more simple and more detailed data can be collected with the aid of computers. Also, the increasing capacity of computers makes estimation of more complex models feasible. Simulation-based estimators (e.g., McFadden and Ruud, 1995) represent contributions in this direction. Another example is resampling techniques like bootstrapping (Efron and Tibshirani, 1993). However, even if data collection and estimation problems in research on complex travel choices are solved, the problem of model evaluation remains. Usually, models are thoroughly formulated and their coefficients, associated standard errors, and model fit are estimated. Still, the assumptions behind the model are rarely tested nor is the sensitivity for departures from model assumptions investigated. These topics are certainly important and need to be addressed in future research. Future research may also benefit from a discussion of the necessary scope of models. If research is exploratory, data and model may be required to grasp the widest possible set of aspects of travel choice. In confirmatory research this may not be so.

CHAPTER SUMMARY

Using their most recent research findings as illustrations, each chapter author has been asked to highlight the theoretical foundations of their research whether it is microeconomic, behavioral, or statistical. In chapters introducing each of the following parts of the book, André de Palma, Ola Svenson, and Daniel McFadden undertook the task of reviewing recent developments of these theoretical foundations. Staffan Algers, John Polak, and Chandra Bhat were asked to comment on the chapters of each section. Below we first briefly summarize the contents of each chapter, then on the basis of the commentatories we highlight what we feel are the most important issues emerging from these chapters.

The microeconomic theoretical foundation constituting the first part of the book is represented by four chapters. In the opening Chapter 2 de Palma addresses general theoretical and methodological concerns. One which is particularly pertinent to this volume is: What should be the level of sophistication in modeling? In discussing the question de Palma focuses on three issues. The first deals with the underlying theory of transportation science, the next concerns the relationship between theory, methods, and models, and the third issue is more policy-oriented, dealing with the social relevance of theory and models.

The traditional microeconomic theory foundation of travel-mode choice modeling is summarized by Jara-Díaz in Chapter 3. The utility of an activity is not ascribed to the activity in itself but how much time is required to perform it including time to travel. On this basis Jara-Díaz proposes a theoretical framework on which travel-activity choice models may be founded.

A common byproduct of discrete-choice models are estimates of the marginal rate of substitution between two attributes such as value of travel time savings (VTTS). Typically, these are point estimates rather than functions. In Chapter 4 Hensher argues that it would be more realistic to model the distribution of VTTS. Empirical tests from a stated-choice study illustrate that this approach leads to different results than the conventional. Applying the results to RP data from commuters furthermore results in a classification of segments of the commuter market by socioeconomic characteristics.

Another frequent assumption in travel choice modeling is that the transportation system is in a state of equilibrium. This assumption is in Chapter 5 questioned by Goodwin who argues that the understanding of behavioral mechanisms may requiring that the concept of a timeless equilibrium is given up. The alternative is to focus on adjustment processes. Dynamic models thus need to be developed based on data on time-dependent changes.

Chapters in the second part of the book discusses and illustrates the potential of BDT to provide a theoretical foundation for travel choice modeling. In Chapter 7 Svenson reviews several behavioral decision theories. He makes a distinction between structural approaches which focus on the relations between decision alternatives and the choices and process approaches which focus on the process leading to the decision. The Differentiation and Consolidation (Diff Con) Theory recently proposed by Svenson is used as an unifying theoretical framework for illustrating many issues associated with decision making in connection with travel.

Another behavioral decision theoretical framework (Decision Field Theory) is discussed in Chapter 8 by Stern who attempts to apply it to accounts of pretrip and en-route decisions that road users make when facing congestions.

In a model of commuters′ repeated decisions concerning departure time and route, Mahmassani and Jou in Chapter 9 show how trip-chaining decisions can be included. The approach is behavioral in that it is assumed that choices are guided by heuristic rules which the model identifies. The utility-maximization principle is abandoned in favor of the assumption of a boundedly rational search for an acceptable outcome. Furthermore, a distinction is in the trip-chaining component made between routine, non-routine, and impulse activities.

A review of the role time has played in travel-choice modeling over the past decades is made by Pas in Chapter 10. Pas believes the role of time has changed to become much more important in current conceptualizations. In the activity-based approach, theorizing about and modeling the determinants of time use is clearly of paramount interest. Pas argues that much of the time-use research can and should be applied to travel choice modeling.

In Chapter 11 Kitamura and Fujii present two computational-process models of activity-travel behavior. By incorporating the concept of a space-time prism, the choice model PCATS builds on and extends previous activity-based theorizing and modeling. The other model, AMOS, simulates adjustments of travel behavior to changes implemented in travel demand management (TDM) strategies.

Kwan, Golledge, and Speigle (Chapter 12) describe how Intelligent Transportation Systems (ITS) and Geographic Information System (GIS) may be developed to become a travel decision support system (DSS). As they point out, much additional theoretical and empirical research is needed for this development. For instance, shortest-path algorithms are not necessarily useful but needs to be replaced by behaviorally more realistic ones.

If the aim is to develop more realistic models of complex travel choices, Axhausen in Chapter 13 notes that the data that are needed will be different. How they will be different is explicated by contrasting current modeling approaches, in particular activity-based such approaches, to existing types of data. Axhausen also discusses the notion of a daily activity stream and our current understanding of it.

The third part of the book includes chapters discussing statistical and data collection issues. In the first chapter (Chapter 15), McFadden discusses new approaches to the measurement and modeling of Willingness-to-Pay (WTP) for transportation improvements.

In Chapter 16 Golob shows how structural equation modeling can be extended to deal with the complex problem of modeling interrelated activity and travel choices which otherwise may be intractable.

Stopher (Chapter 17) reviews the development of travel surveys which are the primary data sources for testing theories and estimating models with the aim of increasing the understanding of travel choices. In addition Stopher identifies several emerging changes in travel surveys which may be extrapolated into the future.

The observation and measurement of complex travel choices may lead to lower data quality. Ways of handling problems of missing data, non-random nonresponses, and panel attrition in travel surveys are discussed by Brownstone in Chapter 18.

Applications of travel choice models are discussed by Daly and Rohr in Chapter 19. Specifically, they focus on the accuracy of forecasts which combine data from SP and RP methods. They also discuss how to overcome problems with representativeness of data samples and the need to update models.

Additional summaries of the chapter contents are found in Chapters 6, 14, and 20. Drawing on many of the constructive critical comments also found in these chapters, we now conclude by highlighting a few important issues.

Let us start by stating that we do not fully agree with one conclusion drawn by Bhat (Chapter 20): it is not our impression that the main task now is to apply the knowledge we have about travel choices. From the chapters in this volume as well as the discussions during the conference of which it is the result, we conclude that progress is being made but that many problems still remain to be addressed. This is, we believe, in particular true if a theoretical foundation of travel choice modeling is sought which has empirical support in behavioral research on attitudes, preferences, and choices. We are afraid that unrealistic claims are not infrequently made concerning, for instance, the predictive ability of travel choice models. Both large systematic and unsystematic deviations from predictions are to be expected. Furthermore, a wider set of outcome variables may need to be taken into account (such as, for instance, measures of consumer satisfaction, physiological indicators, health effects). On the other hand, it is certainly true that applications will not await further research. It is important therefore to continue doing research in application contexts.

The question is then which directions future research should take. Polak (Chapter 14) seems to welcome contributions by BDT. At the same time, he notes that the criteria by which these contributions are evaluated must not be relaxed. We agree although we may disagree about what these evaluation criteria are. We also believe it is not useful to view the contributions from different disciplines as competing with each other. For instance, structural models of travel choice (developed on the basis of MET) must of course be consistent with models of the choice process (developed on the basis of BDT). Sometimes studies of the decision process provide useful tests of assumptions made in structural models (or the reverse) which otherwise would be untestable. If such tests prove that these assumptions are untenable, it is difficult to see how, on the basis of any set of evaluation criteria, the structural model can be retained. If structural models are more feasible in applications, the challenge is to develop such models which incorporate tenable assumptions. We feel this is in the spirit of the final remarks by Algers (Chapter 6) that more truly collaborative efforts are needed. However, we note that modeling a phenomenon is not an arbitrary choice of making a model more or less detailed or realistic. Some models are clearly based on theoretically and empirically untenable assumptions, others are not.

The argument for process approaches is thus not an appeal for more detailed but more valid accounts. Such accounts require that sound methods are available to study the travel-choice process. We agree with Axhausen (Chapter 13) that travel diaries should focus on revealing the decision process rather than its outcome or the resulting behavior (whether called time use, travel, or activity). However, methods need to be developed beyond the diary. Global Positioning Systems (GPS), Geographical Information Systems (GIS), smart cards, and portable computers are all marvelous technologies which have great future promises for the unobtrusive generation of detailed data on travel behavior. Yet, these technologies may not improve our ability to analyze the network of casual determinants. As exemplified in the work by Mahmassani (Chapter 9), an experimental approach in which laboratory and field studies are combined may prove very useful to this end. Of course, this is what we see in studies combining SP and RP methods (Hensher, 1994, Chapter 4; Daly and Rohr, Chapter 19). Given that individuals show great variability (or contingency) across situations, it is however also necessary to systematically pursue extensive series of experimental studies addressing limited sets of theoretically critical questions.

Awareness of the issues discussed here is certainly evident in most of the chapters. Some also bring up several other, equally important issues. One concerns the need to conceptualize travel choices as complex interrelated decisions. As exemplified in several chapters of this volume, here research on travel choice is a forerunner in developing theory. In this case, it is hoped that behavioral researchers will learn. An example is that primarily descriptive research on time use may benefit immensely from developments of theories of activity scheduling.

Both theoretical advancements as well as urgent practical applications may be expected in the future from research on how choices change over time and what factors account for such changes. Humans are adaptive systems. Theoretically it is important to understand the mechanisms of adaptation. Practically, knowledge of the limits and costs of adaptation is clearly essential, more today than ever. As noted by Polak (Chapter 14), key factors are knowledge, learning, and their interrelationships. How do people learn, what do they learn, and how is what they have learnt or know applied in their behavior? These are general questions which more often should be asked in many of the application and academic contexts where travel choice models are developed.

REFERENCES

Ajzen, I. (1987). Attitudes, traits, and actions: Dispositional prediction of behavior in personality and social psychology. In: *Advances in Experimental Social Psychology* (L. Berkowitz, ed.), Vol. 20, pp. 1-63. Academic Press, San Diego, CA.

Algers, S., and S. Widlert (1997). Stockholm model system (SIMS): Application.. In: *World Transport Research: Proceedings of the 7th World Conference on Transport Research: Vol. 2. Modelling Transport Systems* (D. Hensher, J. King and T. Oum, eds.), pp. 345-362. Pergamon, Oxford.

Anderson, J. R. (1993). *Rules of the Mind.* Erlbaum, Hillsdale, NJ.

Arentze, T., A. Borgers and H. P. J. Timmermans (1993). A model of multi-purpose shopping trip behavior. *Papers in Regional Science,* **72**, 239-256.

Axhausen, K. and T. Gärling (1992). Activity-based approaches to travel analysis: Conceptual frameworks, models, and research problems. *Transport Reviews*, **12**, 323-341.

Ben-Akiva, M. and S. R. Lerman (1985). *Discrete Choice Analysis: Theory and Application in Travel Demand.* The MIT Press, Cambridge, MA.

Bhat, C. R. and F. S. Koppelman (1993). A conceptual framework of individual activity program generation. *Transportation Research A*, **27**, 433-446.

Bradley, M. and A. Daly (1994). Use of the logit scaling approach to test for rank-order and fatigue effects in Stated Preference data. *Transportation*, **21**, 167-184.

Burnett, K. P. and S. Hanson (1982). Rationale for an alternative mathematical approach to movement as complex human behavior. *Transportation Research Record, No.* **723**, 11-24.

Busemeyer, J. R. and J. T. Townsend (1989). Decision field theory: A dynamic-cognitive approach to decision making in an uncertain environment. *Psychological Review*, **100**, 432-459.

Camerer, C. F. (1989). An experimental test of several generalized utility theories. *Journal of Risk and Uncertainty,* **2**, 61-104.

Camerer, C. F. (1990). Behavioral game theory. In: *Insights in Decision Making* (R. M. Hogarth, ed.), pp. 311-336. University of Chicago Press, Chicago, Il.

Camerer, C. F. and E. J. Johnson. (1991). The process-performance paradox in expert judgment. In: *Toward a General Theory of Expertise* (K. A. Ericson and J. Smith, eds.), pp. 195-217. Cambridge University Press, Cambridge.

Daly, A. (1987). Estimating "tree" logit models. *Transportation Research B,* **21**, 251-267.

de Palma, A., G. M. Myers and Y. Y. Papageorgiou (1994). Rational choice under an imperfect ability to choose. *American Economic Review*, **84,** 419-440.

Eagly, A. H. and S. Chaiken (1993). *The Psychology of Attitudes*. Harcourt Brace Jovanovich, Forth Worth, FL.

Efron, B. and R. J. Tibshirani (1993). *An Introduction to the Bootstrap*. Chapman and Hall, New York.

Ettema, D. F., A. Borgers and H. J. P. Timmermans. (1993). A simulation model of activity scheduling behavior. *Transportation Research Record, No.* **1493**, 101-109.

Ettema, D. F. and H. J. P. Timmermans (1997). Theories and models of activity patterns. In: *Activity-Based Approaches to Travel Analysis* (D. F. Ettema and H. J. P. Timmermans, eds.), pp. 1-36. Pergamon, Oxford.

Festinger, L. (1957). *A Theory of Cognitive Dissonance*. Stanford University Press, Stanford, CA.

Festinger, L. (1964). *Conflicts, Decision, and Dissonance.* Stanford University Press, Stanford, CA.

Ford, J. K., N. Schmitt, S. L. Schechtman, B. M. Hults and M. L. Doherty (1989). Process tracing methods: Contributions, problems, and neglected research questions. *Organizational Behavior and Human Decision Processes,* **43**, 75-117.

Gärling, T., K. Brännäs, J. Garvill, R. G. Golledge, S. Gopal, E. Holm, and E. Lindberg (1989a). Household activity scheduling. In: *Transport Policy, Management & Technology Towards 2001,* Vol. IV, pp. 235-248. Western Periodicals, Ventura, CA.

Gärling, T. and J. Garvill (1993). Psychological explanations of participation in everyday activities. In: *Behavior and Environment: Psychological and Geographical Approaches* (T. Gärling and R. G. Golledge, eds.), pp. 270-297. Elsevier/North-Holland, Amsterdam.

Gärling, T., R. Gillholm and A. Gärling (1998a). Reintroducing attitude theory in travel behavior research: The validity of an interactive interview procedure to predict car use. *Transportation.* In press.

Gärling, T., R. Gillholm, J. Romanus and M. Selart (1997a). Interdependent activity and travel choices: Behavioral principles of integration of choice outcomes. In: *Activity-Based Approaches to Travel Analysis* (D. F. Ettema and H. J. P. Timmermans, eds.), pp. 135-150. Pergamon, Oxford.

Gärling, T., T. Kalén, J. Romanus, M. Selart and B. Vilhelmson (1998b). Computer simulation of a theory of household activity scheduling. *Environment and Planning*. In press.

Gärling, T., N. Karlsson, J. Romanus and M. Selart (1997b). Influences of the past on choices of the future. In: *Decision Making: Cognitive Models and Explanations* (R. Ranyard, R. Crozier and O. Svenson, eds.), pp. 167-188. Routledge, London.

Gärling, T., E. Lindberg and H. Montgomery (1989b). Beliefs about attainment of life satisfaction as determinants of preferences for everyday activities. In: *Understanding Economic Behavior* (K. G. Grunert and F. Ölander, eds.), pp. 33-46. Kluwer, Dordrecht, The Netherlands.

Gollwitzer, P. M. (1996). The volitional benefits of planning. In: *The Psychology of Action: Linking Cognition and Motivation to Behavior* (P. M. Gollwitzer and J. A. Bargh, eds.), pp. 287-312. Guildford Press, New York.

Hägerstrand, T. (1970). What about people in regional science. *Papers of the Regional Science Association,* **23**, 7-21.

Hanson, S. (1995). Getting there: Urban transportation in context. In: *The Geography of Urban Transportation* (2nd ed.) (S. Hanson, ed.), pp. 3-25. Guildford Press, New York.

Hanson, S. and M. Schwab (1995). Describing disaggregate flows: Individual and household activity patterns. In: *The Geography of Urban Transportation* (2nd ed.) (S. Hanson, ed.), pp. 166-187. Guildford Press, New York.

Hargreaves Heap, S., M. Hollis, B. Lyons, R. Sugden and A. Weale (1992). *The Theory of Choice: A Critical Guide*. Blackwell, Oxford.

Hausman, J. A. and D. A. Wise (1978). A conditional probit model for qualitative choice: Discrete decisions recognizing interdependence and heterogenous preferences. *Econometrica*, **46**, 403-426.

Heckman, J. J. (1981). Statistical models for discrete panel data. In: *Structural Analysis of Discrete Data with Econometric Applications* (C. F. Manski and D. McFadden, eds.), pp. 114-178. MIT Press, Cambridge, MA.

Hensher, D. (1985). Longitudinal surveys in transport: An assessment. In: *New Survey Methods in Transport* (E. S. Ampt, A. J. Richardson and W. Brög, eds.), pp. 77-97. VNU Science Press, Utrecht.

Hogarth, R. M. and H. J. Einhorn (1990). Venture theory: A model of decision weights. *Management Science*, **36**, 780-803.

Huff, J. and S. Hanson (1990). Measurement of habitual behavior: Examining systematic variability in repetetive travel. In: *Developments of Dynamic and Activity-Based Approaches to Travel Analysis* (P. Jones, ed.), pp. 229-249. Gower, Aldershot, UK.

Isen, A. M. (1987). Positive affect, cognitive processes, and social behavior. In: *Advances in Experimental Social Psychology* (L. Berkowitz, ed.), Vol. 20, pp. 203-253. Academic Press, San Diego, CA.

Janis, I. L. and L. Mann (1977). *Decision Making: A Psychological Analysis of Conflict, Choice, and Commitment*. The Free Press, New York.

Jones, P. M., F. S. Koppelman and J.-P. Orfeuil (1990). Activity analysis: State-of-the-art and future directions. In: *Developments of Dynamic and Activity-Based Approaches to Travel Analysis* (P. Jones, ed.), pp. 34-55. Gower, Aldershot, UK.

Kagel, J. H. and A. E. Roth (eds.) (1995). *The Handbook of Experimental Economics*. Princeton University Press, Princeton, NJ.

Kahneman, D., J. L. Knetsch and R. H. Thaler (1990). Experimental tests of the endowment effect and the Coase theorem. *Journal of Political Economy*, **198**, 1325-1348.

Kahneman, D. and A. Tversky (1979). Prospect theory: An analysis of decision under risk. *Econometrica*, **47**, 263-291.

Kitamura, R. (1988). An evaluation of activity-based travel analysis. *Transportation*, **15**, 9-34.

Kleindorffer, P. R., H. C. Kunreuther and P. J. H. Schoemaker (1993). *Decision Sciences: An Integrative Perspective*. Cambridge University Press, New York.

Kühberger, A. (1998). The influence of framing on risky decisions: A meta-analysis. *Organizational Behavior and Human Decision Processes*. In press.

Lind, C. (1998). *Household decision processes in long-range travel*. Doctoral Dissertation, Department of Infrastructure and Planning, Royal Institute of Technology, Stockholm, Sweden.

Levinson, D. and A. Kumar (1995). Activity, travel, and the allocation of time. *Journal of the American Planning Association*, **61**, 458-470.

Loomis, G. (1991). Evidence of a new violation of the independent axiom. *Journal of Risk and Uncertainty*, **4**, 91-108.

Loewenstein, G. (1996). Out of control: Visceral influences on behavior. *Organizational Behavior and Human Decision Processes*, **65**, 272-292.

Luce, R. D. (1959). *Individual Choice Behavior: A Theoretical Analysis*. Wiley, New York.

Manski, C. (1977). The structure of random utility models. *Theory and Decision*, **8**, 229-254.

McFadden, D. (1973). Conditional logit analysis of qualitative choice behavior. In: *Frontiers in Econometrics* (P. Zarembka, ed.), pp. 105-142. Academic Press, New York.

McFadden, D. (1978). Qualitative methods for analyzing travel behavior of individuals. In: *Behavioral Travel Modeling* (D. Hensher and P. Stopher, eds.), pp. 279-231. Croom Helm, London.

McFadden, D. (1981). Econometric models of probabilistic choice. In: *Structural Analysis of Discrete Data with Econometric Applications* (C. F. Manski and D. McFadden, eds.), pp. 198-272. MIT Press, Cambridge, MA.

McFadden, D. (1986). The choice theory approach to market research. *Marketing Science*, **5**, 275-297.

McFadden, D. and P. A. Ruud (1995). Estimation by simulation. *Review of Economics and Statistics*, **76**, 591-608.

Montgomery, H. (1983). Decision rules and the search for a dominance structure: Towards a process model of decision making. In: *Analyzing and Aiding Decision Processes* (P. C. Humphreys, O. Svenson and A. Vari, eds.), pp. 343-369. North-Holland, Amsterdam.

Newell, A. and H. A. Simon (1972). *Human Problem Solving*. Prentice-Hall, Englewood Cliffs, NJ.

Oliver, R. L. (1997). *Satisfaction: A Behavioral Perspective on the Consumer*. McGraw-Hill, New York.

Ortúzar, J. de D. and L. G. Willumsen (1994). *Modelling Transport* (2nd ed.). Wiley, Chichester, U.K.:

Pas, E. I. (1988). Weekly travel-activity behavior. *Transportation, 15*, 89-119.

Pas, E. I. (1990). Is travel demand analysis and modelling in the doldrums? In: *Developments of Dynamic and Activity-Based Approaches to Travel Analysis* (P. Jones, ed.), pp. 3-27. Gower, Aldershot, UK.

Pas, E. I. and F. S. Koppelman (1987). An examination of the determinants of day-to-day variability in individuals′ urban travel behavior. *Transportation, 13,* 183-200.

Payne, J. W., J. R. Bettman and E. J. Johnson (1988). Adaptive strategy selection in decision making. *Journal of Experimental Psychology: Learning, Memory, and Cognition*, **14**, 534-552.

Payne, J. W., J. R. Bettman and E. J. Johnson (1993). *The Adaptive Decision Maker.* Cambridge University Press, New York.

Pipkin, J. S. (1995). Disaggregate models of travel behavior. In: *The Geography of Urban Transportation* (2nd ed.) (S. Hanson, ed.), pp. 188-218. Guildford Press, New York.

Ronis, D. L., J. F. Yates and J. P. Kirscht (1989). Attitudes, decisions, and habits as determinants of repeated behavior. In: *Attitude Structure and Function* (A. R. Pratkanis, S. J. Breckler and A. G. Greenwald, eds.), pp. 213-239. Erlbaum, New Jersey, NJ.

Rumelhart, D. E. and J. L. McClelland (1986). *Parallel Distributed Proccessing: Explorations in the Microstructure of Cognition.* Vol. 1. MIT Press, Cambridge, MA.

Simon, H. A. (1955). A behavioral model of rational choice. *Quarterly Journal of Economics,* **69**, 99-118.

Simon, H. A. (1982). *Models of Bounded Rationality. Vol. 2: Behavioral Economics and Business Organization*. The MIT Press, Cambridge, MA.

Simon, H. A. (1990). Invariants of human behavior. *Annual Review of Psychology,* **41**, 1-19.

Svenson, O. (1979). Process descriptions of decision making. *Organizational Behavior and Human Performance,* **23**, 86-112.

Svenson, O. (1990). Some propositions for the classification of decision situations. In: *Contemporary Issues in Decision Making* (K. Borcherding, O. I. Larichev and D. M. Messick, eds.), pp. 17-31. Elsevier, Amsterdam.

Thift, J.-C. (1992). Choice set formation for discrete choice modeling. *Progress in Human Geography*, **16**, 361-382.

Timmermans, H. J. P. (1983). Non-compensatory decision rules and consumer spatial behavior: A test of predictive ability. *The Professional Geographer*, **35**, 449-455.

Timmermans, H. J. P. (1988). Multipurpose trips and individual choice behaviour: An analysis using experimental design data. In: *Behavioral Modelling in Geography and Planning* (R. G. Golledge and H. J. P. Timmermans, eds.), pp. 356-367. Croom Helm, London.

Tversky, A. (1972). Elimination by aspects: A theory of choice. *Psychological Review,* **79,** 281-299.

Tversky, A. and C. R. Fox (1995). Weighing risk and uncertainty. *Psychological Review,* **102,** 269-283.

Tversky, A. and D. Kahneman (1992). Advances in prospect theory: Cumulative representation of uncertainty. *Journal of Risk and Uncertainty,* **5**, 297-323.

Verplanken, B., H. Aarts and A. Van Knippenberg (1997). Habit, information acquisition, and the process of making travel mode choices. *European Journal of Social Psychology,* **27,** 539-560.

Verplanken, B., H. Aarts, A. van Knippenberg and C. van Knippenberg (1994). Attitude versus general habit: Antecedents of travel mode choice. *Journal of Applied Social Psychology,* **24,** 285-300.

von Neumann, J. and O. Morgenstern (1947). *Theory of Games and Economic Behavior.* Wiley, New York.

PART 1. MICROECONOMIC THEORY

2

INDIVIDUAL AND COLLECTIVE DECISION MAKING: APPLICATION TO TRAVEL CHOICE

André de Palma, University of Cergy-Pontoise, France

INTRODUCTION

The impetus for this chapter comes from discussions with scholars in economics but also in other disciplines (e.g., transportation, psychology, and sociology) regarding research methodology, concepts of rationality and complexity, behavioral versus axiomatic approaches, testing of hypotheses, and the relationship between theory and practice. The explanations I have received vary widely, and are sometimes not satisfactory. I do not claim to have the answers, however. Rather, the main purpose is to illuminate methodological problems that are pertinent to transport research (or even social sciences at large) as well as other issues such as theory production and its practice that have been frequently overlooked by transport researchers.

To begin with, a distinction should be made between scientific and non-scientific and academic and non-academic research. I am concerned here largely with scientific academic research which aims fundamentally to develop high levels of abstractions in the form of concepts or theories, in order to offer explanations of world phenomena or events. Scientific research is an intellectual process which involves a structured way of posing and answering questions to gain knowledge. Such a mode of inquiry proceeds on the assumption that knowledge is best garnered through rational abstractions and empirical observations. This presupposes that while world phenomena or events may appear chaotic or complex, they are governed by regularities and laws, the latter of which awaits to be uncovered by the researcher. Such a positivist orientation means that research methodology is of paramount importance to the researcher because it serves as the *modus operandi* from which he or she is able to infer relationships or laws to represent, explain, describe, and predict the phenomena of interest.

If the object of scientific inquiry is to establish descriptions, explanations, or predictions of observed phenomena or events, then a good way to represent reality is the use of models. Models are relevant because they allow us to illustrate relationships, organize information, or construct coherence from what appears to be a set of unrelated observations and events. But what is an appropriate level of sophistication in modeling? A model that captures a one-to-one, point-to-point correspondence with the characteristics of the phenomena being studied is perhaps the only "true" representation of the real world. This would however be a futile effort. Indeed, there is a tradeoff between a more "realistic" model and increased levels of abstract descriptions. Elegant modeling may be intellectually rewarding but is also less connected to the real world.

May (1976) shows that complex dynamics, such as the evolution of the number of individuals living in a population, can be explained by a simple set of differential equations. This view of the world, which is illustrated, for example, by nonlinear dynamic systems, states that a few equations can display very complex solutions involving stable or unstable branches, single or multiple solutions, single or multiple bifurcations as well as chaotic behavior. A similar approach is also taken by the school of Brussels led by Prigogine (see, Nicolis and Prigogine, 1989). They developed the theory of self-organization and showed that simple reaction-diffusion laws may generate complex spatial and/or temporal patterns such as those observed in chemical systems (chemical clocks, hexagonal structures reminiscent of Christaller's central place patterns, etc.). This approach is in sharp contrast to the Club of Rome type of approach which provides very complex models to forecast the evolution of social and economic systems. Such contrasts beg the following questions: Is a model a simplification of reality which at best suggests some intuition without any capacity to reproduce reality? Or is a model a tool to be used to control our environment? In discussing the questions with respect to transportation, I will focus on three major issues: (1) The nature of transportation science and the underlying economic theory; (2) The relationship between theory, methods, and models; and (3) The social relevance of theory and models in terms of policies.

TRANSPORTATION SCIENCE AND ECONOMIC THEORY

An Economic Tool to Model Demand

A wide class of models that is used in transportation is discrete choice models (DCMs). Although the mechanical derivation of these models is rather straightforward, some attempt has been made to understand the background and intellectual roots of those models (see de Palma and Lefèvre, 1981) as well as their theoretical foundations (Anderson, de Palma, and Thisse, 1992).

Discrete-choice theory dates back to the nineteenth century from the works of Weber and Fechner as well as Helmoltz, Galton, and Wundt who studied the mechanism of perception. To measure sensation, they asked their subjects to detect physical stimuli (lights, sounds, weights, etc.). The key idea here was to relate a continuous physical stimulus to a discrete (binary) reply (YES or NO). The model which describes this process is based on the idea of a threshold. If the stimulus is smaller than the threshold, it is not detected. If the stimulus is larger than the threshold, the stimulus is detected. Psychophysical experimentation shows that the same subject may detect a stimulus on one occasion, but not on another. Therefore, the best representation of the behavioral response of the subject is achieved by using the concept of probability. The threshold would then be defined as the intensity of the stimulus that is detected with a given probability (e.g., 0.5). The key theoretical ingredients of discrete response models are: (1) The relation between a continuous measurable variable and an individual discrete response; (2) The idea of a threshold; (3) The probabilistic ingredient; and (4) The idea that a binary response is the outcome of a comparison between two numbers (i.e., the physical stimulus and the threshold).

Thurstone (1927) provides an extension of the threshold models to (a) a decision process situation and (b) a situation involving a discrete number of alternatives (larger or equal to two). In Thurstone's theory the individual is facing n alternatives and is asked to select the most attractive one. The choice process postulated by Thurstone may be summarized as follows: "The utility is assumed to vary from moment to moment, and the decision process consists of the simple fixed rule of picking the alternative with the largest momentary utility" (Edgell and Geisler, 1980, p. 266). Such a comparison between n alternatives provides an important extension of the threshold approach. It also forms the basis of the standard multinomial DCMs. Mathematically, the two tools - threshold models and discrete choice models - are similar (see Anderson et al., 1992) but conceptually they are different. In the DCMs the individuals are involved in an active decision process while the threshold models test the individuals discriminatory process.

In an analysis of the theoretical foundation of DCMs, we need to consider the following four issues:

1. Although the theory of Thurstone is presented as a direct extension of the theory of threshold, there may be other ways to proceed. Recently, Anderson and de Palma (1997) introduced a direct extension of the idea of a threshold for the case of an individual facing *n* alternatives. The reasoning is as follows. The individual has a discriminatory process which does not allow him or her to perceive differences smaller than a given threshold (just noticeable differences). Those stimuli above the threshold value are chosen with equal probabilities and the others are dominated and rejected. It can be shown that this model is not equivalent to conventional DCMs when the number of alternatives is larger than 2. Such an approach, based on the limited ability of individuals to compare stimuli, is potentially attractive from a theoretical point of view but has not yet been used for empirical estimation.
2. The use of a probabilistic approach may be a good idea, but is also rather tricky. The utility U_i associated with choice i is given by $U_i = V_i + \varepsilon_i$ where V_i is the intensity of the stimulus (a real number) and ε_i a random variable. The underlying hypothesis is that the individual *adds* the objective and measurable aspects of the decision process, V_i, and the irrational or inaccurately measured aspects (of alternatives), ε_i. This assumes that all factors in the decision process can be measured in the same units and added, which does not fully capture reality from a psychological point of view. Though there is a tradeoff between the characteristics of the choice: two routes with different travel times and monetary costs may be equivalent, yet it is not clear that rational (and measurable processes) and irrational processes can be added. Irrationality can be due to the fact that the individual is not willing or unable to accurately evaluate the alternative he or she is facing, or, he or she is not willing (or unable) to compare the alternatives. Such limitations are not biases that are likely to decrease when the individuals learn. Rather, the individual is not a single-task, reliable operator, but a multitask, somewhat unreliable decision maker.
3. The choice procedure assumes that the basic model is deterministic and that the probabilistic components are added to take into account the fact that observed behavior is not fully predictable. An alternative approach was introduced by Luce (1959) who assumed that individual decision processes are genuinely probabilistic. He then introduced some intuitive axioms concerning the choice probabilities. Using this axiomatic approach, Luce derived the multinomial logit model which is the most popular DCM.

4. Finally, DCMs assume that individuals are isolated decision makers. In transportation, individual decisions may be indirectly related since a particular driver will for example select a route as a function of the congestion level created or induced by other drivers. Decisions are also directly interrelated (band wagon or word-of mouth effects).

The first application of DCMs in transportation dates back to Warner (1962) who developed a model of binary choices between car and bus. The immense success of DCMs in transportation (but also in marketing and recently in economics) is due to McFadden (e.g., 1973; 1978) who developed the econometric methods for the estimation of multinomial logit and other more general classes of DCMs. In particular the class of Generalized Extreme Value (GEV) distributions plays an important role by providing a formalized version of the nested model as introduced by Ben-Akiva (1973; see also Ben-Akiva and Lerman, 1985). The recent "theory of mixed models" is rather powerful because it shows that any random utility model can be approximated by a combination of GEV or logit models (cf. Theorem 3, in McFadden, 1996).

The theory underlying DCMs displays strong connections with well-established models used in economics. These formal connections are summarized in Anderson et al. (1992). First, DCMs can be formally represented in a spatial setting, as introduced by Hotelling (1929) in spatial competition models, Lancaster (1971) in industrial organizations, and Hauser and Shugan (1983) in marketing. Second, there is a representative consumer approach such as the constant elasticity of substitution (CES) (see Dixit and Stiglitz, 1977 and Spence, 1976) corresponding to a large class of DCMs (random utility function (RUM) models). A tenable hypothesis is that the utility functions are linear and additive in income. Relaxing this hypothesis is cumbersome and does not guarantee that aggregation is possible. As such, DCMs play a central role given that they provide a missing link which allows us to establish a connection between spatial models and representative consumer models.

The use of DCMs in the context of imperfect competition has been gaining momentum in economics. The researcher in transportation should be interested in those topics since deregulation in transportation implies that demand models will be more and more often embedded in a competitive framework. Recent theoretical developments have shown that existence of a price equilibrium is guaranteed by the logconcavity of the error term. This property appears to be very useful in many theoretical models. Moreover, Anderson, de Palma, and Nesterov (1995) have used the logconcavity of the error terms to prove that the market tends to over-provide variety. Several variants of DCMs have been used in industrial organization and in location theory (de Palma, Ginsburgh, Papageorgiou, and Thisse, 1985), as well as in the automobile and telecommunication industry (see, e.g., the empirical contribution of Berry, Levinsohn, and Pakes, 1995). Finally, we should note that DCMs have also been used as practical tools in experimental game theory (e.g., Anderson, Goeree, and Holt, 1997).

The flexibility of DCMs and their econometrics potential as well as theoretical properties imply that they provide very useful tools to model demand. Combination of disaggregate and aggregate data is also we believe a very fruitful (and important) avenue for further investigation.

Weakness of Economics in Modeling Supply

Economics and econometrics have provided a large body of literature to transportation science to model demand. The research on supply, on the other hand, is much less. For example, despite progress in network analysis, economists still pay very little attention to the modeling of congestion technology in a realistic way (see Arnott, de Palma, and Lindsey, 1993; Arnott and Small, 1994; Small, 1992; Mohring, 1994; Vickrey, 1994). This is, among other things, because (1) the engineering aspects are difficult to model and therefore usually ignored, and (2) the numerical solutions of models are not well received in economics because the discipline is not traditionally associated with numerical analysis. Though economists were active a few decades ago in operations research, it now constitutes a field by itself with few links to economics.

Rationality

Since the ultimate goal of research in social sciences is to develop high-level abstractions or theories, it is necessary to examine their foundations. Economic theory plays an important role and provides a potential tool for transportation science (for the time being more often misused than used). Economic theory relies on the concept of rationality which deserves some further thoughts. Is it an acceptable paradigm for transportation scientists?

Psychology, sociology and marketing science are often based on empirical fact, which may have intuitive explanations rather than theoretical justifications. On the other hand, economics relies much less on common sense and intuition (perceived as non-scientific) and uses a more formal and rigorous, internally consistent description of human behavior. Models also tend to be presented by economists as universal and capable of capturing an increasingly large number of unexplained processes studied in other related disciplines.

The economic literature describes human behavior from the point of view of operators who use the resources available (money, power, information, etc.) in their own interest. Agents are assumed to be selfish and to be able to maximize their objectives: profits for the firms and utility (or intertemporal flow of utility) for consumers. Rationality is often presented by economist as a necessary assumption and even as the only scientific approach in social sciences. The maximization axiom is probably inherited from physics where various phenomena have been explained in terms of the outcome of a maximization principle: the principle of maximum as used in optics (Fermat principle) and mechanics (with the idea of potential), for example. The focus of economics is ex-post behavior of the agents, in terms of simple principles based on static or dynamic optimization and prediction. This view of the world is not neutral and certainly not a tautology. It has been challenged by several psychologists and more and more also, by economists themselves. Probably the most significant attempt to modify the maximization principle is due to Simon (1955) who introduced an alternative way to explain human behavior, based on the concept of bounded rationality and limited ability to process information. (For a discussion of such heuristic choice models, see de Palma, Meyer, and Papageorgiou, 1994).

A behavior is rational if there exists a formal justification explaining why the agent behaves the way he or she does. There are various theories that explain behavior but, those based on maximization and rationality have played a dominant role. The economic approach relies on several strong assumptions. Among those assumptions, the following ones appear to be crucial: (1) The individual knows his or her preferences and the sets of alternatives; (2) The individual is able to measure the variables related to his or her preferences and to the environment; (3) The individual is able to have a global representation of his or her preferences and a global representation of the variables involved in the problem; (4) The individual can grasp the concepts and tools necessary to formulate and solve his or her problem; (5) The individual can optimally choose the coarseness required for measurement and computations to be done; (6) Often it is assumed that the individual has a perfect ability to compute and to store/retrieve information; and (7) The individual is able to explain qualitatively the various dimensions underlying a problem.

All these assumptions have been challenged empirically. Often tasks are far beyond the capability of a human being (and of a computer), that is they are rather unacceptable from a behavioral point of view.

Assumption 1 has been criticized on the grounds that preferences are changing over time. For example, the reply to the question: "do you prefer white wine or red wine?" depends on past consumption, the food that may be consumed, the current mood, the choice of beverage of other people, etc. On the other hand, since the world is changing over time, preferences are also changing constantly. Assumption 2 is dubious because even in the best case, measurements are subject to errors. In the worst case, relevant variables are not even observable. The individual has to deal with this complicated situation on an ′a priori basis. Assumption 3 is also unreasonable because the human mind cannot simultaneously process the various factors related to tastes and preferences; typically some factors will play a more important role than others. Assumption 4 implies that the concepts, algorithms, and various procedures used to solve a problem are all well understood by the problem solver; this assumption is unreasonable from the psychological point of view. Computers, like humans, have a limited ability to grasp abstract concepts. Assumption 5 may be questioned based on the following reasoning: A rational model relies on certain assumptions, namely objective, exogenous and endogenous variables/processes, optimization tools used, etc. The choice of the level of coarseness in the approach has to be determined by the individual. This process is usually either ignored (implicitly treated) or partially discussed. For the time being, there is no way to rationally perform this task since it itself requires cognitive and memory efforts to be computed optimally (see Mongin and Walliser, 1988). Assumption 6 would imply infinite computing ability and unbounded memory which is also unreasonable. Assumption 7 is unreasonable because it is impossible to specify all possible future states of the world.

Rationality has been discussed above in terms of (ex-post) explanations of human behavior. In this approach, the idea is to be able to identify a rational explanation which is compatible with the observed behavior. Does rationalization correspond to a social convention about what is an acceptable explanation by the economist and what is not? If the ingredients of a decision can only be attributed partially to logic and could not be used within a preset rational framework, the decision process itself is and will never be rational. The rational decision maker would have to compute the optimal rule to use and the optimal amount of information to collect (before this, he or she would have to determine the optimal set of search rules from which to select, etc.). For the time being, the decision-making process is still an open process, in the sense that one cannot provide a closed description of it. Rational models cannot be produced rationally: any model relies on implicit or explicit assumptions outside the scope of the model (see the argument of Goedel).[4] The human brain, in contrast to the computer, makes mistakes which require that humans constantly adapt themselves to new situations. In other words, the decision-making process uses logical elements but at the same time has emotional

---

[4] Goedel's theorem published in 1931 shows that any system of axioms is essentially incomplete. For example, Goedel showed that the meta-mathematical statement: "calculus is consistent" cannot be proved within calculus. In other words, for a given set of axioms, there always exists statements which cannot be derivable for those set of axioms. (Interestingly, Goedel's proof relies on the fallacious but fruitful "Richard Paradox" first discussed by Richard in 1905). For a excellent nontechnical presentation of this theorem, see Nagel and Newman (1958). See also Ladrière (1957) for a philosophical discussion of the scope of an axiomatic approach.

components which are typically irrational. Hence, there is a mix of rational and irrational ingredients. The nature of the boundary between the rational and the irrational one remains somewhat obscure.

Rationality can also be used to construct a predictive model. But here again, the limitations discussed above play an important role. The results are less satisfactory at the predictive level as they are usually too ambitious: no economist can predict the price to be set for a new product for example. Practitioners in marketing, who use a body of knowledge that is less formalized can probably do a better job, combining cumulative knowledge, past experience, and judgment. Moreover, behavior can be explained using other approaches, which attribute a larger role to emotion, motives, cultural and social constraints or simply trial and error. It is clear that a stock market is mainly possible because the expectations of the agents differ: some agents sell their stocks because they believe that the price of those stocks will decrease, while others buy because they have the opposite expectation.

Though the rationality approach can be criticized on many grounds, it does provide a reasonable (though inocmplete) way to describe individual behavior. While rationality provides a safe framework for modeling, it cannot be regarded as the panacea of modeling. Claiming for example that there is one way to build rational models and many ways to build irrational models, is just not true and not enough. Such traditional defense by the economics profession is somewhat undermined because, as we have seen, there is a set of implicit assumptions which defines the scope of the model. There is insufficient justification to dismiss all other descriptions of human behavior, although any departure from rationality should be based on strong empirical and theoretical foundations. Moreover, various economic models, especially those in game theory, may display multiple equilibria which restrict the scope of rational models. Criteria for equilibrium selection have been proposed but are often based on some arguable requirements.

Clearly, the discussion on rationality is still open. For the time being we propose the following agenda for further discussion:

1. Rationality is a possible ex-post explanation of human behavior.
2. It is not possible to have a closed and rational model. Assumptions underlying a rational model are in fact irrational, that is, model production and external consistency are outside the scope of rationality.
3. Rationality can be justified in specific repetitive situations, as the outcome of a heuristic procedure which involves trial and error strategies as well as the imitation and transmission of knowledge between agents.

4. Competing with rationality are various alternate models based, for example, on impulse actions, emotion, routine, and implicit behavior, etc. In real-world situations, decisions are the result of a subtle mix of emotional and rational ingredients. Rationality provides a possible, albeit narrow, description of human behavior.

A model may be regarded as an analog which allows us to develop better intuitions about the world. A model is constructed from a set of assumptions. These assumptions provide the framework from which to operate or estimate the model which can then be compared with observable phenomena or data. It may be difficult to test the assumptions of the model. It may be difficult to know what to test. For example, there are various models that describe the behavior of investors. It is unlikely that we will be able to recognize ourselves in these investors models. Indeed, many factors may influence an investor's decisions: habit, information provided by a friend, positive or negative past experience, etc. This is not to say that financial markets are inefficient. As can be shown theoretically, a market may behave as if all investors were rational even if a fraction of them are not. Similar reasoning applies in transportation. Take the example of two routes in parallel with one origin and one destination. Assume that a fraction of the drivers select either route randomly and that the rest of the drivers select the shorter route. Aat the user equilibrium, the travel time on the two routes will be the same if the number of random drivers is not too large. Similarly, random variations are likely to cancel out. Given the law of large numbers, a system may behave *as if* all agents behave in a deterministic and rational way[5] even if many or furthermore even all agents display irrational behaviors. That is, individual irrationality does not necessarily translate into aggregate irrationality.

Complexity

Even if economic models have a well defined structure, their approach could be easily criticized on the ground that they are unrealistic. The "as if" argument provides one line of defense but we may wish to have an explanation of human behavior that are at least compatible with observations. Interestingly, the theoretical models and some business

[5] A similar example can be found in the study of ants' behavior (see in particular Deneubourg, Pasteels, and Verhaegue, 1983. Ants exhibit a probabilistic behavior which looks almost random from the human point of view. The ant society in its entirety is nevertheless able to perform well-structured tasks such as foraging, the building of nests with periodic structures, and keeping the nest's temperature stable up to 0.1 degrees. This is even more remarkable since the queen in the ant colony plays only a marginal role, and does not rule the society. In those cases structure and orders are the outcome of a self-organizing phenomenon and not the outcome of any planned strategies or outcome "decentralized" by information on prices which is indeed available to *all* agents.

practices have been criticized even by the business community on the grounds that they are too complex. Complexity is a construct which is intuitively easy to understand but difficult to formalize[6]. Some attempts have been proposed to tackle this problem in economics (see Mongin and Walliser, 1988). Roughly speaking, the complexity of a model is a function of the number of elementary computations required and amount of memory required to solve the problem. This definition is incomplete since it does not take into account the difficulty of the concepts involved. Additionally, many concepts are difficult to grasp when they are still in the process of taking shape, although they may wrongly appear to be trivial for some. Finally, complexity depends on the context of the phenomena in question and is therefore difficult to quantify.

We may wonder how economic or social agents are able to solve problems that researchers are themselves unable to describe with a formal model. For example, drivers know how to drive their car (usually) and how to select their itinerary. Yet, researchers are still unable to model driver. That is not to say that when aggregated, the movement of cars would not be possible to model (indeed, even very simple cellular automata may be displaying realistic properties at the aggregate level). Developing more complex models is not the solution. Rather, a new paradigm is likely to be necessary.

Models have two purposes. First, they are to be seen as black boxes which relate inputs to outputs. This is the cybernetic approach introduced by Wiener, von Neumann, and Ashby (see, e.g., von Neumann, 1958). This approach can then accommodate a formal description inside the box since the outcome is only evaluated in terms of input and output and not as a function of the processes used to bridge the gap between these two ends. The second approach is more phenomenological and is in line with the reasoning developed by the philosopher Husserl (1950). It relies on the direct observation of the behavior (or phenomenon) and tries to derive a close description of models which can be compared to the real world at each step of the proposed description.

---

[6] In more formal fields, such as operations research, "complexity" is fairly well-defined (cf. e.g., the definition of the complexity of an algorithm).

DIFFICULTIES IN IMPLEMENTING THEORIES

Change in Scale

A theoretical model in social sciences usually only provides qualitative predictions. There is still a long way to go between a model's capacity to predict and its applications. One main problem lies in the change in scale. For example, simple networks are found in various analytical models: one route, one intersection, one bus line, etc. (see, e.g., de Palma, 1992). It is not clear how the results can be extended to an entire network with several interrelated intersections (see de Palma, Marchal, and Nestorov, 1997). The insights provided by a simple model may just not suffice when more realistic situations are considered.

Theory has received some recognition among practitioners because it provides useful guidelines for reasoning. Such a process, combined with non-rigorous, non-rational, implicit rules, can increase the relevance of modeling to reality. However, given the limited focus of theoretical ideas, the relationship between the theory and its application remains an open question.

In social sciences, there appears to be a trade-off between the internal structure, or the formal level of a theory, and its domain of application. A theory which is more formal and dictated by greater rigorous formalism is likely to have a lower degree of application. On the other hand, relaxing assumptions and allowing for more realistic behaviors may widen the scope of the theory. This is the case for DCMs. One starting point of the theory is that the assumptions of the theory are relaxed: for example, dismissing the assumption of transitivity as well as the belief that one can determine with certainty the behavior of an individual. Some scholars may be disappointed by the fact that this approach is less ambitious. However, such relaxation, while imposing less structure, also has a larger predictive power. This idea has been elaborated with the concept of approximate reasoning, as introduced by Zadeh (1978) and has led to the theory of fuzzy sets (see also the applications of fussy sets by Ponsard, 1983, in spatial economics). Unfortunately, there are few such examples so that the rigorous and rational approach remains the bench mark to evaluate a theory. Thus, theoretical works based on a "science of approximate reasoning" (Moles, 1990) remain to be developed. Similarly, approximate reasoning continues to be an open field to be further explored with tool specifically designed for social science.

General Equilibrium Effects

Transportation analysis in isolation provides a potentially biased view of the real world. While the researcher normally defines the boundary of the system that is being studied, it is not clear what the nature of the boundary should be. Is location theory a separate field from transportation science for example?

Some interests in interdisciplinary work have surfaced such as the attempt in economics to integrate concepts from various other disciplines in the study of the organization and processes of economic activities. Krugman (1991), in his new economic geography, tries to integrate geography and economics,by establishing microeconomics foundations for the geographic and spatial distribution of human activities. Similarly, the new urban economics has tried to integrate the areas of land use, location, housing, and local public finance, treated previously as separate topics. Such attempts are important because they question the robustness of general equilibrium effects which may play an important role in the long run. For example, the construction of a new road cannot be evaluated using a conventional cost-benefit analysis alone, while ignoring the change in the price of land and the relocation of residential and economic activities that it induces.

A growing concern in the EC relates to the interaction between land use and transportation. The decentralization of business and residential activities, for instance, will mean that such trends should increasingly influence the future of public transportation research in Europe. Clearly any long-run cost-benefit analysis which ignores the impact of relocation will strongly bias the results. In this respect, several softwares now exist which address questions of interfaces between various areas of study, such as the integration of transportation and land use. TRANUS, for example, which was developed in Venezuela, has now been applied in the EC context. (The model is however not without its limitations, such as treating transportation as a static phenomenon.) Because of the integration of transportation and the location of economic activities, such an approach should become more useful for long-run cost-benefit analysis. The bigger challenge remains, however, to narrow the gap between theory and day-to-day decision making in transportation.

Political Constraints

One reason for the gap between theory and practice lies in political constraints. To take an example, a cost-benefit analysis may demonstrate that it improves welfare to retain certain projects. Yet these projects may be politically unacceptable to parties. One way to circumvent such a problem is to try to incorporate the behavior and reactions of relevant parties, namely voters, into the model. Assuming that voters decide on the basis of their best interests, we could simulate and compute voting behavior thereby allowing us to define a set of politically acceptable projects. Specific questions such as what the true preferences are with respect to individual valuations of public goods (i.e., roads) still have to be studied more extensively.

Coalitional Conflicts

The society is made up of a complex of various actors and coalitions whose receptiveness or resistance to transport proposals or policies will need to be taken into account. In this case, their inputs during the research process may help to construct a more realistic model. In a recent study, a team of us solicited the views of six actors/coalitions in the formulation of regulations pertaining to the taxi cab industry in Geneva. The six actors/coalitions were (1) taxi drivers, (2) the company running taxi cabs, (3) the provider of information, (4) customers, (5) the transportation department, and (6) pressure groups. The program involved an iterative process where the various actors listed were consulted before and after the formulation of the proposed regulations. Within the constraints of local and state laws, an economic-engineering model was developed which incorporated the implicit negotiating process. Such a program expends more time and energy but may also be more successful if the recommended policies derived from the theoretical model are to be palatable to the affected parties.

The above suggests that the role of pressure groups and what Olson (1965) has termed distributional coalitions deserve special attention. Interest-group politics are widespread in many societies and the acceptance of a policy may well depend on the resolution of state and coalitional tensions. For example, road pricing may be socially beneficial yet rejected by various groups because of redistribution problems. In this regard, models in the public-choice theoretic tradition offer an understanding of state-coalitional behavior but are yet to be adopted in the transport literature.

Socio-Cultural Context

The increased complexity of transport phenomena implies that any study of the latter cannot be contextually divorced as if it occurs in a socio-cultural vacuum. Implicit norms and interactions within a socio-cultural setting also affect the behavior of various economic agents. For instance, driving habits differ from country to country. Such implicit norms have received scant attention in transport modeling, yet they are important in that they could lead to multiple equilibrium regimes.

TRANSPORTATION PLANNING AS A HEURISTIC PROCESS

Understanding and practicing transportation can be learned but relies also on experience. As early as the seventeenth century, the French philosopher Descartes proposed a systematic way to approach scientific problems in his famous "Discours de la méthode" (Descartes, 1826). His emphasis on rigor is widely practiced in transportation research, yet it is not true at all that the search for a solution associated with an algorithm is completely driven by rigor. This may be true of a medical doctor attempting to diagnose the nature of an illness. But a researcher often has to work within constraints beyond his or her control such as time constraints, budget constraints, data availability, team performance, etc. Intuition seems also to be outside the scientific methodology. Furthermore, there is no guarantee that a more rigorously derived model yields superior results or outcomes.

CONCLUDING REMARKS

More work that establishes a bridge between theory and practice is necessary. But empirical models which have a strong theoretical background do not automatically translate into "added value" compared to models that are developed on an ad-hoc basis. Packaging and communicating our research to the pertinent audience does have a more and more important role. The government has to provide the right incentives to universities and consulting companies in order to be informed about the consequences of potential investment for private and public transportation. This question is related to the generic problem of costly monitoring which is being studied in banking, in the audit industry, and the like. Such problems seem to be mostly neglected in the transportation field.

Furthermore, researchers need to spend enough time understanding the complexity of real-life problems. Unfortunately, a theoretical paper which addresses the essential tradeoffs in a practical problem does not guarantee publication. Prevailing publication culture continues to emphasis mathematical elegance over such considerations, and the trend within various disciplines is to move towards greater specialization rather than integration.

Many of the issues raised in this chapter remain unsolved and may leave the reader with a bitter taste. It seems to me that while transportation research is application driven, it cannot ignore the underlying epistemological processes that influence and rule it. On the contrary, because the goal of transportation research is to explain and provide useful tools for understanding and managing the flow of passengers and goods in society, the researcher needs to be aware of his or her historical evolution and patterns. Perhaps more discourse will ensue from the concerns raised in this chapter.

ACKNOWLEDGEMENTS

I would like to thank Richard Arnott, Moshe Ben-Akiva, Jacques Drèze, Daniel McFadden, Robin Lindsey, Philippe Mongin, Emile Quinet, as well as Thomas Laitila, Tommy Gärling and Kerstin Westin for comments. The theoretical foundations of discrete choice models were jointly developed with Simon Anderson and Jacques-Francois Thisse. The intellectual influence of Abraham Moles and Isabella Stengers should also be acknowledged. Finally, I would like to thank the organizers as well as the participants of the Stockholm conference on theoretical foundations of travel choice modeling for many stimulating discussions.

REFERENCES

Anderson, S. A. and A. de Palma (1997). *Firm pricing when consumers have imperfect discriminatory power.* Unpublished manuscript, Department of Economics, University of Virginia.

Anderson, S. A., A. de Palma and Y. Nesterov (1995). Oligopolistic Competition and Optimal Provision of Products. *Econometrica*, **63**, 1281-1301.

Anderson, S. A., A. de Palma and J.-F. Thisse (1992). *Discrete Choice Theory of Product Differentiation.* The MIT Press, Cambridge, MA.

Anderson, S., Goeree, J. and C. Holt (1997). *Rent Seeking with Bounded Rationality: An Analysis of the All-Pay Auction.* Discussion paper 283, Department of Economics, University of Virginia, Charlottesville, VA.

Arnott, R., A. de Palma and R. Lindsey (1993). A structural model of peak-period congestion: A traffic bottleneck with elastic demand. *American Economic Review*, **83**, 161-179.

Arnott, R. and K. Small (1994). The economics of traffic congestion. *American Scientist*, **82**, 446-455.

Ben-Akiva, M. (1973). *Structure of Passenger Travel Demand Models.* Doctoral dissertation, Department of Civil Engineering, Massachusetts Institute of Technology MIT, Cambridge, MA.

Ben-Akiva, M. and S. Lerman (1985). *Discrete Choice Analysis: Theory and Application to Travel Demand.* The MIT Press, Cambridge, MA.

Berry, S., J. Levinsohn and A. Pakes (1995). Automobile prices in market equilibrium. *Econometrica,* **63**, 841-890.

de Palma, A. (1992). A game-theoretic approach to the analysis of simple congested networks. *American Economic Review, Paper and Proceedings*, **82**, 494-500.

de Palma, A., V. Ginsburgh, Y. Papageorgiou and J.-F. Thisse (1985). The principle of minimum differentiation holds with sufficiently heterogeneity. *Econometrica,* **53**, 763-781.

de Palma, A. and C. Lefèvre (1981). Simplification procedures for a probabilistic choice model. *Journal of Mathematical Sociology*, **6**, 43-60.

de Palma, A., F. Marchal and Y. Nesterov (1997). METROPOLIS a modular architecture for dynamic urban simulation. *Transportation Research Board, Paper and Proceedings*. In press.

de Palma, A. Meyers and Y. Papageorgiou (1994). Rational choice under imperfect ability to choose. *American Economic Review,* **84**, 419-440.

Descartes, R. (1826). *Règles pour la direction de l'esprit, in Oeuvres complètes*. V. Cousin, t. XI.

Deneubourg, J. L., J. M. Pasteels and J. C. Verhaegue (1983). Probabilistic behavior in ants: A Strategy of errors? *Journal of Theoretical Biology*, **105**, 259-271.

Dixit, A. and J. E. Stiglitz (1977). Monopolistic competition and optimum product diversity. *American Economic Review,* **69**, 297-308.

Edgell, S. E. and W. S. Geisler (1980). A set-theoretic random utility model of choice behavior. *Journal of Mathematical Psychology*, **21**, 265-278.

Hauser, J. R. and S. M. Shugan (1983). Defensive marketing strategies. *Marketing Science*, **2**, 319-360.

Hotelling, H. (1929). Stability in competition. *Economic Journal,* **39**, 41-57.

Husserl, E. (1950). Idées directrices pour une phénoménologie. *Nouvelle Revue Francaise*, Paris.

Krugman, P. (1991). *Geography and Trade*. The MIT Press, Cambridge, MA.

Ladrière, J. (1957). *Les limitations internes des formalismes*. Université Catholique de Louvain, Belgium.

Lancaster, K. (1971). *Consumer Demand: A New Approach.* Columbia University Press, New York.

Luce, R. D. (1959). *Individual Choice Behavior: A Theoretical Analysis*. Wiley, New York.

May, R. (1976). *Stability and Complexity in Model Ecosystems*. Princeton University Press, Princeton, NJ.

McFadden, D. (1973). Conditional logit analysis of qualitative choice behavior. In: *Frontiers in Econometrics* (P. Zarembka, ed.), pp. 105-142. Academic Press, New York.

McFadden, D. (1978). Modelling the choice of residential location. In: *Spatial Interaction Theory and Residential Location* (A. Karlquist, L. Lundqvist, F. Snickars and J. W. Weibull, eds.), pp. 75-96. North-Holland, Amsterdam.

McFadden, D. (1996). *On Computation of Williness-To-Pay in Travel Demand Models.* Unpublished manuscript, Department of Economics, University of California, Berkeley, CA.

Moles, A. (1990). *Les Sciences de l'imprécis.* Collection Seuil. Paris.

Mongin, Ph. and B. Walliser (1988). Infinite Regressions in the optimizing theory of decision, In: *Risk Decision and Rationality* (B. R. Munier ed.), pp. 435-457. Reidel, Dordrecht, The Netherlands.

Nicolis, G. and I. Prigogine (1989). *Exploring Complexity.* Freeman, New York.

Olson, M. (1965). *The Logic of Collection Action.* Harvard University Press, Cambridge.

Ponsard, C. (1983). *Fuzzy Data Analysis in a Spatial Context.* Research Paper, IME, University of Bourgogne, Dijon, France.

Simon, H. (1955). A behavioral model of rational choice. *Quarterly Journal of Economics,* **69**, 99-118.

Small, K. (1992). Urban transportation economics. In: *Fundamental of Pure and Applied Economics, 51.* Harwood, Chur.

Spence, M. (1976). Product selection, fixed costs, and monopolistic competition. *Review of Economic Studies,* **43**, 217-235.

Thurstone, L. L. (1927). A law of comparative judgement. *Psychological Review* **34**, 273-286.

Vickrey, W. (1994). *Public Economics, Selected papers* (R. Arnott, K. Arrow, A. Atkinson and J. Drèze, eds.). Cambridge University Press.

Warner, S. L. (1962). *Strategic Choice of Mode in Urban Travel: A Study of Binary Choice.* Northwestern University Press, Evanston, IL.

Weber, S. (1992). On hierarchical spatial competition. *Review of Economic Studies,* **59**, 407-425.

Zadeh, L. (1978). Fuzzy sets as a basis for a theory of possibilities. *Fuzzy Sets and Systems,* **1**, 3-28.

3

TIME AND INCOME IN TRAVEL CHOICE: TOWARDS A MICROECONOMIC ACTIVITY-BASED THEORETICAL FRAMEWORK

Sergio R. Jara-Díaz, University of Chile, Santiago, Chile

INTRODUCTION

The main objective of this chapter is to summarize the microeconomic foundations of discrete mode-choice models emphasizing the role of time and income. The central idea is to highlight both the genesis and properties of such models, using economically meaningful constructs such as the marginal utility of income or the subjective value of time. This will be shown to have important consequences for their correct specification.

The second objective of the paper is to expand the theoretical framework in order to encompass all travel decisions. To achieve this, the evolution of time-related theories of consumer behavior is described; the need to understand travel as part of a general activity framework is thereby highlighted.

DISCRETE CHOICES OF TRAVEL MODE

Disaggregate choice models are the most popular type of models to apply to travel mode choice. The most important element is the (mode-specific) utility, usually represented as a linear combination of cost and characteristics of each mode, and socio-economic variables for each group of individuals. In this approach it is assumed, for each individual type, which variables determine the level of non-random utility associated with each discrete alternative. This poses many questions regarding model specification: the structure of decisions, the distribution of the unknown portion of utility, functional form of the observable part, type and form of variables that should be used, and criteria to decide which group of individuals will be regarded as "alike."

The choice of the word utility to describe the equation that represents the level of satisfaction associated with each mode, is not arbitrary. It is in fact borrowed from the terminology of microeconomics, a discipline that provides a theoretical framework to understand and specify choice models. The objective in this section is to expose this theoretical foundation in order to understand the role of income, time, characteristics, preferences, etc. on choice of travel mode. Two caveats should be noted. First, the primary sources of utility will not be examined (i.e., the psychological mechanisms that make consumption or activities pleasurable). Secondly, in order to avoid confusion it is important to stress that what is called utility of a mode in discrete choice models is in fact a conditional indirect utility function that already includes the role of the constraints faced by the individual as well as a first level of decisions. Thus, although the genesis of direct preferences are not analyzed, the formation of mode specific utility is the focus of the following synthesis.

Quality and Income in Discrete Choice Models

The traditional microeconomic theory of consumer behavior is stated in terms of a bundle of continuous goods $\boldsymbol{X}$ which are chosen by the individual in an attempt to obtain the maximum level of satisfaction, within all possible bundles allowed by his or her purchasing power. After Lancaster (1966), who introduced the notion of goods characteristics as the primary source of utility, the level of satisfaction can be stated in terms of those characteristics (e.g., flavor, nutrient, warmth, beauty); accordingly, the problem of choice can be understood properly accounting for the fact that characteristics can be obtained through the purchase of market goods which in turn require money.

In order to account for discrete choices, the theoretical framework needs to be slightly modified. Choices are discrete when decisions are made to acquire one unit of a certain good (e.g., a trip) followed by the choice of a specific type (e.g., a mode). Then the consumer can be viewed as choosing both the amount of the continuous good and one of the discrete alternatives (travel mode), each one described by a vector $\boldsymbol{Q}_j$ with its qualitative characteristics. Formally (adapting from McFadden, 1981), an individual is assumed to behave as if

$$\underset{x,j}{\text{Max}}\ U(\boldsymbol{X},\boldsymbol{Q}_j) \tag{1}$$

subject to

$$\Sigma P_i X_i + c_j \leq I \qquad j \in M \tag{2}$$

where P_i and X_i are the price and quantity of good i respectively, c_j is the cost of using travel mode j, I is income, and M is the set of alternative travel modes.

The problem can be solved for $\boldsymbol{X}$ conditional on the discrete choice j, obtaining conditional demands $X_i(\boldsymbol{P}, I - c_j, \boldsymbol{Q}_j)$. Once these are replaced in U, the resulting conditional indirect utility function $V(\boldsymbol{P}, I - c_j, \boldsymbol{Q}_j) \equiv V_j$ represents the maximum utility the individual can obtain if travel mode j is chosen. Then the preferred travel mode will be that which fulfills $V_j > V_i \;\; \forall\, i \neq j$. This means that not necessarily all arguments in V will actually influence travel-mode choice. The portion of V_j that determines the result of the discrete comparison is a truncated utility $\overline{U}_j$. From the derivation of the conditional indirect utility function V_j, it follows that

a) Marginal utility of income (MUI) $= \dfrac{\partial V}{\partial I} = -\dfrac{\partial V}{\partial c_j}$ (3)

b) Subjective value of characteristic k (SV_k) $= \dfrac{\partial V / \partial q_{kj}}{\partial V / \partial c_j}$ (4)

c) Travel-mode utility is in fact a truncated conditional indirect utility function.

To illustrate this, let us represent V_j with a linear function, that is

$$V_j = \alpha + \sum_i \beta_i P_i + \sum_k \gamma_k q_{kj} + \lambda(I - c_j) \tag{5}$$

Which one is the largest value for V_j among all $j \in M$ will depend only on the characteristics in $\boldsymbol{Q}_j$ and the cost c_j (all other terms cancel out when comparing V_i and V_j). Thus, the relevant part of V_j for discrete-choice modeling is

$$\overline{U}_j = -\lambda c_j + \sum_k \gamma_k q_{kj} \tag{6}$$

Equation (6) justifies the usual linear in cost and time (and other variables) specification of travel-mode utility, so frequent in applications. According to equation (3), the MUI is minus the coefficient of cost, and the SV_k is simply the ratio of the corresponding quality coefficient over the MUI. The simplicity of the model has a cost: income plays no role in the discrete choice. In fact, McFadden's (1981) AIRUM (additive income random utility maximizing) model structure yields choice probabilities that are independent of income.

A very simple extension of the usual linear utility model permits a much better understanding of the role of income (Jara-Díaz and Videla, 1989). For simplicity only, assume that the utility function U in (1) is separable in $\boldsymbol{X}$ and $\boldsymbol{Q}_j$. This implies that the level of satisfaction attained from consuming a bundle $\boldsymbol{X}$ is independent of travel mode characteristics, that is

$$\frac{\partial^2 U}{\partial x_i \partial q_{jk}} = O, \ \forall i, \ \forall j,k \tag{7}$$

Under the separability assumption, we can write the utility function as

$$U\{\boldsymbol{X},\boldsymbol{Q}_j\} = U_1(\boldsymbol{X}) + U_2(\boldsymbol{Q}_j) \tag{8}$$

The maximization problem on $\boldsymbol{X}$ has a solution that is conditional on c_j alone, yielding a set of functions $X^*(\boldsymbol{P}, I-c_j)$; once they are replaced in $U_1(\boldsymbol{X})$, a partial indirect utility function is obtained, that is

$$\underset{X \in x}{\text{Max}} \ [U_1(\mathbf{X}) | \mathbf{PX}^T \le I - c_j] \equiv V_1(\mathbf{P}, I-c_j) \tag{9}$$

Thus, the conditional indirect utility function is

$$V(\boldsymbol{P}, I-c_j, \boldsymbol{Q}_j) = V_1(\boldsymbol{P}, I-c_j) + U_2(\boldsymbol{Q}_j) \tag{10}$$

The role of income involves V_1 only. Assuming that prices of continuous goods are constant, V_1 can be approximated by a Taylor expansion around $(\boldsymbol{P}, I)$, that is

$$V_1(\boldsymbol{P}, I-c_j) = V_1(\boldsymbol{P}, I) + \sum_{i=1}^{n} \frac{1}{i!} V_1^i(\boldsymbol{P}, I)(-c_j)^i + R_{n+1} \tag{11}$$

where V_1^i denotes the ith derivative of V_1 with respect to $I-c_j$ evaluated at I, and R_{n+1} represents terms of order $n+1$ and higher. If a Taylor expansion to the order n is assumed to be sufficiently accurate, then R_{n+1} is close to zero; therefore, V_1^n is a function of $\boldsymbol{P}$ only. Thus, V_j is given by

$$V_j = V_1(\boldsymbol{P}, I) + \sum_{i=1}^{n-1} \frac{1}{i!} V_1^i(\boldsymbol{P}, I)(-c_j)^i + \frac{1}{n!} V_1^n(\boldsymbol{P})(-c_j)^n + U_2(\boldsymbol{Q}_j) \tag{12}$$

This shows that travel-mode choice depends on the level of individual income for $n \geq 2$, since at least one term of the form $V_I^i(\boldsymbol{P},I)$ will appear. Comparing $V(c_i,\boldsymbol{Q}_i)$ to $V(c_j,\boldsymbol{Q}_j)$ may yield a different result for different levels of income. In other words, if the best specification for V_j involves terms in c_j of order two or higher, then income influences travel-mode choice.

This theoretical framework has been applied to a middle-low income corridor in Santiago, Chile. Travel-mode utility was specified using linear and quadratic terms in cost for the whole sample; as the quadratic term was statistically significant, the sample was divided into four homogeneous income groups and travel-mode choice models were estimated using the second-order specification. Within each subsample, the quadratic cost term was significant only for the three poorest groups, with its level diminishing with income, which implies that the influence of income on choice was reduced as purchasing power increased.

In general, the *MUI* can be calculated at an individual level as

$$\lambda = \frac{\partial V^*}{\partial I} = V_I^1(\mathbf{P},I) + \sum_{i=1}^{n-1} \frac{1}{i!} V_I^{i+1}(\mathbf{P},I)(-c_d)^i \tag{13}$$

where d stands for the chosen mode. Applying this to the described data, the *MUI* was as expected found to diminish with income.

It is worth mentioning that, although weakly justified, specifications including travel-mode cost and income can be found in the literature. The discrete choice theoretical framework suggest that they should be linked but usually the treatment of units makes this a fuzzy point (for a discussion of related matters, see Viton, 1985).

THE GOODS/LEISURE THEORETICAL FRAMEWORK

The preceding approach to modeling discrete choices is fairly general, that is it applies to most type of purchasing decisions when the choice has to be made among a family of qualitatively different goods. The transport-specific dimensions enter the picture when variables like the components of travel time (in-vehicle, waiting, and access times) are included in $\boldsymbol{Q}_j$. An obvious alternative for the modeling of an activity like travel is to include time in the framework from the beginning.

The analysis of travel choices within a theoretical framework of consumer behavior explicitly including time is a fairly natural extension of the early theoretical attempts to account for time as a "requisite" for goods consumption (reviewed in the next section). By 1970, Gronau adapted Becker's (1965) theory to model travel-mode choice including both time and income constraints, showing that the (discrete) decision depended on something that we now would call travel-mode utility, defined as a weighted sum of cost and travel time (see Gronau, 1986).

One of the most popular microeconomic approaches specifically aimed at understanding travel-mode choice, introduced modal travel time t_i and cost c_i as variables that influence utility through the impact on goods consumption G and leisure time L. This goods/leisure trade-off approach (Train and Mc Fadden, 1978) can be formalized as follows for the case of a single trip:

$$\text{Max } U(G,L) \tag{14}$$

subject to

$$G + c_i = wW + E \tag{15}$$

$$L + W + t_i = \tau,\ i \in M \tag{16}$$

where W is working time, w is wage rate, E is income from other sources and τ is total available time. By virtue of equation (15), working more (increasing W) means consuming more (G) reducing leisure (L), and vice versa. Thus, the trade-off between goods and leisure is captured by W. As in the previous problem, represented by equations (1) and (2), this can be solved in two steps, using W as a "pivot," replacing G and L as functions of W from (15) and (16) in (14). Then the optimal value of W can be found conditional on travel-mode choice (i.e., on c_i and t_i), which yields a conditional demand for working time W^* as a function of w, $E - c_i$ and $\tau - t_i$. If this is replaced back in the utility function, a conditional indirect utility V_i is obtained. Giving U the Cobb-Douglas form $KG^{1-\beta}L^{\beta}$, the result is

$$V_i = K(1-\beta)^{1-\beta}\beta^{\beta}[w^{1-\beta}(\tau - t_i) + w^{-\beta}(E - c_i)] \tag{17}$$

Again, travel-mode choice is made by comparisons among V_i, $\forall i \in M$. For a given individual, this approach yields choices governed by the maximum value of $-(c_i / w) - t_i$ or $-c_i - wt_i$.

It should be noted that what we have called the truncated conditional indirect utility function $\overline{U}_i$ corresponds in the case of equation (17) to an expression of the form

$$\overline{U}_i = K'w^{1-\beta}t_i - K'w^{-\beta}c_i \tag{18}$$

which is again linear in cost and time. Note that when $\beta \rightarrow 0$, then $K' = K$ and choice is determined by $-c_i - wt_i$; when $\beta \rightarrow 1$, then $K' = K$ and choice follows the maximum of

$-t_i - (c_i / w)$. This is the origin of the popular cost over income specification of travel-mode utility in which income is in fact a proxy for the wage rate.

The preceding model includes a rather strong assumption, that is that the individual can choose working time freely at a pre-specified wage rate. Nothing essential changes, however, if a fixed working schedule and a fixed income is introduced provided the individual works extra time at a marginal wage rate w. On the other hand, if no additional work is produced, then a model with exogenous income I is obtained. Under this specification, the trade-off between goods and leisure no longer depends on assigning more or less time to work but on choosing faster (and more expensive) travel-modes or slower and cheaper ones. In this model, the conditional indirect utility function is directly obtained replacing G and L from the constraints on U. If utility with a Cobb-Douglas form is approximated to a second-order Taylor expansion around (*I,T-W*), replacing G and L and rearranging terms, a truncated conditional indirect utility function is given by (Jara-Díaz, 1990)

$$\overline{U}_i = -\theta(1-\beta)\frac{c_i}{g} - \theta\beta t_i + \frac{1}{2}\theta\beta(1-\beta)(S_T - S_I)(\frac{c_i}{g} - t_i) \tag{19}$$

where g is an expenditure rate $I/(\tau - W)$, S_I and S_T are the share of income $\left(\frac{c_i}{I}\right)$ and free time $\left(\frac{t_i}{\tau - W}\right)$ spent in transport respectively, and θ is $Kg^{1-\beta}$ (note that $S_I \neq S_T$).

The expression for the travel-mode utility represented by equation (19) involves a number of novelties. First of all, if either S_I or S_T were significantly different from zero, then second-order terms in travel cost, travel time, or both, should be included in the specification. This is consistent with a previous observation regarding the role of income in travel-mode choice captured by second-order terms in cost, because (as should be consistent with standard theory of consumer behavior) a high share of income in the consumption of a particular good is indicative of the presence of an income effect. Secondly, if both S_I and S_T were negligible, a linear specification would be appropriate, keeping some resemblance with the previous version of $\overline{U}_i$ in equation (18) which involves the travel-mode cost over the wage rate; however, in this fixed income case cost is divided by an expenditure rate which represents the amount of money to be spent in a $(\tau - W)$ period (Jara-Díaz and Farah, 1987). We have labeled these specifications the wage rate and expenditure rate models respectively. Such specifications were empirically explored by Jara-Díaz and Ortúzar (1989). Note that a constant working schedule across the population in a sample of fixed income travelers would provide a clear case for the cost over income specification.

The generalized expenditure rate model represented by equation (19) helps clarifying an important point regarding the stratification of travelers for model estimation. Imagine that a traditional travel-mode choice model with linear utility is specified with the usual cost over income and time variables; assume as well that individuals in the sample have similar preferences (i.e., same K and β) but trips involve a variety of travel distances (or times). This means that individuals in the sample would have different values for S_T and, therefore, different coefficients for cost and time according to equation (19). Thus, different linear models should be estimated for individuals traveling short and long distances. In other words, the sample should be stratified according to travel distance.

The goods/leisure theoretical framework can be used to explore the relation between income and "pure" or unrestricted preferences, represented by the parameter β in direct utility. If second-order effects are assumed negligible in equation (19) and the first-order terms are conveniently manipulated, one obtains

$$\overline{U}_i = -Ag^{-\beta}c_i - Bg^{1-\beta}t_i \tag{20}$$

Travel-mode choice models can be estimated to obtain A, B, and β for populations with different income in order to examine possible monotonicity between the income level and the estimated β values. This approach was used by Jara-Díaz (1991) in a study involving two income groups within each of two corridors. The results did not show a correlation between income and the parameter representing unrestricted preferences.

Finally, the appeal of the goods/leisure theoretical framework goes beyond its simplicity. It can be adapted to cases like interurban travel or vacation trips to a resort area. Imagine an individual that is self-employed and whose vacations are planned, including destination, length of stay, and travel mode. In this case the existence of earnings per unit time and the endogenous decision on the length of time out of work play a key role in the specification of utility; the resulting model will be similar to a wage rate model. On the other hand, if the individual has a pre-specified vacation period, the expenditure rate approach (properly adapted) could be used, making the vacation budget play the role of fixed income.

Extensions and Discussion

One might be tempted to include here a discussion of the value of time but, although related, the emphasis is on the formation of the truncated conditional indirect utility function. Yet, one may observe that the goods/leisure theoretical framework yields a value of time equal to the (marginal) wage rate under its endogenous income version, but the result is different (and more interesting) if income is regarded as exogenous. In fact, for the Cobb-Douglas form of direct utility, the subjective value of time (*SVT*) is given by

$$SVT = \frac{\beta}{1-\beta} \cdot \frac{I - c_i}{\tau - W - t_i} \tag{21}$$

which is nearly proportional to the expenditure rate when c_i and t_i are negligible compared to income and leisure time respectively; in fact, to a first-order approximation, *SVT* is equal to $g\beta / (1-\beta)$ from the first part of equation (19). Note that, for a given income level, a person that works less has a lower value of time. This is consistent with empirical results like those obtained by Bates and Roberts (1986) regarding the low *SVT* for retired individuals. Also, note that *SVT* increases with t_i which means that the (marginal) subjective valuation of travel time increases with trip length. This is an important point as some claim that one more minute of a short trip would be perceived as more valuable than one more minute of a long trip; this ignores however the fact that what is valuable to an individual is leisure time, which is the complement of t_i. Thus, what matters is the perception of one minute relative to leisure, which diminishes as leisure increases, or *increases* with t_i.

In applications, any version of the alternative-specific utility function introduced here includes c_i divided by some form of income (e.g. wage rate, income itself or expenditure rate), all components of travel time, other (travel-mode) characteristics, socio-economic indexes, etc. Each variable has a parameter such that the *MUI* and SV_j are easily calculated using equations (3) and (4) respectively. As discussed, the *SVT* under the original version of the goods/leisure trade-off theoretical framework is equal to the wage rate, but this is rarely the result in empirical work in which the ratio of the travel time coefficient over the cost/wage coefficient is usually less than one (a result theoretically supported by arguments provided by Gronau, 1986). This is related to the formulation of the trade-off model in which the absolute perception of time is captured by the multiplier of the corresponding constraint that is the same for all activities included in *L*. Thus, the price of time is equal for all activities and equal to the wage rate. The case would be different if restrictions regarding time were identified beyond equation (16). One possibility is that of minimum time requirements, like those identified by Truong and Hensher (1985). On the other hand, a ratio significantly greater than one has also been obtained (Jara-Díaz and Ortúzar, 1989). In this latter case, an expenditure rate approach would accept such values as a possibility, as shown in equation (21) where $\beta / (1-\beta)$ can take any positive value for β lying within the interval $0 \le \beta \le 1$. Note that β represents the importance of time in direct utility which means that individuals with a large absolute perception of time would reveal a high *SVT* if the fixed income and fixed working schedule is the relevant one.

So far, it seems as if the main issue for the correct specification of (travel-mode) utility is the role of income, its endogeneity or exogeneity, depending on whether paid working hours are decided or not by the individual. In fact, time plays a key role which will be exposed in the next section.

From Goods To Activities: The Development of Time-Related Microeconomic Theory

Modeling travel choice introduces time and space in consumer theory. For a given location pattern, an individual has to choose what goods to buy and what activities to perform, including leisure, work and travel. The role of time began to be discussed between 1965 and 1972 in the microeconomic literature. The traditional theoretical framework of consumer behavior sees individuals as trying to achieve the highest level of satisfaction given the constraints they face. As the level of satisfaction was assumed to depend only on the amount of goods consumed, the natural constraint was that of a limited ability to purchase (budget constraints). The need to understand the labor market made it necessary to introduce time as an important element in that framework, as the consumer was assumed to face a choice between work and non-work time. The emphasis was on the relation between wage and the willingness to work (labor supply). In this context, an activity which is essentially time consuming as travel was also of interest. In this section, the microeconomic theories involving the assignment of time are reviewed and discussed, in order to make the way for the integration of the theory of travel choice with the theory of time allocation.

The Allocation of Time

Becker (1965) brought attention to the fact that market goods $\boldsymbol{X}$ are not consumed as they are bought but have to be transformed into "basic commodities" $\boldsymbol{Z}$ which require time to be prepared. Thus, as satisfaction comes from $\boldsymbol{Z}$, and each Z_i depends upon goods consumption X_i and preparation time T_i, then utility should be defined as $U(\boldsymbol{X},\boldsymbol{T})$. In Becker's theory, income is essentially an endogenous variable as the individual decides how many hours W to work at a pre-specified wage rate w. Thus, two constraints appear originally in his formulation: the traditional budget constraint which relates expenses in market goods to income wW, and a time constraint which states that working hours plus $\sum_i T_i$ should be equal to total available time. However, Becker turns the two constraints into one by noting that "time can be converted

Table 3.1
Time-Related Microeconomic Theories of Consumer Behavior

Framework Element	Traditional	Becker (1965)	Johnson (1966)	Oort (1969)	De Serpa (1971)	Evans (1972)
Utility	$U(X)$	$U[Z(X,\overline{T})]^a$	$U(B,G,W,L)$	$U(W,L,tt,I)$	$U(X,T)^b$	$U(T)$
Income Constraint	$P'X < I$	$\sum_{i,j}(P_j b_{ij} + T_i w) = \tau w$	$G + cB = wW$	$I = wW$	$P'X < I$	$P'OT = O$
Time Constraint	—	included in income	$W + L + tB = \tau$	$W + L + tt = \tau$	$\sum T_i = \tau$	$\sum T_i = \tau$
Technical Constraint	—	implicit in income	—	—	$T_i \geq a_i X_i$	$JT \leq O$ $(X = OT)$

a) $\overline{T}$ does not include work time.
b) X can include "pure time" commodities.

into goods by using less time at consumption and more at work" (pp. 496-497). The resulting constraint is stated in money terms where a full price for each generalized good Z_i appears; this full price encompasses the expenses on the necessary market goods plus a time cost given by wT_i which represents the value of forgone income, that is, the amount of money the individual would earn if he or she assigned T_i to work more (see Table 3.1). Both Johnson (1966) and Oort (1969) used the new time constraint in order to model and understand trip generation and the role of travel time respectively; both, however, included work time in utility.

DeSerpa (1971) developed a theory that resembles Becker's, as both goods and time are included as arguments in the utility function; however, the approach has also important differences. The first one is with the inclusion of all time components in the utility function in addition to goods, particularly working time which was explicitly excluded in Beckers's theory. The second difference is the addition of a series of constraints reflecting minimum time requirements to consume each good. DeSerpa's notation is not the most appropriate, as reflected by the need to introduce a number of concepts to explain potential limitations of the model (e.g., the concept of pure time commodities, a work commodity, and negative prices). Although he makes no reference to the concept of "basic commodities," consumption time is introduced in the utility function together with goods; later the consumption of each good is called an "activity." The income and time constraints are presented as independent equations, and the role of the minimum required consumption time is highlighted as the source of the positive valuation of a reduction in T_i only if the corresponding constraint is active, implying that the individual would have liked to spend less time on it.

The first microeconomic theory focusing on activities was formulated by Evans (1972). Here, the primary source of satisfaction is the type of activity performed, and its measure is the amount of time T_i assigned to the particular activity within a period. Thus, in essence, Evans introduced $U(\boldsymbol{T})$ as an apparently simple departure from the classic goods-consumption theory; however, activities are costly because to be performed they require that goods are purchased. Thus, market goods are inputs which are necessary to develop activities and, in turn, goods are the source of the activity cost. What DeSerpa called pure time activities are allowed to exist in Evans' framework simply as a special case; their cost can be either positive (the individual pays) or negative (the individual is paid). If $\boldsymbol{Q}$ is a matrix containing the input of goods at a certain rate per unit time which are required for each activity, then $\boldsymbol{QT}$ is the vector of goods that has to be bought in order to be able to perform the activities in $\boldsymbol{T}$. Thus, the budget constraint is in fact related to $\boldsymbol{QT}$. On the other hand, activities will be interdependent. This is taken into account by introducing a matrix $\boldsymbol{J}$ representing links among activity times.

The relation $\boldsymbol{X} = \boldsymbol{QT}$ is the first explicit introduction of a transformation function that turns activities into goods, which was implicitly expressed both in Becker's theory (the b_{ij} coefficients in Table 3.1) and in DeSerpa's (the a_i coefficients). For Evans, the amount of time to be assigned to each activity is the basic variable, that is, the source of direct utility and the original source of both expenses and income. Accordingly, his theory is stated in terms of the vector $\boldsymbol{T}$ only, as shown in Table 3.1. For completeness, it should be mentioned that Michael and Becker (1973) further elaborated the role of the "household production function" $Z(\boldsymbol{X},\boldsymbol{T})$ along different lines.

Discussion

Time evolved in consumption theory from a secondary role to a central one in a short period. However, today the basic approach to model consumer behavior still rests on the idea of goods consumption as the primary source of direct utility. From our brief account, it seems however as if the fundamental role assigned by Evans (1972) to activity time allocation generates a more general and meaningful theoretical framework. If one looks at Table 3.1 trying to make a synthesis, there are three key issues to discuss. The first one is the relation between goods and time. Such a relation is fairly general in both Becker's and Evans' theories through a matrix of fixed coefficients (b_{ij} in the first one and $\boldsymbol{Q}$ in the second). The second issue is the presence (or absence) of working time W in the direct utility function. Unlike Becker (1965), who explicitly leave them out, both DeSerpa (1971) and Evans (1972) include working hours as a direct source of utility. This is an important issue, as including working hours in the utility function would make Becker's synthesis of income and time constraints into one a mistake. The reason is that W could not be used as a pivot variable since the utility level would be affected. This is specifically pointed out by Evans (1972), and it has previously been highlighted by Johnson (1966) and Oort (1969) in their independent departures from Becker's theory. In fact, Evans criticizes both Johnson and Oort for not introducing other time-related restrictions as, for example, minimum time required to perform an activity (DeSerpa's theory is not mentioned by Evans). If no time restrictions are accounted for, the value of time would be equal for all activities because time is adjusted accordingly. And this leads to the third issue, which is more ample than specific minimum time requirements: the interrelation among activities.

TOWARDS A SYNTHESIS: A MICROECONOMIC TRAVEL/ACTIVITY THEORY

Travel Choice and Time Allocation Theory

In their original form, both the goods/leisure approach (Train and McFadden, 1978) and Becker's time allocation theory yield the same value of time, the wage rate. This should be no surprise, as in both cases three conditions concur: income is endogenously determined by freely choosing working hours, work time does not affect direct utility, and no constraints besides income and time budgets are included. Although they look different and their utilities have different rationales, both models are in fact conceptually the same; it should be recalled, however, that $\boldsymbol{X}$ and $\boldsymbol{T}$ in Becker's theory are the inputs to obtain the basic commodities $\boldsymbol{Z}$, and both are *vectors*, as opposed to G and L in Train and McFadden's theory where they are aggregates.

The preceding argument makes Gronau's (1986) extension a relevant one, as he includes a work commodity in the direct utility as well as a potentially given labor supply, making Becker's theory a special case. By association, a generalized version of the goods/leisure theory can be constructed by simply replacing W in (16) by $W_F + W_V$, representing fixed and variable (endogenously decided) working hours respectively, and set $W = W_V$ and $E = I$ (fixed income) in equation (15). Such a model would still not have work in direct utility. However, both the wage rate and expenditure rate specifications could be obtained as special cases using W_V as pivot; if W_V is positive the wage rate approach holds, whereas a zero (corner solution) implies an expenditure rate model.

There are some attempts to formulate travel-mode choice models according to the general frameworks developed to account for time allocation. One example is Truong and Hensher (1985), later improved by Bates (1987). They try to translate both Becker's and DeSerpa's general theories into (discrete) travel-mode choice models. Due to DeSerpa's constraints regarding minimum time requirements, they show that the conditional indirect (travel-mode) utility should have a mode-specific time coefficient; this coefficient should be generic if travel-mode choice is derived from Becker's theory. This difference is also influenced by the fact that travel time does not enter direct utility in the so-called Becker-type model, while it does appear explicitly in DeSerpa's counterpart. In both cases the Truong-Hensher-Bates model follows the goods-leisure approach which, as we have seen, is in fact Becker's. However, as goods and "activities" in DeSerpa's theory are also vectors explicitly written as such, as working time is not adjustable, and as additional time constraints appear, interpreting DeSerpa's utility arguments $\boldsymbol{X}$ and $\boldsymbol{T}$ as goods and leisure seems to be wrong.

In the preceding paragraphs, the possibility of building a more general framework for travel decisions linked to theories of time allocation has been highlighted. A basic issue to be solved is however what the arguments in direct utility should be. Here, the so-called "final commodities" $\boldsymbol{Z}$ seems to be an excuse to plug $\boldsymbol{X}$ and $\boldsymbol{T}$ in utility than anything else. In fact, $\boldsymbol{Z}$ is never defined with enough precision with the exception of Gronau (1986) who calls them "activities." On the other hand, Evans (1972) argues in favour of *time devoted to activities* as the basic quantifiable source of utility.

For Becker (1965) $\boldsymbol{T}$ is time to prepare the final commodities (which is the reason why W is left out of utility); for DeSerpa (1971) $\boldsymbol{T}$ is consumption time; for Train and McFadden (1978) the aggregate source of utility is leisure; Truong and Hensher (1985) include travel time in direct utility in the so-called DeSerpa model. As particularly emphasized by Evans (1972), Bates (1987), and Gronau (1986), including or not including an activity time in direct utility plays a key role in the interpretation of a model. In analytical terms, the behavior represented by the corresponding first-order conditions for optimality might include or not include a marginal utility of time assigned to the particular activity. The basic point is whether the individual level of satisfaction can change only because of transfers among leisure, work, and travel, through the time constraint with an impact on purchasing ability, or also due to direct pleasure or displeasure.

It seems that there has been an emphasis on keeping as arguments in the utility function only those elements which are believed to increase satisfaction (e.g., leisure, goods). Somehow the idea of non-leisure activities as direct arguments has been ignored, in spite of the previous examples and discussions. To test whether a variable should enter U, the problem can be restated as satisfaction. All variables that act through other variables in U should not belong to it (as, for instance, income). Thus, work and travel time should indeed be included; goods should not, as they require the assignment of time to their use. Even if some goods are bought for the pleasure of acquiring, satisfaction is realized in the act of buying; if it is a piece of art, satisfaction is experienced by the act of admiring or by enhancing an action (either at work or during free time). Thus, all identifiable activities should enter U as separate entities. The only justification for $\boldsymbol{X}$ in U would be a generic description of the activities (e.g. having a more comfortable bed increases the satisfaction of sleeping).

In sum, what to use as an argument in direct utility, what constraints should be considered, and what is fixed or what is variable, are key decisions in proposing a theoretical framework aimed at the modeling and understanding of travel decisions.

A Unified Theory of Travel and Activities[7]

Based on the microeconomic theories of travel-mode choice and time allocation, a unified theory may be proposed. It has three basic elements. First, the source of utility is the time devoted to each activity, including *all* activities (sleep, eat, talk, travel, work, and so on); second, market goods and services are ingredients of the different activities, and they are the sources of expenses; finally, besides time and income budget constraints, there are relations among $\boldsymbol{T}$ and $\boldsymbol{X}$ (feasible $\boldsymbol{T}$ for given $\boldsymbol{X}$, necessary $\boldsymbol{X}$ for given $\boldsymbol{T}$), and among the T_i's themselves (e.g., activities which require other activities) which need to be taken account.

A model of travel choices may be viewed as a time allocation problem, recognizing that utility is directly derived from what the individual does (activities) requiring goods which incur costs. Formally,

$$\max_{T, W_v, \{m\}, B} \quad U(\boldsymbol{T}, W_F, W_v, \boldsymbol{t}) \tag{22}$$

subject to

$$\sum_i T_i + W_v + W_F + \sum_{j=1}^{B} \sum_{i \in M_j} \delta_{ij} t_{ij} = \tau \tag{23}$$

$$F(\boldsymbol{X}, \boldsymbol{T}, W_F, W_v, \boldsymbol{t}) \geq 0 \tag{24}$$

$$\sum_i \sum_d P_{id} X_{id} + \sum_{j=1}^{B} \sum_{i \in M_j} \delta_{ij} c_{ij} = I_F + wW_v \tag{25}$$

---

[7] This subsection summarizes the essence of Jara-Díaz (1998).

$$B = B(\boldsymbol{X}) \tag{26}$$

where

- $\boldsymbol{T}$: vector of activity times T_i in period τ
- t: vector of travel times t_{ij} in period τ
- B: number of trips in period τ
- δ_{ij}: 1 if travel-mode i is used in trip j; 0 otherwise
- F: transformation function that converts activities into goods and vice versa; it includes the interrelations among activities
- X_{id}: amount of good i bought in zone d in period τ
- P_{id}: price of good i in zone d
- M_j: set of travel-modes available for trip j

It is assumed that goods can be bought in different locations at potentially different prices. As residence and work places are given, the number of trips is only sensitive to the choice of $\boldsymbol{X}$ (equation (26)). This can be viewed as the result of a network-related sub-problem (e.g., optimal number of trips given $\boldsymbol{X}$). Note that the transformation function in equation (24) is not Becker's $Z(\boldsymbol{X},\boldsymbol{T})$, but Evans' implicit functions $\boldsymbol{X}\text{-}\boldsymbol{QT}\text{=}\boldsymbol{O}$ and $\boldsymbol{J}'\boldsymbol{T} \leq \boldsymbol{O}$.

The given variables are W_F, I_F, t_{ij}, c_{ij}, τ, P_{id} and w, while the decision variables are $\{T_i\}$, $\{\delta_{ij}\}$, $\{X_{id}\}$, W_v and B. The solution for B is the generation model, the solution for $\boldsymbol{X}$ is the distribution model, and the solution for δ is the travel-mode choice model. Note that this theory is not compatible with the goods-leisure framework. As discussed earlier, $\sum T_i = L$ and $\sum P_{id} X_{id} = G$. Because of the relation between goods and activities, there is an implicit relation between G and L which has the straightforward interpretation that goods consumption requires L and vice versa. This assumption is missing in both the Becker and Train-McFadden theories.

Let us see how the theory works when analyzing travel-mode choice in the case of one trip k, which is the prevailing modeling approach in the field. All other trip decisions will be assumed to be given, that is, the number of trips B, destinations (which are one of the dimensions in $\boldsymbol{X}$), and all other travel-mode choices. The problem is

$$\underset{X,T,W_v,i \in M_k}{\text{Max}} \quad U(\boldsymbol{T}, W_F, W_v, t_1, \ldots, t_{ik}, \ldots, t_B) \tag{27}$$

subject to

$$\sum_i T_i + W_v + W_F + \sum_{j \neq k} t_j + t_{ik} = \tau \tag{28}$$

$$F(\boldsymbol{X}, \boldsymbol{T}, W_F, V_r, \boldsymbol{t}) \geq 0 \tag{29}$$

$$\sum P_i X_i + \sum_{j \neq k} c_j + c_{ik} = I_F + W_v \tag{30}$$

and the non-negativity constraints. For simplicity the relation between B and $\boldsymbol{X}$ has been dropped, which means that the amount of goods does not affect the number of trips. As usual in the discrete choice approach, (27) - (30) can be solved conditional on travel-mode choice yielding the conditional solutions for $\boldsymbol{T}$, $\boldsymbol{X}$ and W. Formally,

$$T_i^* = T_i^*[\tau - W_F - \bar{t} - t_{ik}, W_F, \vec{\boldsymbol{t}}, t_{ik}, \frac{1}{w}(I_F - \bar{c} - c_{ik})] \geq 0 \tag{31}$$

$$X_i^* = X_i^*[\%] \geq 0 \tag{32}$$

$$W_v^* = W_v^*[\%] \geq 0 \tag{33}$$

where $\bar{t}$ and $\bar{c}$ are obviously defined, and $\vec{\boldsymbol{t}}$ is the vector of travel times except t_{ik}. Then the conditional indirect utility function corresponds to (if $W_v > 0$)

$$U(\mathbf{T}^*, W_F, W_v^*, t_{ik}, \vec{\boldsymbol{t}}) \equiv V[\tau - W_F - \bar{t} - t_{ik}, W_F, t_{ik}, \vec{\boldsymbol{t}}, \frac{1}{w}(I_F - \bar{c} - c_{ik})] \tag{34}$$

This expression discloses explicitly some key aspects in the specification of travel-mode utility. First, unlike travel-mode cost, travel time plays a dual role in the indirect utility: it provides direct dissatisfaction, as a survivor from U in equation (27), and it affects available time to do other activities, as a consequence of constraint in equation (28). This latter role of travel time deals with the trade-off with pleasurable activities. The second key aspect is that both roles cannot be distinguished if V in equation (34) is approximated linearly. If this is accepted as a reasonable representation of (indirect) utility, the conditional comparisons would be based upon an expression like

$$V_i = k + \alpha(\tau - W_F - \bar{t} - t_{ik}) + \beta W_F + \gamma t_{ik} + \sum \theta_h t_h + \varepsilon \frac{1}{w}(I_F - \bar{c} - c_{ik}) \tag{35}$$

and the only terms that would influence mode choice would be

$$\overline{U}_i = (\gamma - \alpha) t_{ik} - \varepsilon \frac{c_{ik}}{w} \tag{36}$$

from which one can only estimate $(\gamma - \alpha)$ and ε. However, neither γ nor α can be estimated. Note also that a first-order approximation like (35) would make all other variables but travel time and cost irrelevant (e.g., income). As explained earlier, this would not happen if a second-order expansion was believed to be a better model.

The third key aspect is the role of the wage rate w. In this theory, the relevant value of w is the hourly payment the individual is offered to do extra work. It is true that this might have a relation to I_F / W_F but w represents the opportunity cost of activities performed outside the (fixed) working schedule. According to this, if the individual has a fixed salary and fixed working time, he or she should be asked the value of the wage rate for additional work, as this is the value that should enter travel-mode utility, provided $W_v > 0$.

The conditional indirect travel-mode utility in equation (34) can be interpreted in terms of "goods and leisure." The first argument is in fact total time available to perform $\boldsymbol{T}$ (which can be associated with L) or to keep on working, and the last argument is the time equivalent to buy $\boldsymbol{X}$, that is, G / w minus the actual extra time worked. Formally,

$$V_i = V(L + W_v, W_F, t_{ik}, \bar{\boldsymbol{t}}, \frac{G}{w} - W_v) \tag{37}$$

which explicitly highlights the difference to the goods-leisure approach.

Comments

The proposed theory to account for travel behavior rests on Evans' (1972), and also on the goods/leisure version of the discrete-choice approach. Accordingly, it should be no surprise that a wage rate type specification for travel-mode utility is recovered when a travel-mode choice is derived under the appropriate assumptions, provided that variable working hours exist. At this point, it seems fairly clear that the role of labor supply is highly relevant: if it is fixed (exogenous income, at least in the short run) what matters is the time available to spend the money, while if it is variable (endogenous income) marginal adjustments make the wage rate a key variable. Some additional properties of the theory are

a) decisions to allocate time to travel and activities pertain to the same class;
b) the interrelations among activities, and those between activities and goods, make it difficult to accept continuous analytical solutions because of minimum required durations and the presence of durable goods;
c) the subjective value of each activity can be different;
d) if income is relatively small, choices in the time space can be very limited because of the relations between goods and activities, which can make the time constraint irrelevant;

e) if income is relatively large, a number of activities are open for consideration because the necessary goods and services could be acquired. This could make the income constraint irrelevant.

An approach like the one presented here emphasizes time allocation and, therefore, the perception of time. The decisions about what to do within a time frame become the relevant phenomenon to investigate. Part of this deals with the analysis of labour supply (how much to work), but understanding individual time allocation as a whole requires a more extensive analysis of human activities. Analyzing travel decisions may not require understanding the profound motives, but the influence of dominant social values is indeed relevant when studying the structure of daily activities. This should redirect research towards the identification of socially induced activities, telecommunication, or the relations between prices and uses of goods (e.g., add to the "Do I have money?" question, "Do I have extra time to use it?" and "What will I stop doing in order to use this?"). Thus, acquiring cable TV, having a compact disc player in the car or playing soccer with the neighbors become something relevant to understand and model. On the other hand, there is a need to understand activity choice when income is small enough to make impossible the acquisition of leisure goods (e.g., toys, gadgets) or the performance of leisure activities (e.g., movies, sports). It might well be that we are facing the emergence of two social “classes:“ those that still have money when the day ends, and those that still have time when they run out of money.

Needless to say, the aggregate trends of social behavior, the role of technology and social values, or social idiosyncracy, seem essential for understanding travel. Along these lines, the high subjective valuation of time in Santiago, Chile, previously mentioned, has been examined from the viewpoint of the absolute perception of time. A detailed survey indicated that Chilean students’ perceptions are closer to individuals in the U.S. than in Brazil regarding punctuality, coordination of activities, synchronization, and so on. The conclusion was that the high valuation of time relative to income revealed by the travel demand models could well be the result of time perceptions which are highly influenced by the life style in the developed world (Jara-Díaz and Romero, 1992).

The activity-travel theory presented here has also provided a microeconomic basis for understanding residential location (Jara-Díaz, Martínez and Zurita, 1994). One of the results was the analytical deduction of a term that represents accessibility, associated to each location, which combines the utility obtained from performing activities in different locations with the generalized cost to reach those places.

CONCLUSIONS

Consumer theory essentially provides a framework to describe economic behavior. Within this theoretical framework, the concept of utility function has been instrumental to model demand for goods and services and to model labor supply. It is assumed that the individual behaves as if such a function is maximized. Although travel choice modeling has also benefited from this theoretical framework, it seems necessary to make a revision of the manner in which it has been adapted to understand and model urban transport users' behavior. In this article, travel choices have been examined from the perspective of consumer theory, in an attempt to unveil the specific role of the different elements which are part of users' decisions.

Discrete choices, the goods/leisure framework and time-related theories of behavior have been exposed and examined. From this analysis of the microeconomic foundations of models related to trip decisions, some issues have been clearly established. First is the question about the sources of direct utility; starting from goods consumed and going through the concept of basic commodities, consumption time appeared as a necessary item to realize utility. After this modest beginning, time devoted to activities emerged as the basic source of satisfaction, and goods should be looked at as means to an end. Once this is accepted, every part of a period should be considered. This means, among other things, that both working and travel times are variables that should enter utility just as all other activities. Thus, time cannot be converted into money (through more work) without altering utility, which makes the fusion of income and time constraints a mistake.

On the other hand, the traditional time and income budget constraints are not enough to make the picture complete of individual behavior since market goods and activity times are interrelated (as well as activities themselves). The addition of a set of technical constraints is necessary to strengthen the fact that certain activities which would be omitted otherwise are performed. This is a point raised originally by DeSerpa (1971) and Evans (1972), introduced later in the discrete choice modeling by Troung and Hensher (1985). However, no explicit reference to a transformation function has been made within the context of travel-mode choice. This needs revision and discussions, and Evans' contribution seems to be the best departure point.

It is somewhat surprising to realize that little discussion has taken place regarding the variables in direct utility. In fact, goods and services seemed a reasonable choice until the recognition of a time constraint. The introduction of such a constraint implies a relation between goods and activities that cannot be overlooked. Furthermore, once this has been firmly established, identifying the assignment of time to activities as the basic source of satisfaction seems evident. However, this gives travel a different status.

According to Gronau (1986) and Jara-Díaz and Romero (1992), activities related to personal care (eating, sleeping, and other biological needs) consume on average a little more than eleven hours daily. A normal working schedule would leave something like four hours for discretionary activities in a working day. Time assigned to mandatory travel can consume a relevant part of this potentially uncommitted time. Thus, understanding travel choice relies on an understanding of activity participation.

This suggests the need to augment the travel-choice theoretical framework with the elements and analysis of time allocation coming from the home economics literature. In fact, in this literature the role of travel has been already highlighted. "The shadow price of time affects customer's choice of the optimum combination of time and market inputs and the decision whether to participate in market work or not. The imputation of this shadow price is therefore based on the observation of choices where time is traded for goods, and the choice concerning labor force participation. Unfortunately, most often in situations where goods are traded for time, the amount of time saved is unrecorded"... "One of the few exceptions is the field of transportation" (Gronau, 1986, pp. 292). In this quote, emphasis reveal a goods-leisure point of view; the explanation lies in the type of problems addressed, particularly those pertaining to (domestic) daily life. The individual can make a choice among buying frozen food (which requires a microwave oven), cooking, hiring somebody to cook, or dining out; in essence, this is a choice involving quality, money and time. And it is true that the trade-offs are difficult to establish empirically because of lack of recorded information. It would be desirable indeed to collect and analyze such information in order to be able to model and understand travel choice as well as goods consumption and activity patterns.

Although a theoretical framework does not necessarily translates immediately into an operational model, implementation should be kept in mind. For example, an activity-travel theory as the one proposed here yields conditional demands for goods, work and activities as intermediate results when modeling travel-mode choice (see equations 31 to 33) which turn into unconditional ones if choice is observed. All variables are potentially known, so that a system of equations could be estimated. Undoubtedly, there has been a historical emphasis on market demand for goods, which has blurred the activity-oriented approaches. Maybe the present universal trend towards the "I have no time" syndrome will reverse the situation.

ACKNOWLEDGEMENTS

This research was partially funded by FONDECYT, Chile. The first version of the activity-travel model was presented at the 7th conference of the International Association of Travel Behavior Research (IATBR), Valle Nevado, Chile, 1994. Comments and feedback by the participants in this conference are gratefully acknowledged.

REFERENCES

Bates, J. (1987). Measuring travel time values with a discrete choice model: a note. *The Economic Journal*, **97,** 493-498.

Bates, J. and M. Roberts (1986). Value of time research: summary of methodology and findings. Presented at the 14th PTRC summer annual meeting.

Becker, G. (1965). A theory of the allocation of time. *The Economic Journal,* **75,** 493-517.

DeSerpa, A. (1971). A theory of the economics of time. *The Economic Journal,* **81,** 828-846.

Evans, A. (1972). On the theory of the valuation and allocation of time. *Scottish Journal of Political Economy*, **19**, 1-17.

Gronau, R. (1986). Home production - a survey. In: *Handbook of Labor Economics* (O. Ashenfelter and R. Layard, eds.), Vol. 1, pp. 273-304. North Holland, Amsterdam.

Jara-Díaz, S. R. (1990). Valor subjetivo del tiempo y utilidad marginal del ingreso en modelos de partición modal (Subjective value of time and marginal utility of income in mode choice models). *Apuntes de Ingeniería*, **39**, 41-50.

Jara-Díaz, S. R. (1991). Income and taste in mode choice models: are they surrogates? *Transportation Research,* **25B,** 341-350.

Jara-Díaz, S. R. (1998). A general micro-model of users' behavior: The basic issues. In *Travel Behavior Research: Updating the State of Play* (J. de D. Ortúzar, D. Hensher, and S. Jara-Díaz, eds.). Pergamon, Oxford. In press.

Jara-Díaz, S. R. and M. Farah (1987). Transport demand and user's benefits with fixed income: The goods/leisure trade-off revisited. *Transportation Research,* **21B,** 165-170.

Jara-Díaz, S. R., F. Martínez and I. Zurita (1994). A microeconomic framework to understand residential location. In: *Proceedings of 22th PTRC Summer Annual Meeting - Seminar H*, pp. 115-128. London

Jara-Díaz, S. R. and J. de D. Ortúzar (1989). Introducing the expenditure rate in the estimation of mode choice models. *Journal of Transport Economics and Policy,* **23,** 293-308.

Jara-Díaz, S. R. and C. Romero (1992). Valoración subjetiva del tiempo: transporte, trabajo y ocio en Santiago. (Subjective valuation of time: transport, work and leisure in Santiago). In: *Tecnología y Modernidad en Latinoamérica* (E. Sabrovsky, ed.), pp. 35-48. Hachette, Santiago.

Jara-Díaz, S. R. and J. Videla (1989). Detection of income effect in mode choice: Theory and application. *Transportation Research,* **23B,** 393-400.

Johnson, B. (1966). Travel Time and the Price of Leisure. *Western Economic Journal,* 135-145.

Juster, F. T. (1990). Rethinking utility theory. *The Journal of Behavioural Economics,* **19,** 155-179.

Lancaster, K. (1966). A new approach to consumer theory. *Journal of Political Economy,* **74,** 132-157.

McFadden, D. (1981). Econometric models of probabilistic choice. In: *Structural Analysis of Discrete Data with Econometric Applications* (C. Manski and D. McFadden, eds.), pp. 198-272. MIT Press, Cambridge, MA.

Michael, R. and G. Becker (1973). On the new theory of consumer behaviour. *Swedish Journal of Economics,* **75,** 378-396.

Oort, C. (1969). The evaluation of travelling time. *Journal of Transport Economics and Policy,* **3,** 279-286.

Train, K. and D. McFadden (1978). The goods/leisure trade-off and disaggregate work trip mode choice models. *Transportation Research,* **12,** 349-353.

Truong, P. and D. Hensher (1985). Measurement of travel time values and opportunity cost from a discrete - choice model. *The Economic Journal,* **95,** 438-451.

Viton, P. (1985). On the interpretation of income variables in discrete choice models. *Economic Letters,* **17,** 203-206.

Winston, G. C. (1987). Activity choice: A new approach to economic behaviour. *Journal of Economic Behavior and Organization,* **8,** 567-585.

4

EXTENDING VALUATION TO CONTROLLED VALUE FUNCTIONS AND NON-UNIFORM SCALING WITH GENERALISED UNOBSERVED VARIANCES

David A. Hensher, The University of Sydney, Australia

INTRODUCTION

Transportation researchers have progressed our understanding of individual travel behavior within a set of modeling paradigms of an essentially economic nature, as represented by the rule of utility maximisation. With rare exception, the mainstream empirical emphasis has been on the application of revealed-preference (RP) data in a static discrete-choice framework with linear additivity assumed for the functional form of the indirect utility expression associated with each alternative in the choice set. In recent years interest has grown in extending the RP data paradigm to incorporate stated-choice (SC) data as an enrichment tool in contexts where the attribute levels and choice sets extend beyond the utility space observed in real markets (see Hensher, 1994, in Special Issue of Transportation, as indicative of the state of art). However this extension has stayed essentially within a linear-additive framework and assumed the constant variance assumption of the simple logit model within each of the RP and SC data sub-sets (e.g., Bradley et al., 1995; Bradley and Daly, 1992, 1994; Bradley and Gunn, 1990; Hensher and Bradley, 1993; Morikawa, 1989; Suzuki et al., 1995; Swait et al., 1994).

Two restrictions of current practice are the emphasis on deriving marginal rates of substitution between attributes such as the value of travel time savings as *point* estimates rather than as functions, and the *uniform* scaling of RP data when combined with SC data. The constant variance assumption normalises the location parameter which scales the taste weight parameter for each attribute. Relaxing the constant variance assumption enables identification of the unique variances and hence scale parameters associated with each alternative in a choice set. The combined extensions into valuation functions and variable scaling of parameters adds further richness to our understanding of human behavioral response to travel opportunities. A comparison with the traditional logit form highlights the loss of information associated with the more restrictive empirical discrete-choice studies.

This paper is organised as follows. The next section discusses the structure of valuation functions and the implications of relaxing the constant variance assumption. In recognizing the non-linearity of valuation functions, special consideration must be given to the data form for introducing higher-order attributes such as quadratics and interactions. A series of four discrete-choice models are estimated - two basic logit models with constant variance with/without quadratic and interaction terms for time and cost, and two heteroskedastic logit models with/without quadratic and interaction terms. The distribution of values of travel time savings (VTTS) based on the models with quadratic and interaction terms estimated on a stated choice commuter mode data set for 6 capital cities in Australia are compared between constant and non-constant variance models, highlighting the potential loss of information from the more restrictive assumption. The valuation functions are then applied on revealed preference data from the same sample of commuters to derive a distribution of market-based VTTS, which are used in a classification procedure (CARTS) to segment the commuter market by socioeconomic characteristics with respect to VTTS. Some important implications of using point estimates of VTTS are highlighted.

VALUATION FUNCTIONS INSTEAD OF POINT ESTIMATES

It is typically assumed that the marginal rates of substitution between any two attributes vary by a number of market segments such as mode, trip purpose, trip length and personal income; indeed in the limit we can treat each individual in the sampled population as a segment with unique empirical valuation properties. Within each aggregated market segment, point estimates such as the value of travel time savings (VTTS) are routinely derived and applied as mean estimates of an unknown (but assumed) distribution. A strictly linear interpretation of the marginal rate of substitution between travel time and travel cost is assumed within the segment. Little understanding of the true profile of VTTS within each segment is known; indeed the set of segment-specific values are treated as threshold values with limited appreciation of the best set of cut-off points if segmented single values are required. For example, VTTS for short, medium and long trips are preconditioned on an exogenous definition of each trip length. Bradley and Gunn (1990), for example, selected a series of travel speeds with arbitrary ranges. Whether these arbitrary cut-off points are appropriate or not was not considered. Valuation functions enable us to seek out a continuous distribution of values and to partition this distribution in a more useful way if a set of point estimates are required for practical analysis. Alternatively, the richness of the continuous distribution can be preserved and implemented (see Bradley and Gunn, 1990; Hensher and Truong, 1984).

We propose a distribution of VTTS derived from the non-linearity inherent in the attributes of alternatives (such as travel time and travel cost), and then map these into socio-economic space to identify any systematic variation between VTTS and individual characteristics. This is more appealing and richer than relying on socioeconomic characteristics as both proxies for the non-linearity of travel time and travel cost and the role that socioeconomic characteristics play as proxies for other unobserved influences on choice and hence value. A *value function* is by our definition a functional relationship between VTTS and levels of time and cost both in respect of higher-order (e.g., quadratic) and interaction (e.g., two-way product of time and cost). A *segmented valuation function* is a value function conditioned on profiles of characteristics of the individual traveller and/or other external influences.

To begin, assume that the theoretical parameter κ_i from an indirect utility function of the linear additive form (DeSerpa, 1971; Truong and Hensher, 1985):

$$V_i = \alpha_i - \lambda C_i - \kappa_i T_i \qquad (1)$$

is a function of C_i and T_i:

$$\kappa_i = \kappa(T_i, C_i) \qquad (2)$$

A Taylor series expansion of (2) around the mean levels $\overline{T}$ and $\overline{C}$ for each alternative i (neglecting second-order terms) results in equation (3):

$$\kappa_i = \overline{\kappa} + (\partial\kappa / \partial T)_i (T_i - \overline{T}) + (\partial\kappa / \partial C)_i (C_i - \overline{C}) \qquad (3)$$

Substitution of (3) into (1) and some rearrangement of terms gives:

$$V = \alpha_i - \lambda C_i - \overline{\kappa}\, T_i + (\beta T_i^2 + \gamma C_i T_i + \omega) \qquad (4)$$

where

$\beta = (\partial\kappa / \partial T)_i$, $\gamma = (\partial\kappa / \partial C)_i$, and $\omega = -\beta\overline{T} - \gamma\overline{C}$

By neglecting second-order terms in (3) we implied that $\partial\kappa / \partial T$ and $\partial\kappa / \partial C$ are constants, independent of alternative i. Equivalently, the parameters ω, β and γ are unsubscripted. The VTTS can now be derived from (4) as follows (Truong and Hensher, 1985; Hensher, 1997):

$$\text{VTTS}=\frac{\partial V / \partial T_i}{\partial V / \partial C_i}\Big|V_i = \text{constant}$$

$$=\frac{\bar{\kappa} + \gamma C_i + 2\beta T_i}{-\lambda + \gamma T_i} \qquad (5)$$

Thus VTTS is dependent on the levels of travel time and cost. This formula can be generalised to account for the disaggregation of travel time (see Truong and Hensher, 1985, for more details). We can introduce interactions between each travel time and between travel time and other attributes of alternatives. The ability to enrich the valuation function to test for a richer specification is conditioned by the quality of data.

RP data are usually somewhat limiting in its ability to offer sufficient richness in both variability and correlation structure to enable each potential influence to be included without producing confoundment. This is particularly true when accounting for non-linearity and when "new" alternatives are assessed for market share. Data derived from a SC experiment, however, increases the opportunity to account for the independent (i.e., additive) contribution of each source of variability in the valuation function in an expanded choice set. It is for this reason, amongst other reasons, that the SC paradigm has evolved to become an important feature of a preferred empirical approach to obtaining behavioral values of travel time savings (see Bradley et al., 1995).

DEVELOPING A VALUATION FUNCTION

To identify the underlying structure of a SC experiment, and to understand how the valuation function is derived from the information in this experiment, we begin by defining the information potential of each attribute to be evaluated in the choice experiment. Assume that each attribute in a mode choice experiment has three levels, represented two ways - as nominal or mean centred values (-1, 0, 1) or as actual levels (e.g., \$1, \$2 and \$3). The non-linearity of an attribute can be accounted for by specifying higher-order dimensions such as quadratics. A three-level attribute contains information up to the quadratic dimension (or order two). A quadratic term is constructed by squaring the value of the attribute and replacing the nominal mid-point (0) with the nominal -2.

Let $v(X)$ denote the marginal utility which arises from attribute X alone (e.g., travel time or cost). If $v(X)$ is non-linear in X, a second-order polynomial approximation for $v(X)$ can be written as

$$v(X) = \alpha X + \beta X^2 \tag{6}$$

We can derive relationships between $v(X)$ and X that show the role of the quadratic term and why it is important to be able to test for non-linearity by separating out the independent contribution of X^2, and hence determining if both X and X^2 are required to capture the full effect due to attribute X (given the limiting assumption of a maximum order 2 with only 3 levels of an attribute) (Hensher 1997):

$$\alpha = (1/2)(v_3 - v_2) \tag{7}$$

$$\beta = (1/6)[\,(v_3 - v_2) - (v_2 - v_1)] \tag{8}$$

where α measures the mean slope between $X = -1$ and $X = 1$ and β measures the change in slope between ($X = -1, X = 0$) and ($X = 0, X = 1$).

In addition to the non-linearity of each attribute's influence on the value of travel time savings (equation 5), the possibility of two-way interactions between pairs of attributes needs to be considered. In fractional factorial designs the number of two-way interactions which are independent (i.e., orthogonal) from the main effects (i.e., X and X^2) are determined by the actual fraction selected and the degrees of freedom ensuing from the design. For VTTS functions, we are interested in the interaction between travel cost and travel time. With a quadratic term in the main effect for travel time and one interaction term for cost and time, we are able to evaluate the empirical form of the valuation function. A valuation function arises from a specification that includes quadratic and/or interaction terms.

Since the quadratic of an orthogonally coded nominal attribute is not equivalent to the quadratic of the actual value of an attribute, we require a mapping from one metric to the other. We estimate the logit model using the orthogonal codes but determine the value of travel time savings based on the actual levels of the attributes shown to the sampled population.

Define the nominal level as N and the actual level as A. For a three-level attribute, define:

$$N = \theta_0 + \theta_1 A \tag{9}$$

$$N^2 = \delta_0 + \delta_1 A + \delta_2 A^2 \tag{10}$$

Denote the three levels for A as low (L), medium (M), and high (H) and L^2, M^2, and H^2 for A^2. Given the orthogonal codes for N (-1, 0, 1) and N^2 (1, -2, 1), substitution into equations (9) and (10) gives the following transformation functions:

$$-1 = \theta_0 + \theta_1 L \tag{11}$$

$$0 = \theta_0 + \theta_1 M \tag{12}$$

$$1 = \theta_0 + \theta_1 H \tag{13}$$

where $\theta_1 = 1/\Delta_i$, $\theta_0 = -M/\Delta_i$, and $\Delta_i = M - L = H - M$ (assuming equal spacing for attribute i), and

$$1 = \delta_0 + \delta_1 L + \delta_2 L^2 \tag{14}$$

$$-2 = \delta_0 + \delta_1 M + \delta_2 M^2 \tag{15}$$

$$1 = \delta_0 + \delta_1 H + \delta_2 H^2 \tag{16}$$

and $\delta_2 = 3/\Delta_i^2$, $\delta_1 = -6M/\Delta_i^2$, and $\delta_0 = -2 + 3M^2/\Delta_i^2$.

The utility expression associated with each alternative as estimated using nominal attribute levels can be specified as:

$$V_i = \eta + \eta_1 C + \eta_2 T + \eta_3 T^2 + \eta_4 CT \tag{17}$$

which can be translated from nominal to actual levels of attributes as follows:

$$V_i = \omega + \omega_1 c + \omega_2 t + \omega_3 t^2 + \omega_4 ct \tag{18}$$

Defining M_i as the actual mean level of attribute i and Δ_i as the step change in the level of attribute i, we derive the following conversion relationships:

$$\omega_1 = [\eta_1 / \Delta_c - M_t \eta_4 / (\Delta_c \Delta_t)] \tag{19}$$

$$\omega_2 = [\eta_2 / \Delta_t - 6M_t \eta_3 / \Delta_t^2 - M_c \eta_4 / (\Delta_c \Delta_t)] \tag{20}$$

$$\omega_3 = [3\eta_3 / \Delta_t^2] \tag{21}$$

$$\omega_4 = [\eta_4 / \Delta_c \Delta_t] \tag{22}$$

RELAXING THE CONSTANT VARIANCE ASSUMPTION

Each attribute in the indirect utility expression associated with an alternative has a beta parameter identifying its contribution to variation in the level of relative utility which is the product of two components - a location or scale parameter and a taste weight parameter. In the simple logit form with constant variance in the unobserved effects, it is well known that the scale parameter, as an index of variability in the unobserved effects, can be arbitrarily set equal to one. When the interest centres on valuation, the ratio of any two or more attributes within the utility expression for a single alternative cancels the scale effect. However the simple logit form assumes that this constant scalar is independent of the alternatives in the choice set and hence has no effect on comparing values *across* alternatives (McFadden, 1981). Assuming uniqueness may be unsatisfactory. If the scale parameters vary between alternatives, then the relative valuation between alternatives will be distorted. The extent is unknown and is essentially empirical (Bhat 1995). This, however, is an important issue since we should be interested in both the absolute values *and* the relative values. For example, if there are noticeable unobserved effects with high variance associated with the public transport alternatives in contrast to the automobile alternatives, then the constant variance assumption can overvalue the VTTS for public transport use relative to automobile use.

To relax the constant variance assumption requires a more complex choice model, called the heteroscedastic extreme value logit (HEVL) model. Allenby and Glinter (1995) and Bhat (1995) have recently implemented the HEVL model on RP data, but have not considered the benefits of the HEVL specification for relative valuation. Hensher et al., (1998) review the HEV model and other variants of logit models.

Define the behavioral choice rule as

$$P_i = \text{Prob}[U_i > U_j] \text{ for all } j \neq I \tag{23}$$

$$= \int_{-\infty}^{+\infty} \pi(j \neq i) F(\lambda_j)\{V_i - V_j + \varepsilon_i\} \lambda_i f(\lambda_i \varepsilon_i) d\varepsilon_i \tag{24}$$

where $f(t)$ is the density function, $f(t) = \exp(-t)*\exp(-\exp(-t)) = -F(t)*\log(F(t))$. The probabilities are evaluated numerically as there is no closed-form solution for the integral. The scale parameter is $1/\lambda$, with the square of lambda proportional to the inverse of the variance of the unobserved random component of utility, ε_i. The integral can be approximated using Gauss-Laguerre quadrature (Bhat, 1995). Thus we can replace equation (23) with equation (24), where w is the weight and z(l) is the abscissa of the Gauss-Laguerre polynomial. A 32 point approximation provides sufficiency (Greene, 1996). The derivatives can be used in constructing the log-likelihood and for calculation of the full matrix of direct and cross share elasticities:

$$\int_{-\infty}^{+\infty} \pi(j \neq i) F[t(j|i)] \exp[-u(i)] du(i) \approx \Sigma(1) w(1) F(z(1)) \tag{25}$$

SPECIFICATION OF AN EMPIRICAL INQUIRY TO EVALUATE THE BENEFITS OF GENERALIZATION

As part of a study investigating the impact of transport policy instruments on reductions in greenhouse gas emissions in six Australian capital cities, Sydney, Melbourne, Brisbane, Adelaide, Perth, and Canberra (Hensher et al., 1995), a SC experiment centred on commuter mode choice was designed (Louviere et al., 1994). The universal choice set comprised the currently available modes plus two "new" modes - light rail and busway.

Each respondent considered a context in which the offered set of attributes and levels represent the available ways of undertaking a commuter trip from the current residential location to the current workplace location. The purpose is to establish their coping strategies under these circumstances. Four alternatives appeared in each travel choice scenario - car (no toll), car (toll), bus or busway, and train or light rail. There were 12 types of showcards, with three trip lengths and four combinations of public transport: bus vs. light rail, bus vs. train (heavy rail), busway vs. light rail, and busway vs. train. The public transport combinations appeared in each showcard with incidence determined by the statistical design. The range of attribute levels are summarised in Table 4.1; the contextual statement is given in Appendix A, an illustrative showcard in Appendix B.

The five attributes of the public transport alternatives are total in-vehicle time, frequency of service, closest stop to home, closest stop to destination, and fare. The attributes for the car alternatives are travel time, fuel cost, parking cost, travel time variability, and for the toll road - departure time and toll charge. Three levels were selected for each attribute. Detailed definitions of each attribute are given in Appendix C. The design allows for six alternativespecific main effect models for car no toll, car toll road, bus, busway, train, and light rail. Linear by linear interactions are estimable for both car models, and generically for the bus/busway and train/light rail models. While cross effects have been assumed negligible, the four-alternative design is perfectly balanced across all attributes. The design has been blocked into 27 versions of size three also balanced for every attribute such that each respondent sees each level of each attribute exactly one time. The base design for the travel choice task was a 27 x 3^{27} orthogonal fractional factorial in 81 runs. The 27 level factor was used for creating the versions, and the three level factors selected and arranged to meet the design requirements. Two factors each at two levels were constructed for the bus/busway and train/light rail modes. In the design, bus and train appear in 36 scenarios while busway and light rail appear in 45 scenarios.

Table 4.1
The Set of Attributes and Attribute Levels in the Travel Choice Experiment
(all cost items are in $s, all time items are in minutes)

SHORT (<30)	**Car no toll**	**Car toll road**	**PUBLIC TRANSPORT**	**Bus**	**Train**	**Busway**	**Light Rail**
Travel time to work	15,20,25	10,12,15	Total time in the vehicle (one-way)	10,15,20	10,15,20	10,15,20	10,15,20
Pay toll if you leave at this time otherwise free)	None	6-10, 6:30-8:30, 6:30-9	Frequency of service	Every 5,15,25	Every 5,15,25	Every 5,15,25	Every 5,15,25
Toll (one-way)	None	1,1.5,2	Time from home to your closest stop	Walk 5,15,25 Car/Bus 4,6,8	Walk 5,15,25 Car/Bus 4,6,8	Walk 5,15,25 Car/Bus 4,6,8	Walk 5,15,25 Car/Bus 4,6,8
Fuel cost (per day)	3,4,5	1,2,3	Time to your destination from the closest stop	Walk 5,15,25 Bus 4,6,8	Walk 5,15,25 Bus 4,6,8	Walk 5,15,25 Bus 4,6,8	Walk 5,15,25 Bus 4,6,8
Parking cost (per day)	Free,$10,$20	Free,$10,$20	Return fare	1,3,5	1,3,5	1,3,5	1,3,5
Time variability	0, ±4,±6	0,±1,±2					
MEDIUM (30-45)							
Travel time to work	30,37,45	20,25,30	Total time in the vehicle (one-way)	20,25,30	20,25,30	20,25,30	20,25,30
Pay toll if you leave at this time (otherwise free)	None	6-10, 6:30-8:30, 6:30-9	Frequency of service	Every 5,15,25	Every 5,15,25	Every 5,15,25	Every 5,15,25
Toll (one-way)	None	2,3,4	Time from home to your closest stop	Walk 5,15,25 Car/Bus 4,6,8	Walk 5,15,25 Car/Bus 4,6,8	Walk 5,15,25 Car/Bus 4,6,8	Walk 5,15,25 Car/Bus 4,6,8
Fuel cost (per day)	6,8,10	2,4,6	Time to your destination from the closest stop	Walk 5,15,25 Bus 4,6,8	Walk 5,15,25 Bus 4,6,8	Walk 5,15,25 Bus 4,6,8	Walk 5,15,25 Bus 4,6,8
Parking cost (per day)	Free,$10,$20	Free,$10,$20	Return fare	2,4,6	2,4,6	2,4,6	2,4,6
Time variability	0, ±7, ±11	0, ±2, ±4					
LONG (>45)							
Travel time to work	45,55,70	30,37,45	Total time in the vehicle (one-way)	30,35,40	30,35,40	30,35,40	30,35,40
Pay toll if you leave at this time(otherwise free)	None	6-10, 6:30-8:30, 6:30-9	Frequency of service	Every 5,15,25	Every 5,15,25	Every 5,15,25	Every 5,15,25
Toll (one-way)	None	3,4.5,6	Time from home to your closest stop	Walk 5,15,25 Car/Bus 4,6,8	Walk 5,15,25 Car/Bus 4,6,8	Walk 5,15,25 Car/Bus 4,6,8	Walk 5,15,25 Car/Bus 4,6,8
Fuel cost (per day)	9,12,15	3,6,9	Time to your destination from the closest stop	Walk 5,15,25 Bus 4,6,8	Walk 5,15,25 Bus 4,6,8	Walk 5,15,25 Bus 4,6,8	Walk 5,15,25 Bus 4,6,8
Parking cost (per day)	Free,$10,$20	Free,$10,$20	Return fare	3,5,7	3,5,7	3,5,7	3,5,7
Time variability	0, ±11, ±17	0, ±7, ±11					

EMPIRICAL RESULTS

Four travel-mode choice models have been estimated as the basis for deriving a valuation function in the presence and absence of constant variance in the unobserved effects. The four models are: Basic logit model (BL), heteroscedastic extreme value logit model (HEVL), basic non-linear logit model (BnlL) and heteroscedastic extreme value non-linear logit model (HEVnlL). Model estimation is limited to only two car design attributes (fuel and time) and four public transport attributes (fare, main mode time, access time, and egress time). The car alternatives have been pooled across toll and no toll and segmented by drive alone and ride share.

A profile of the data set is given in Table 4.2, with full model results in Tables 4.3-4.6. A direct comparison of all four models is not informative - rather we need to consider the derivative outputs in the form of values of travel time savings. Two models produce point estimates of VTTS for each mode (i.e., BL, HEVL); two models provide valuation functions for VTTS (i.e., BnlL, HEVnlL), and two models provide point estimates of direct and cross share elasticities (i.e. BL, HEVL). The BnlL and HEVnl can be applied to obtain a distribution of share elasticities as a function of price and time, but we do not report them here. It is sufficient to compare the point estimates of weighted aggregate elasticities to establish the potential loss of information from a constant location scale parameter.

Table 4.2

Stated Choice Profile: Means and Proportions for each Attribute in Base Model

Attribute	Drive Alone	Ride Share	Bus	Train	Light rail	Busway
In choice set	2016	2016	902	889	1127	1114
Chosen	559	325	252	235	340	305
Mode shares	.277	.161	.125	.117	.168	.151
One-way fuel/fare cost ($)	2.815	2.815	1.784	1.769	1.776	1.766
Main mode time (mins.)	27.794	27.794	18.359	20.332	20.546	18.456
Access time (mins.)	-	-	11.032	11.965	11.827	10.764
Egress time (mins.)	-	-	11.062	12.120	12.068	11.048
Daily parking cost ($)	8.507	8.507	-			
Personal income ($000s)	33.487	-	-			
Car availability index	1.090	-	-			

Note: car availability index is defined as the ratio of number of cars in household to number of workers.

Table 4.3
Base Stated Choice Model: Simple Logit for 6 Cities (5 Iterations)

Attribute	Alternative(s)	Acronym	Estimate	T-value
Drive alone constant	DA	DASC	.39918	1.22
In-vehicle cost	All	CST	-.55209	-13.73
Main mode time	DA, RS	TMCAR	-.059778	-10.00
Parking cost (all day)	DA,RS	PRK	-.09745	-13.71
Personal income	DA	PINC	.00784	2.39
Car availability index	DA	CARAV	.34451	3.80
Ride share constant	RS	RSASC	.51850	2.82
Bus specific constant	BS	BSASC	.051506	0.46
Main mode time	BS,TN,LR,BWY	TMPT	-.069077	-9.38
Access mode time	BS,TN,LR,BWY	ACT	-.04253	-7.53
Egress mode time	BS,TN,LR,BWY	EGT	-.020825	-3.74
Train specific constant	TN	TNASC	.23524	2.22
Light rail specific constant	LR	LRASC	.32225	3.40
Log-likelihood at convergence	-2241.16			
Pseudo-R^2	0.37			

Table 4.4
Basic Stated Choice Model: HEV Logit for 6 Cities (70 Iterations)

Attribute	Alternative(s)	Acronym	Estimate	T-value
Drive alone constant	DA	DASC	1.7449	19.71
In-vehicle cost	All	CST	-.051265	-3.60
Main mode time	DA, RS	TMCAR	-.005719	-3.35
Parking cost (all day)	DA	PRK	-.013878	-3.45
Personal income	DA	PINC	.0016314	2.14
Car availability index	DA	CARAV	.062650	2.45
Ride share constant	RS	RSASC	1.7054	17.97
Bus specific constant	BS	BSASC	1.7981	27.09
Main mode time	BS,TN,LR,BWY	TMPT	-.00294	-2.61
Access mode time	BS,TN,LR,BWY	ACT	-.003857	-2.95
Egress mode time	BS,TN,LR,BWY	EGT	-.000818	-1.44
Train specific constant	TN	TNASC	1.7692	22.79
Light rail specific constant	LR	LRASC	1.7691	24.17
Lambda	DA		4.3109	3.29
Lambda	RS		4.0631	3.34
Lambda	BS		17.357	3.76
Lambda	TN		9.2403	2.90
Lambda	LR		10.107	3.80
Lambda	BWY		1.000	-
Log-likelihood at convergence	-2186.76			

The overall goodness of fit of the model set varies from 0.32 to 0.38. The valuation function models are more complex and produce a slightly smaller reductions in the log-likelihood towards zero at convergence. The additional attributes introduced into the valuation function models vary in significance across the six travel-mode alternatives. The quadratic in time contributes to the bus and light rail modes but with opposite signs. A negative sign suggests

Table 4.5
Valuation Function Stated Choice Model: Basic Logit for 6 Cities (6 Iterations)

Attribute	Alternative(s)	Acronym	Estimate	T-value
Drive alone constant	DA	DASC	-.344484	-3.29
In-vehicle cost	All	CST	-.57166	-14.15
Main mode time	DA, RS	TMCAR	-.068657	-10.86
Parking cost (all day)	DA	PRK	-.097506	-13.54
Personal income	DA	PINC	.007868	2.39
Car availability index	DA	CARAV	.35052	3.83
Ride share constant	RS	RSASC	-.80128	-6.99
Bus specific constant	BS	BSASC	.16340	1.28
Main mode time	BS,TN,LR,BWY	TMPT	.060689	-8.08
Access mode time	BS,TN,LR,BWY	ACT	-.042895	-7.5
Egress mode time	BS,TN,LR,BWY	EGT	-.020764	-3.70
Train specific constant	TN	TNASC	.028564	0.26
Light rail specific constant	LR	LRASC	-.01880	-.16
Quadratic time	BS	TMPTQB	-.002184	-2.69
Quadratic time	LR	TMPTQL	.0009425	1.65
Interactive time and cost	DA	TCMDA	.012865	4.73
Interactive time and cost	RS	TCMRS	.009216	2.66
Interactive time and cost	TN	TCMPT	-.016104	-1.51
Interactive time and cost	BWY	TCMPTW	-.014921	-1.68
Log-likelihood at convergence	-2239.67			
Pseudo-R^2	0.380			

Table 4.6
Valuation Function Stated Choice Model: HEV Logit for 6 Cities (114 Iterations)

Attribute	Alternative(s)	Acronym	Estimate	T-value
Drive alone constant	DA	DASC	1.7707	7.45
In-vehicle cost	All	CST	-.05765	-1.04
Main mode time	DA, RS	TMCAR	-.006877	-1.43
Parking cost (all day)	DA	PRK	-.008829	-1.53
Personal income	DA	PINC	.0005549	1.4
Car availability index	DA	CARAV	.052777	1.6
Ride share constant	RS	RSASC	-.03264	-.05
Bus specific constant	BS	BSASC	1.7904	8.19
Main mode time	BS,TN,LR,BWY	TMPT	-.007243	-1.53
Access mode time	BS,TN,LR,BWY	ACT	-.004343	-1.76
Egress mode time	BS,TN,LR,BWY	EGT	-.0018195	-1.54
Train specific constant	TN	TNASC	1.7993	8.60
Light rail specific constant	LR	LRASC	1.7884	8.11
Lambda	DA		9.4223	1.32
Lambda	RS		.9783	1.75
Lambda	BS		7.7896	1.43
Lambda	TN		9.4361	1.54
Lambda	LR		9.3416	1.76
Lambda	BWY		1.000	-
Quadratic time	BS	TMPTQB	-.0002086	-1.53
Quadratic time	LR	TMPTQL	.00006717	1.83
Interactive time and cost	DA	TCMDA	.001289	1.32
Interactive time and cost	RS	TCMRS	.0003655	1.42
Interactive time and cost	TN	TCMPT	-.002344	-1.78
Interactive time and cost	BWY	TCMPTW	.0004882	1.03
Log-likelihood at convergence	-2206.48			

that the rate of change in relative utility associated with a travel-time increase decreases at an increasing rate as travel-time increases. A positive sign suggests that the rate of change in relative utility associated with a travel-time increase decreases at a decreasing rate as travel time increases. The two-way interactions between time and price are statistically significant in drive alone and ride share (positive), and in train and busway (negative). A positive sign indicates that, as travel time increases, the change in relative utility is less as price increases. A negative sign indicates that, as travel time increases, the change in relative utility is greater as price increases.

An important result emerges from a comparison of the BL and HEVL models (Table 4.7). The mean VTTS for public transport is substantially lower when we relax the constant variance assumption. In contrast the mean VTTS for drive alone and ride share is slightly higher. The reasoning lies in the variance estimates. The mean estimate of variance varies substantially across the four public transport modes, given busway is set equal to 1.00. The large variance associated with unobserved effects which clearly are more prominent for public transport appear to have confounded the contribution of time and cost in public transport more than for the automobile. Consequently the inflated VTTS for public transport in the basic logit model is the result of failure to account for some unobserved influences on relative utility which are suppressed through the constant variance assumption and consequently distributed to the observed effects. However, when we introduce non-linearity into the VTTS estimate expressed as a valuation function we find that at the mean of the sample travel time and costs the higher VTTS associated with the basic linear model is "reinstated" for public transport. The fact that we have a similar value at the mean of the sample for HEVnlL does not imply that the basic linear model is correct; rather it suggests in this single empirical study that the downward valuation for public transport VTTS consequent on allowing for non-constant variance in a linear specification is approximately offset by an upward valuation after allowing for non-linearity in the non-constant variance model, producing a result at the sample mean very close to that of a constant variance model with linearity. What is of more importance is the distribution of values across the sample summarised in Table 4.8 for a range of travel times and costs for each of the 6 travel-modes.

Table 4.7
Mean Estimates of Commuter Values of Travel Time Savings ($ per person hour)

	BL	HEVL	BnlL *	HEVnlL *
Drive Alone	6.50	6.69	7.21	7.12
Ride Share	6.50	6.69	7.21	7.12
Bus	7.51	3.44	6.37	7.54
Train	7.51	3.44	6.37	7.54
Light Rail	7.51	3.44	6.37	7.54
Busway	7.51	3.44	6.37	7.54

note: * columns are based on mean time and cost from Table 4.1

Table 4.8
VTTS Distribution in Travel Time and Cost Space for each Commuter Mode based on a Synthesised Range of Times and Costs - BnlL Model (HEVnlL in parenthesis)

	DA	RS	BS	TN	LR	BWY
c=1, t=20	8.22 (8.25)	7.96 (7.56)	7.12 (8.34)	5.12 (5.81)	6.48 (7.70)	4.97 (8.13)
c=1, t=25	9.09 (9.13)	8.58 (7.80)	9.41 (10.54)	4.48 (4.80)	5.49 (6.99)	4.42 (8.49)
c=1, t=30	10.16 (10.21)	9.29 (8.05)	11.71 (12.74)	3.98 (4.10)	4.50 (6.29)	3.97 (8.90)
c=1, t=35	11.53 (11.59)	10.14 (8.32)	14.00 (14.93)	3.59 (3.57)	3.51 (5.58)	3.61 (9.34)
c=1, t=40	13.31 (13.40)	11.16 (8.61)	16.29 (17.13)	3.26 (3.17)	2.52 (4.87)	3.31 (9.83)
c=1.5, t=20	7.64 (7.67)	7.53 (7.38)	7.12 (8.34)	5.97 (7.06)	6.48 (7.70)	5.72 (7.86)
c=1.5, t=25	8.45 (8.49)	8.11 (7.61)	9.41 (10.54)	5.23 (5.84)	5.49 (6.99)	5.08 (8.22)
c=1.5, t=30	9.45 (9.50)	8.79 (7.86)	11.71 (12.74)	4.65 (4.98)	4.50 (6.29)	4.57 (8.61)
c=1.5, t=35	10.72 (10.78)	9.59 (8.12)	14.00 (14.93)	4.19 (4.34)	3.51 (5.58)	4.16 (9.04)
c=1.5, t=40	12.38 (12.46)	10.56 (8.40)	16.29 (17.13)	3.81 (3.85)	2.52 (4.87)	3.81 (9.52)
c=2, t=20	7.07 (7.09)	7.10 (7.19)	7.12 (8.34)	6.82 (8.31)	6.48 (7.70)	6.48 (7.60)
c=2, t=25	7.82 (7.85)	7.65 (7.42)	9.41 (10.54)	5.97 (6.87)	5.49 (6.99)	5.75 (7.95)
c=2, t=30	8.74 (8.78)	8.29 (7.66)	11.71 (12.74)	5.31 (5.86)	4.50 (6.29)	5.18 (8.33)
c=2, t=35	9.91 (9.97)	9.05 (7.92)	14.00 (14.93)	4.78 (5.11)	3.51 (5.58)	4.70 (8.74)
c=2, t=40	11.45 (11.53)	9.95 (8.19)	16.29 (17.13)	4.35 (4.53)	2.52 (4.87)	4.31 (9.20)
c=2.5, t=20	6.49 (6.52)	6.67 (7.01)	7.12 (8.34)	7.68 (9.56)	6.48 (7.70)	7.23 (7.34)
c=2.5, t=25	7.18 (7.21)	7.19 (7.23)	9.41 (10.54)	6.72 (7.91)	5.49 (6.99)	6.42 (7.68)
c=2.5, t=30	8.03 (8.07)	7.79 (7.47)	11.71 (12.74)	5.98 (6.75)	4.50 (6.29)	5.78 (8.04)
c=2.5, t=35	9.11 (9.16)	8.50 (7.72)	14.00 (14.93)	5.38 (5.88)	3.51 (5.58)	5.25 (8.44)
c=2.5, t=40	10.52 (10.59)	9.35 (7.99)	16.29 (17.13)	4.89 (5.21)	2.52 (4.87)	4.81 (8.89)
c=3, t=20	5.92 (5.94)	6.24 (6.83)	7.12 (8.34)	8.53 (10.81)	6.48 (7.70)	7.98 (7.08)
c=3, t=25	6.54 (6.57)	6.72 (7.04)	9.41 (10.54)	7.47 (8.94)	5.49 (6.99)	7.09 (7.41)
c=3, t=30	7.32 (7.35)	7.29 (7.27)	11.71 (12.74)	6.64 (7.63)	4.50 (6.29)	6.38 (7.76)
c=3, t=35	8.30 (8.35)	7.95 (7.52)	14.00 (14.93)	5.98 (6.65)	3.51 (5.58)	5.80 (8.14)
c=3, t=40	9.59 (9.65)	8.75 (7.78)	16.29 (17.13)	5.44 (5.89)	2.52 (4.87)	5.31 (8.57)
c=3.5, t=20	5.34 (5.36)	5.81 (6.64)	7.12 (8.34)	9.38 (12.06)	6.48 (7.70)	8.73 (6.82)
c=3.5, t=25	5.91 (5.93)	6.26 (6.85)	9.41 (10.54)	8.22 (9.98)	5.49 (6.99)	7.76 (7.13)
c=3.5, t=30	6.61 (6.64)	6.78 (7.08)	11.71 (12.74)	7.31 (8.51)	4.50 (6.29)	6.98 (7.47)
c=3.5, t=35	7.50 (7.54)	7.40 (7.31)	14.00 (14.93)	6.58 (7.42)	3.51 (5.58)	6.35 (7.84)
c=3.5, t=40	8.66 (8.72)	8.15 (7.57)	16.29 (17.13)	5.98 (6.58)	2.52 (4.87)	5.82 (8.26)
Range:						
BnlL	5.34 - 13.31	5.81 - 11.16	7.12 - 16.29	3.26 - 9.38	2.52 - 6.48	3.31 - 8.73
HEVnlL	5.36 - 13.40	6.64 - 8.61	8.34 - 17.13	3.17 - 12.06	4.87 - 7.70	6.83 - 9.83

Empirical Valuation Functions

There are two empirical forms of equation (5), the VTTS function. For drive alone, ride share, train and busway we have:

$$\text{VTTS}_{(\text{DA,RS,TN,BWY})} = \frac{\beta + \eta C}{\alpha + \eta T} \tag{25}$$

For bus and light rail we have:

$$\text{VTTS}_{\text{(BS,LR)}} = \frac{\beta + 2\gamma T}{\alpha} \tag{26}$$

where

β = linear time parameter

γ = quadratic time parameter

η = interaction of time and cost parameter

α = linear cost parameter

C and T are nominal levels

The range of VTTS BnlL and HEVnlL in Table 4.7 may appear to be similar, but in fact the differences are quite substantial, especially for public transport. The only travel-mode with consistently similar VTTS is car drive alone. For the other modes, the difference can be as high as 90%, with the HEVnlL model producing higher VTTS. The implications for forecasts of user benefits in transport project evaluation is clear - there is the potential for substantial differences in the net benefits to different segments of the travelling market; currently simple mean point estimates have the following properties in the current study: they over-estimate the time benefits for train, light-rail, and busway trips, and under-estimate them for the shorter train trips, longer car drive alone, and bus trips. As will be shown below, these findings translate to the distribution of VTTS by travel-mode derived from distributions of actual market times and costs.

Mapping Valuation Function Outputs with Socioeconomic Segments

To meaningfully evaluate the relationship between VTTS and socioeconomic characteristics of the commuter and their household, we derived VTTS for each sampled commuter for each alternative in their choice set for both BnlL and HEVnlL models (Figures 4.1-4.4). We then implemented a method known as CARTS (classification and regression trees - Breiman et al., 1993) to partition the data into relatively homogeneous terminal nodes (i.e., classes), taking the mean value of VTTS observed in each node as its predicted value. The mean square error is used to rank trees of different sizes. In addition, a resubstitution relative cost measure is generated to measure the misclassification rate for the data used to grow the classification tree. After extensive testing we were able to establish the role of socio-economic characteristics in partitioning the sample, namely age of commuter, personal income, and car availability. A simple linear generalised least-squares regression confirmed the statistical significance of these three influences on variation in VTTS across the sampled population. Other socioeconomic variables - occupation, employment status, sex, city of residence, drivers licence status, and highest level of education - had no significant influence at all.

The VTTS for all modes were pooled in the CARTS analysis for each of BnlL and HEVnlL derived VTTS. The BnlL analysis produced 17 nodes and the HEVnlL analysis produced 21 nodes, in both cases at a depth of 5 (i.e., a 5-level tree). Age of the commuter, followed by car availability and then personal income defines the relative importance of socio-economic discrimination in the two analyses. The concept of important in a CART application involves calculation of the improvement measure attributable to each socio-economic variable in its role as a surrogate to the primary split. The value of these improvements are summed over each node and totalled, and then scaled relative to the best performing variable. The most important variable, age, is given an importance score of 100 in both analyses; car availability has a value of 60.22 in BnlL and 31.6 in HEVnlL; personal income has importance values of 34.7 in BnlL and 16.24 in HEVnlL. The dominating influence of age in stratifying VTTS is clear, although much more so for the HEVnlL analysis.

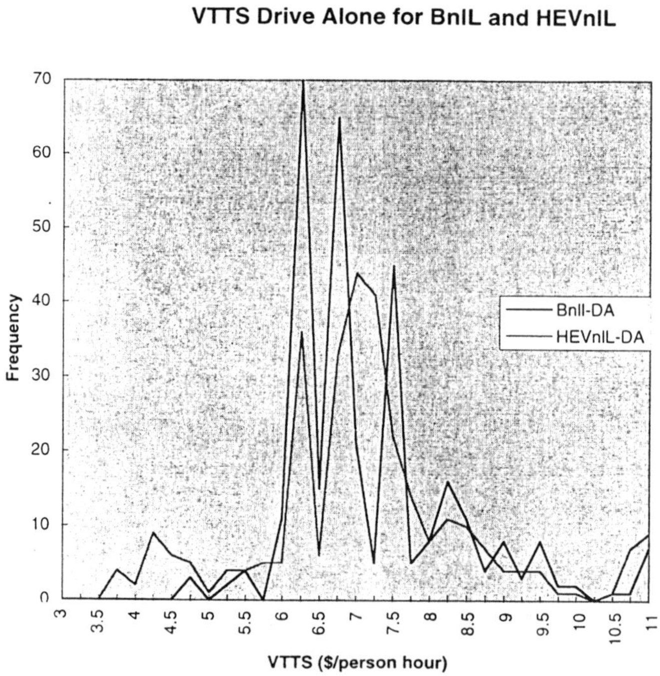

Figure 4.1
Distribution for VTTS drive alone for BnIL and HEVnIL

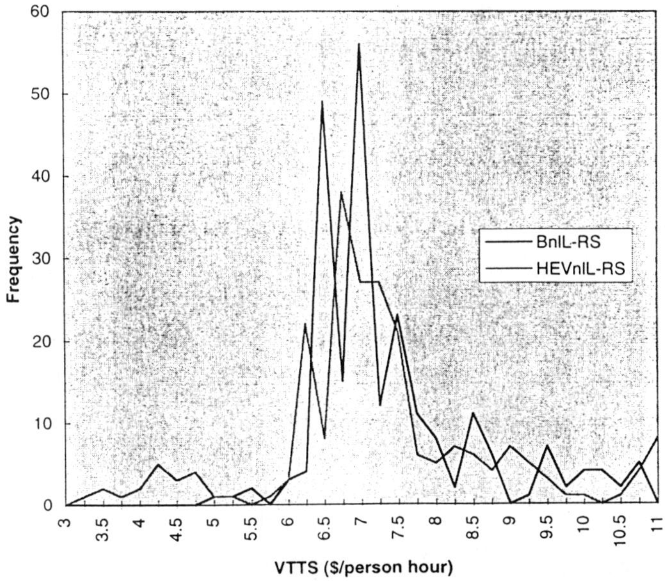

Figure 4.2
Distribution for VTTS ride share BnIL and HEVnIL models

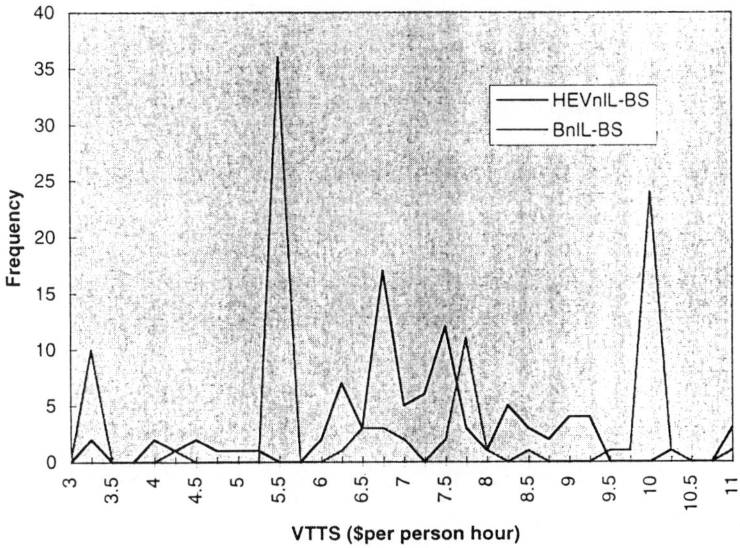

Figure 4.3
Distribution for VTTS bus BnIL and HEVnIL

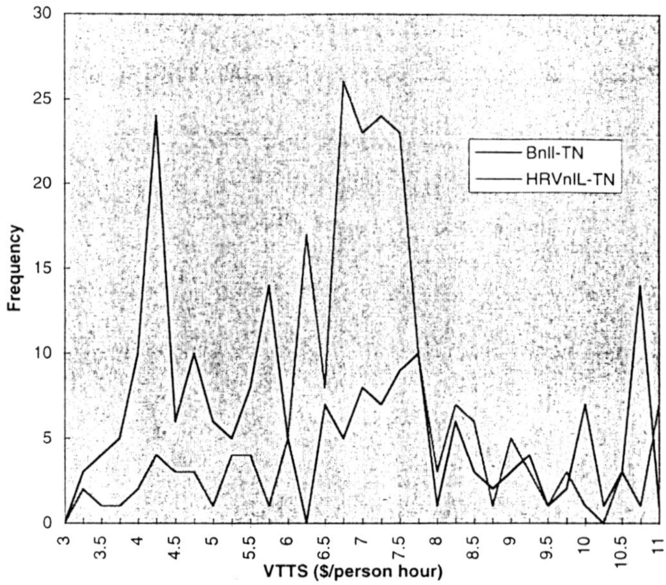

Figure 4.4
Distribution for VTTS train BnIL and HEVnIL

The two trees derived from CARTS are summarised in Figures 4.5 and 4.6, together with the mean VTTS for the sub-sample, the number of observations in each part of the tree, and the socio-economic partitioning variable. The variation in VTTS between socio-economic segment is quite substantial, ranging from $5.89 to $9.97 around a mean of $7.29 for the BnlL model, and, from $6.34 to $10.2 (excluding one clear outlier of $3.86) around a mean of $7.36 for the HEVnlL model. The combined impact of personal age and personal income in identifying the market segments with different mean VTTS is clear. The further discriminatory role of car availability (defined by the ratio of number of cars to number of workers) is a very strong influence on splitting a mean VTTS by as much as 15%.

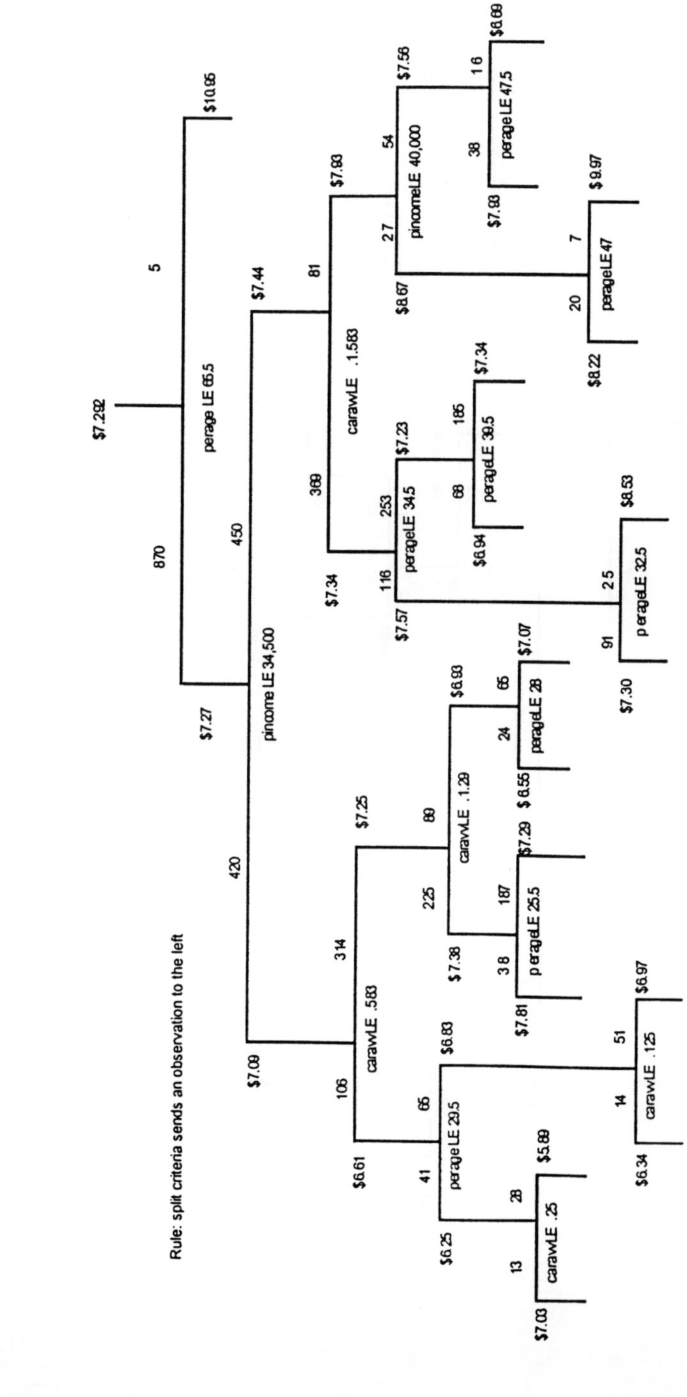

Figure 4.5. Socio-economic classification for BnIL model

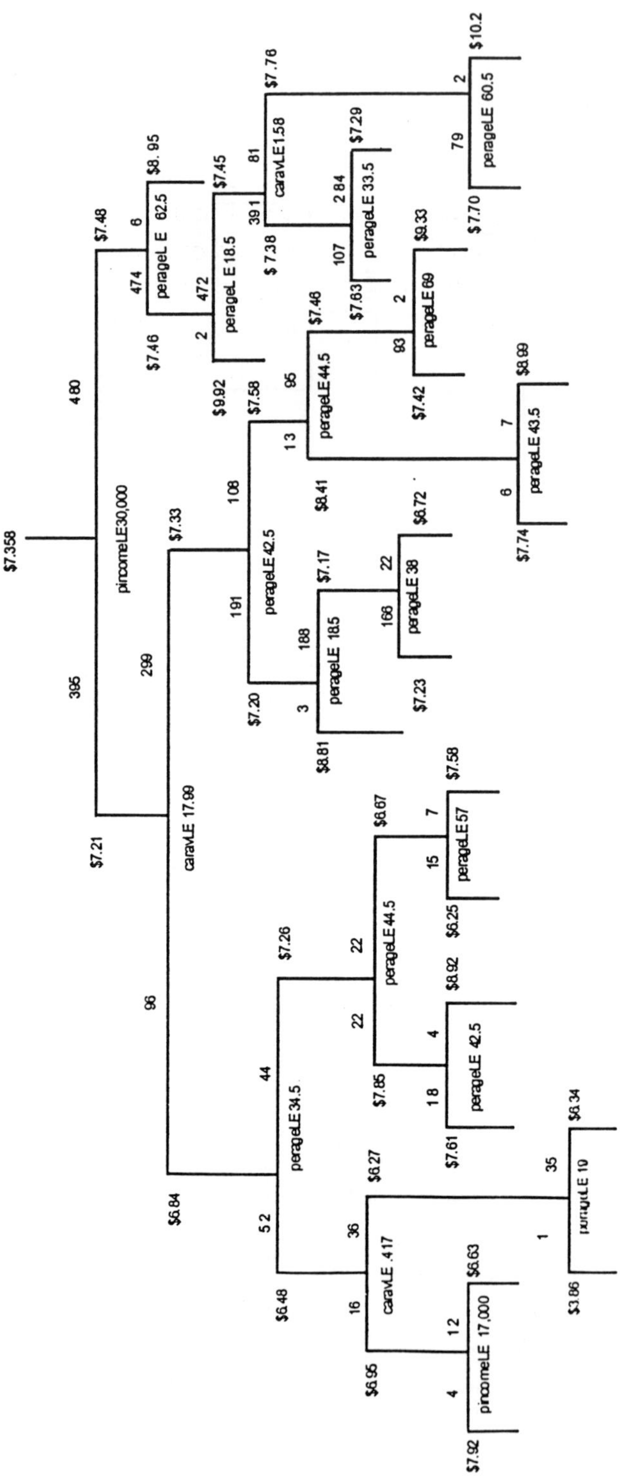

Figure 4.6. Socio-Economic classification for HEVnIL model

CONCLUSIONS

This chapter offers a number of important insights into the benefits of generalising valuation from point estimates to functions, and the implications that relatively restrictive assumptions on the nature of variation in unobserved yet potentially important influences on choice might have in the definition of the distribution of values of attributes in a population. The approach has obvious merit in the valuation of the wider set of influences on individual behavior, especially the growing set of environmental considerations (Daniels, 1997).

ACKNOWLEDGEMENTS

Discussions with Bill Greene, Chandra Bhat, Mark Bradley, Jordan Louviere, Don Anderson, Michael Hanemann, Richard Carson, Truong T. Truong, and Andrew Daly are appreciated. Permission from Chris Thorpe, Director, Bureau of Transport and Communication Economics (BTCE), to use a subset of the data collected by the Institute of Transport Studies for the BTCE's Greenhouse Gas Emissions study is gratefully acknowledged. The comments of Thomas Laitila have assisted greatly in the revision of the chapter.

REFERENCES

Allenby, G. and J. Ginter (1995). The effects of in-store displays and feature advertising on consideration sets. *International Journal of Research in Marketing*, **12**, 67-80.

Bhat, C. (1995). A heteroscedastic extreme value model of intercity travel mode choice. *Transportation Research B*, **29**, 471-483.

Bradley, M. A. and A. J. Daly (1992). *Uses of the logit scaling approach in stated preference analysis*. Paper presented at the 7th world conference on transport research, Lyon, France.

Bradley, M. and A. Daly (1994). Use of the logit scaling approach to test rank-order and fatigue effects in stated preference data. *Transportation*, **21**, 167-184.

Bradley, M. and H. Gunn (1990). Stated preference analysis of values of travel time in the Netherlands. *Transportation Research Record, No.* **1285**, 78-89.

Bradley, M., C. Rohr and C. Heywood (1995). *The value of time in passenger transport: a cross-country comparison*. Paper presented at the 7th world conference of transport research, Sydney, Australia.

Breiman, L., J. Friedman, R. Olshen and C. Stone (1993). *Classification and Regression Trees*. Chapman and Hall, New York.

Daniels, R. (1997). *Combining stated choice methods and discrete choice models in the development of valuation functions for environmental attributes influencing choice behavior.* Doctoral thesis, Institute of Transport Studies, The University of Sydney, Australia.

DeSerpa, A. C. (1971). A Theory of the economics of time. *Economic Journal*, **81**, 828-45.

Greene, W. (1996). *Heteroskedastic extreme value model for discrete choice.* Unpublished manuscript, Econometric Software Inc, New York and Sydney.

Hensher, D. A. (1994). Stated preference analysis of travel choices: The state of practice. *Transportation*, **21**, 107-134.

Hensher, D. A. (1997). Behavioral value of travel time savings in personal and commercial automobile travel. In: *The Full Costs and Benefits of Transportation* (D. Greene, D. Jones and M. Delucchi, eds.), pp. 245-280. Springer-Verlag, Berlin.

Hensher, D. A. and M. Bradley (1993). Using stated response data to enrich revealed preference discrete choice models. *Marketing Letters,* **4**, 139-152.

Hensher, D. A and T. P. Truong (1984). Valuation of travel time savings from a direct experimental approach. *Journal of Transport Economics and Policy,* **19**, 237-261.

Hensher, D. A., F. W. Milthorpe and M. Lowe (1995). *Greenhouse Gas Emissions and the Demand for Urban Passenger Transport: Final Report: Summary of Approach and Selective Results from Application of the ITS/BTCE Simulator,* Report 8, Institute of Transport Studies, The University of Sydney, Australia.

Hensher, D. A., J. J. Louviere and J. Swait (1998). Combining Sources of Preference Data: The Case of the Lurking Lambdas. *Journal of Econometrics.* In press.

Louviere, J. J., D. A. Hensher, D. A. Anderson, T. Raimond and H. Battellino (1994). *Greenhouse Gas Emissions and the Demand for Urban Passenger Transport: Design of the Stated Preference Experiments,* Report 3, March, Institute of Transport Studies, The University of Sydney, Australia.

McFadden, D. (1981). Econometric models of probabilistic choice. In: *Structural Analysis of Discrete Data with Econometric Applications* (C. F. Manski and D. McFadden, eds.), pp. 198-272. The MIT Press, Cambridge, MA.

Morikawa, T. (1989). *Incorporating Stated Preference Data in Travel Demand Analysis.* Doctoral Dissertation, Department of Civil Engineering, Massachusetts Institute of Technology MIT, Cambridge, MA.

Suzuki, S., N. Harata and K. Ohta (1995). *A study on the measurement methods of the value of time.* Paper presented at the 7th world conference of transport research, Sydney, Australia.

Swait, J., J. Louviere and M. Williams (1994). A sequential approach to exploiting the combined strengths of SP and RP data: application to freight shipper choice. *Transportation*, **21**, 135-152.

Truong, T. P. and D. A. Hensher (1985). Measurement of travel times values and opportunity cost from a discrete-choice model. *Economic Journal,* **95**, 438-451.

APPENDIX A

THE CONTEXTUAL STATEMENT ASSOCIATED WITH THE TRAVEL CHOICE EXPERIMENT

We would now like to ask some questions about your trip TO work. We need to speak to the person in the household who completed the Commuter Questionnaire.

IF THERE IS NO COMMUTER IN THE HOUSEHOLD GO TO QUESTION 14

How long does it take you to travel to work, door to door, on a normal day (ie. without any abnormal traffic or public transport delays) READ OPTIONS

Less than 30 minutes... 1
30 to 45 minutes.... 2
Over 45 minutes.. 3

SELECT THE RELEVANT SET OF CHOICE CARDS FOR THE RESPONDENT'S TRAVEL TIME

We are going to show you 3 possible choices of transport options in your area. We are not suggesting that these changes will happen in your area, we are just using these options to understand how individuals and households choose to cope with possible changes to transport. We need your help to try to understand how transport facilities can best service your needs under a variety of possible conditions.
We would like you to consider each choice with reference to your current trip TO work.

TRAVEL CHOICE 1.

CHOOSE A SET OF THREE CARDS AT RANDOM FROM THE TRAVEL TIME SET WHICH IS RELEVANT FOR THE RESPONDENT. TAKE ONE OF THOSE CARDS. WHAT IS THE NUMBER OF THE CARD ___________

This is the first choice. (SHOW THE RESPONDENT THE CARD AND EXPLAIN THE FEATURES OF THE OPTIONS)
If these were the options available for your trip to work, which one would you choose?

Car toll route................. 1
Car no toll route................. 2
Bus............... 3
Train 4
Light rail (Tram)................ 5
Busway................ 6
(A busway is a dedicated lane for buses which extends for most of your trip)

Which set of times did you consider when you were thinking about getting to/from the public transport options (regardless of whether you chose public transport)?

From home:	To your destination:
Walk.............. 1	Walk.............................1
Car/bus 2	Bus..............................2

If you were to travel by private vehicle on either a toll or a no toll route (regardless of whether you chose these options), would you

Drive alone..1
Carpool or share a ride as driver..............2
Carpool or share a ride as passenger.......3

If these were the set of travel choices that were available for your trip to work, do you find them so unattractive that you would seriously consider

	Yes	No
Changing your work place	1..................	2
Changing where you live..........................	1	2

IF THE RESPONDENT CHOSE EITHER OF THE CAR OPTIONS CONTINUE WITH THE FOLLOWING QUESTION, IF NOT GO TO CHOICE 2.

Given the choice that you have made to travel by private vehicle on a (TOLL/NO TOLL ROUTE) how would this affect the time that you leave home compared with now. Would you leave

Earlier, if so by how many minutes............1 ________mins
Later, if so by how many minutes..............2 ________mins
At the same time..3

APPENDIX B

EXAMPLE OF THE FORMAT OF A TRAVEL CHOICE EXPERIMENT SHOWCARD

SA101	1. CAR TOLL ROAD	2. CAR NO TOLL
Travel time to work	**10 min.**	**15 min.**
Time variability	**None**	**None**
Toll (one way)	**$1.00**	**free**
Pay toll if you leave at this time (otherwise free)	**6-10 am**	—
Fuel cost (per day)	**$1.00**	**$3.00**
Parking cost (per day)	**Free**	**Free**

	3. BUS		4. TRAIN	
Total time in the vehicle (one way)	**10 min.**		**10 min.**	
Time from home to your closest stop	Walk **5 min.**	Car/Bus **4 min.**	Walk **5 min.**	Car/Bus **4 min.**
Time to your destination from the closest stop	Walk **5 min.**	Bus **4 min.**	Walk **5 min.**	Bus **4 min.**
Frequency of service	**Every 5 min.**		**Every 5 min.**	
Return fare	**$1.00**		**$1.00**	

APPENDIX C

DESCRIPTION OF VARIABLES IN THE TRAVEL CHOICE EXPERIMENT

Travel Time to Work

There are three different sets of showcards representing short (under 30 minutes), medium (30-45 minutes) and long (45 minutes and over) commutes. They are matched to the commute times currently experienced by respondents. Within each set of showcards, there are three levels of travel times. All public transport options have the same levels as each other, allowing for different combinations across the public transport pairs in each replication. Having the levels the same enables us to investigate the influence that image (through the mode-specific constants) plays in determining preferences within the public transport modes after allowing for the influence of the balanced set of attributes and levels in the design. Travel times on the tolled road have been selected so that it is never worse than the time on the non-tolled route.

Pay Toll if You Leave at this Time (otherwise free)

The tolled route option only has a toll at peak congestion times. The peak over which the toll applies is varied to see what impact a short, medium, and long toll period have on mode and departure time decisions.

Toll (one-way)

The toll only applies to the tolled routes when the respondent's commute trip commences within the times specified by the previous variable. There are three levels of toll for each travel time set, with toll levels increasing for the longer travel time sets. Tolls in excess of current tolls in Sydney have been included to assess the impact of increases beyond the current levels in one City. The toll on the M4 in Sydney is $1.50 for a car; the M5 toll is $2 on the section, currently in place, but is likely to increase up to $4 when the second section is open. Tolls in the experiment vary from $1 to $6.

Fuel Cost (per day)

Fuel cost is varied from current levels to a tripling of current levels to assess possible changes commuters will make as a response to large increases in fuel prices. The daily fuel cost for the commute trip on the tolled road is assumed to be equal to or lower than that experienced on the non-tolled route. Fuel costs can be as high as $15 per day for trips in excess of 45 minutes on a free route. This represents a tripling of fuel prices.

Parking Cost (per day)

Another method for reducing the attractiveness of private vehicle use, particularly in central city areas, is to increase parking charges. Three levels of parking charge are used in the experiment to see how sensitive respondents are to parking costs. A fixed set of charges ranging from free to $20 in evaluated.

Travel Time Variability

This variable is calculated for private vehicle modes only, with levels based on 0, ±20% and ±30% of the average trip time on "no toll" roads, and 0, ±5%, and ±10% of the average trip time for "tolled routes". Toll roads will always be equal to or better than non-tolled roads on trip reliability.

Total Time in the Vehicle (one-way)

For public transport only, this variable refers to the time spent travelling on a train, bus, light rail (LRT) or busway. There are three travel time sets to match those of private vehicles. Only two public transport systems are compared or traded off at once to make the experiment more realistic for the respondent. Thus, there are 4 sets of public transport combinations, listed above. Any other combinations are not meaningful. All public transport options share the same experimental levels enabling the investigation of the role of image in respondent's preferences.

Frequency of Service

This variable gives the number of minutes between each service, and has three levels. The frequency for all modes has a range from a low of 5 minutes to a high of 25 minutes.

Time from Home to Your Closest Stop

The distance from the respondent's home to the public transport stop, in minutes, is measured in both walk time and time travelling by a motorised form of transport. The respondent will be asked to indicate which means of access they would us if they were to use public transport. There are three levels: 5, 15, and 25 minutes walk time, and 4, 6, and 8 minutes by a motorised mode. This same logic is applied to the *Time to Your Destination from the Closest Stop* except that the only motorised mode available is bus. It is very rare that a commuter will use a car to complete a trip after alighting from public transport. The taxi option is excluded.

Return Fare

This variable gives the return fare in dollars. This has three levels, with the same fare sets being used for all public transport modes for each trip length.

5

THE END OF EQUILIBRIUM

Phil Goodwin, ESRC Transport Studies Unit, University College London, U.K.

INTRODUCTION

> "The great attraction of equilibrium is that it is essentially timeless, so it is not necessary to look in detail at behavioural mechanisms, such as *how* or indeed *when* decisions are taken."
>
> Bell and Iida (1997)

> "Higher day-to-day variation in activity than in travel is found. Substantial year-to-year changes in activity and travel are also observed, with travel exhibiting more regularity than activity confirming past research on habitual behaviour...Models of activity behaviour cannot be specified without considering time explicitly...using time-scales such as short-term versus long-term. This will allow (us) to separate components of behavioural variation that are due to day-to-day, week-to-week, month-to-month and year-to-year scheduling and/or habits from structural effects due to changes in lifestyle and activity/transportation system changes."
>
> Ma and Goulias (1997)

This chapter explores the tensions between those two statements, arguing that they represent two long established analytical traditions in which the historic dominance of the former is now giving way to the future pre-eminence of the latter. The key question at stake is: if we *do* want to understand the how and when of behavioural mechanisms, are we able to sustain the concept of a timeless equilibrium?

The concept of equilibrium has exerted a commanding authority over the development of economic theory in general, and travel behaviour modelling in particular. It underpins the construction of elaborate and elegant mathematical chains of reasoning, inferences from observed data, and analytical structures, which are by any standards impressive intellectual achievements. It is at the heart of computer programmes, used by many of the great cities of the world, which forecast the traffic flow on every street, and the number of trips by each of hundreds of combinations of method of transport, type of person, and journey purpose, between a million or more origin-destination pairs - and all built on a tractable set of surveys to be carried out in a single base year.

However, many of the chapters in this book, as of the best of recent research, are implicitly moving away from the equilibrium tradition, focusing on the *processes* of adjustment, of individuals and systems, rather the states they happen, temporarily, to be passing through. This chapter argues that the concept of equilibrium is itself no longer "in equilibrium" with prevailing trends in the analysis of travel demand and the evaluation of transport policy. It acts as a barrier to understanding how choices are made, and can offer dangerously misleading bias in the assessment of policy.

The current situation may be described as a cognitive dissonance among transport modellers. On the one hand, equilibrium is recognised as an unlikely description of real behaviour and systems. On the other hand, it is thought to be a necessary condition for workable models. Both views are expressed on the same page of Ortuzar and Willumsen's (1994, p. 350) textbook:

> "There are good reasons to suspect that equilibrium does not happen in practice, not even at the simplest network level. The real systems are in a permanent state of change, with travellers experimenting with new routes, modes and destinations. Families change residences, jobs, shopping and social patterns and lifestyles."

But

> "The 'bottom line' is that one needs to use models to provide advice on transport decisions and this requires comparing alternative ways of intervening in the transportation system. Consistency in the use of models is then of capital importance as we wish to compare 'like with like'. Casting the transport modelling effort into a general equilibrium framework seems a prerequisite for ensuring this consistency... the state of the art of equilibrium modelling is such that one seldom has to sacrifice too much realism to achieve it."

This proposition depends on the view that equilibrium is, for practical purposes, a reasonably unbiased approximation to the dynamic reality. The argument in this chapter is that this is, in general, not the case: the difference between static and dynamic modes of analysis is not to be understood simply as alternative legitimate ways of seeking insight into a complex world (as is the case, for example, when we might choose to compare the advantages of revealed and stated preference surveys, or aggregate and disaggregate data, or trip-based and activity-based modelling frameworks). Rather, the claim is made that - in general - equilibrium foundations for travel choice modelling are likely to be systematically wrong, while dynamic ones at least have the possibility of being right[8].

THE EQUILIBRIUM TRADITION

As an informal starting point, the condition of equilibrium applies in the relationship between a dependent variable Y and an independent variable X which influences it, when there is no incentive for Y to change. For example, suppose that income is the only influence on car ownership. Then car ownership is in equilibrium with income if it has already adjusted fully to the present level of income. If however income has recently changed, but car ownership has not yet adjusted to the new level, it is not yet in equilibrium. By extension, a whole system is in equilibrium if all its variables have fully adjusted to each other, such that there will be no change unless a perturbation comes from outside.

Random movements, or repeated cyclical movements, around a stable base position and with no overall systematic change, may be included as equilibrium without damage to the concept. Complicated systems with many interactions can be handled, at least conceptually, by suitable use of ceteris paribus assumptions. There is no condition that an equilibrium has to be unique, or to give a social optimum, though these cases have been much analysed.

For mainstream microeconomics, there are two classic uses of equilibrium. The first relates to

[8] The most frequent criticism made by some commentators on an earlier draft of this paper vigorously defended the *concept* of equilibrium as valid and useful: they redirected my criticisms towards inadequate *use* of it by imperfect human agencies. Thus one might distinguish between Platonic equilibrium theory, and real equilibrium practice. I find this philosophical stance awkward, but suspect that it does not make much difference to the underlying solutions to the problems discussed: the changes to practice that would need to be made to bring it into line with true equilibrium theory would be in substance the same as are here described as "dynamic." In either case we would need to abandon current interpretation of cross-section data, choice without process, outcomes without mechanisms, and forecasts without lags. But if the reader prefers to recast my argument as a defence of "proper" equilibrium analyses, I do not think I mind very much.

individual choice, and supposes that the individual will arrange expenditures and activities in response to prevailing prices and other conditions in such a way as to maximise their personal utility: since they gain no further improvement by altering any choice, the individuals are in a stable equilibrium. The second relates to the market as a whole, and is represented by the point of intersection of an upward-sloping supply curve and a downward-sloping demand curve, showing that at this point, and no other, there is a price which "clears the market". Since markets do, manifestly, clear, the existence of such an equilibrium price is seen as axiomatic. The text-book commentary on such a diagram typically notes that if price is not set at this point, either demand or supply or both will adjust *towards* the intersection due to the actors' own actions to maximise utility.

In transport applications, the role of price is usually replaced either by travel time[9], or by a suitable definition of generalised cost consisting of a weighted sum of time, money, and sometimes other cost-like variables (i.e., sharing the property that their expenditure may be counted as a disutility to offset the utility to be gained from the journey's end). The use of generalised cost enriches the idea of utility, but does not otherwise alter the classical framework.

The defining empirical relationship in transport studies is the observation that the speed on a road reduces if more traffic uses it, hence the journey time increases with the volume of traffic. Wardrop (1952) provided the theoretical core of modern transport modelling practice in suggesting that drivers choose among alternative routes for the same journey by picking that route which gives the minimum journey time. However, if too many chose the most direct route, it would become congested and less attractive than an initially rejected longer route. From any initial position, there would be an incentive for drivers to seek quicker routes. Adjustments would continue until equilibrium is reached when no individual can make a further improvement as a result of any individual choice[10]. This would obtain when all routes

---

[9] It is helpful to spell out that the word "time" is used in two different senses. Here, travel time is being treated as analogous to a currency or resource, to be spent or saved. This is conceptually quite distinct from its use as a dimension, analogous to space, which specifies the coordinates at which activities take place. The former is measured in numbers of minutes, hours, or days, the latter by times of day or dates. Dynamic analysis (like any postulate of causal relationships) is marked by the use of time as a dimension, but ordered to preserve sequence, that is, so that earlier and later are not reversible. The distinction made here is somewhat similar to the distinction between area and location in spatial analysis, where confusion in less likely because the words are different.

[10] The effect of traffic flow on speed has the result that each additional vehicle imposes delays on others, in which case an equilibrium achieved through the sum of individual decisions does not necessarily provide a welfare optimum for those individuals in total. This justifies interventions such as road pricing, bus priority systems, etc.

used between any origin and destination have the same journey time, any other possible route being slower, and not used.

Small and Chu (1997) give a graphic description of the importance of the next step in the argument:

> "It has been a third of a century since A. A. Walters (1961) established what is now the standard way economists think about congestion...It was the genius of Walters to recognise that by identifying flow as quantity, the engineering relationship could be viewed as an average cost curve (with a suitable transformation from travel time to cost) and then combined with a demand curve to analyse equilibria and optima."

Wardrop's rules, in their earliest and simplest form, are most elegant if it may be assumed that the total number of journeys, and their distribution among available origins and destinations, are fixed, so that the only dimension of choice open to the driver is which route to take. Although initially only a simplifying assumption, the economy, efficiency, and mathematical rigour of traffic assignment models which discount any behavioural response other than route choice gave them an overwhelming advantage as a practical, though thoroughly misleading, tool which could be applied quickly and cheaply to assess traffic measures on a street or a small part of the network. This is very common in small scale local traffic modelling, and has been applied even for large scale road construction projects (e.g., for trunk roads in the UK in the period 1970 to 1995). The intuitive ex post justification was that for a small scheme, or certain classes of large scheme, one can ignore changes in trip rates, destination choice, mode choice, and other possible responses.

In fact there is no known empirical evidence to support the proposition that route choice is the only behavioural response people make to changes in network conditions, whether small or large. It is by now common ground among researchers (including, as may be seen in their chapters, every author in this volume), academics, the better consultants, and virtually all policy-makers who are sensibly engaged with their field of action, that a wider range of responses than simply adjustment in route choice must be taken into account if travel behaviour is to be understood, or if any transport policy of importance is to be sensibly assessed.

This acceptance of a wider range of behavioural response has not generally been seen as implying criticism of the equilibrium approach of assignment models - quite the reverse. By far the greatest research effort into travel behaviour since the mid 1960s has been devoted to developing equilibrium models which would encompass ever-wider dimensions of choice:

mode choice and destination choice in distribution models in the late 1960s and early 1970s; time of day choice in the late 1980s and early 1990s; recently revisiting trip frequency in policy-sensitive trip generation models. As a parallel track, activity-based models have sought to embed travel choices within a more realistic (though typically less practically-used) context of activity participation, in the early 1980s and again recently.

In many cases, these choices are treated as responses to external influences, rather than being influenced by the behaviour itself - except via the speed-flow curve which then in effect provides the feedback mechanism for all travel choices. In modelling terms, this means that the assignment model most frequently provides the equilibration mechanism for the model as a whole[11]. In other cases, overcrowding on public transport, or market prices of land in an interactive land-use/transport model, provide analogous endogenous feedback mechanisms leading to equilibrium, and Mogridge (1997) reviews analyses in which public and private transport are treated as an interacting equilibrium-seeking mechanism.

The continuing importance of the concept for both practical and theoretical work is expressed in a recent textbook on network modelling by Bell and Iida (1997 p. 2):

> "The techniques presented here are based on the notion of an *equilibrium* between the demand for, and supply of, transportation. While a transportation system may never actually be in a state of equilibrium, it is assumed that it is at least near equilibrium, tending towards equilibrium, and only prevented from attaining equilibrium by changes in external factors...At equilibrium, the transport system reduces to a *fixed point* (the equilibrium costs and flows), and powerful analytical techniques...exist for finding the fixed point. Proponents of equilibrium theory take it as an article of faith that, given the existence of an equilibrium, there are behavioural mechanisms which push the transport system to this fixed point"

It may easily be seen that this "article of faith" is vulnerable. That an equilibrium *can* exist as a potential state of an individual or a system, does not necessarily mean that equilibrium actually *does* apply at the moment of an observation. However, this assumption is ubiquitous. All models estimated using cross-section data assume that equilibrium is not only reachable, but already reached.

---

[11] It is intriguing that the very short-term dynamics of this equilibrating mechanism are rather well-explored, notably the build-up of queues over periods of minutes or, say, a single peak period, leading to the odd result that behavioural equilibrium is often assumed within a model using an elaborate dynamic formulation at its heart. Small and Chu (1997) helpfully resolve some resulting confusion.

The practice of equilibrium analysis following from this may be summarised as follows.

1. Observed patterns of behaviour and choices are, at the time of observation, already in equilibrium with the factors that influence them. (This applies equivalently whether the 'observations' are revealed choices or stated preferences).
2. Relationships estimated from these observations accurately represent the strength of the various influences.
3. Those relationships may then be used to describe an equilibrium position that would obtain as a result of different values of the influencing variables than have been observed.

However, this practice is not always apparent. This is because virtually all discussion, justification, interpretation, and use of equilibrium analysis is carried on using the *language* of dynamics, with verbs that are about changes rather than states ("increase" and "reduce" rather than "higher" and "lower"), and ubiquitous use of the adverb "when". For example, observation that higher income groups own more cars than lower income groups is almost invariably written up in terms such as "when income increases, so does car ownership." Analysis of different life-cycle groups is introduced by "when people get married..." or "when people retire..." When outputs are compared of a model in which higher and lower fuel prices have been used as an input, the analyst will report something like "when prices go up, such and such a proportion of people will switch from car to train."

Use of such language is almost never avoided, as it reflects a firm human commitment to the understanding of the universe in terms of process, change, and cause-and-effect, but it does create great confusion - sometimes in the minds of the modellers themselves - about which dynamic processes are genuinely included in the model structure, and which are imposed as a creative (but inconsistent) interpretation.

AN ALTERNATIVE TRADITION

This section starts with a more formal statement of current thinking on the difference between equilibrium and dynamic formulations, and with this in mind then reconsiders a rich but under-valued tradition of empirical studies from the 1970s, which may be seen as the proper precursors to current work on panel data.[12]

---

[12] Equations (1) to (9) are due largely to a series of papers written by Kitamura, the notation and sequence of argument here being that of his contribution to Goodwin, Kitamura, and Meurs (1990).

Consider a measure of travel behaviour, $Y(t)$, and a vector of exogenous variables, $\boldsymbol{X}(t)$, that influences it. Assume that $Y(t)$ is a differentiable continuous variable in $0 \leq t$, and that both $Y(t)$ and $\boldsymbol{X}(t)$ are observable over a continuous time axis.

A behavioural unit is said to be in equilibrium if its travel behaviour has fully adjusted to the prevailing values of the contributing factors. Formally:

Let, for some s, $\boldsymbol{X}(r) \neq \boldsymbol{X}(s)$ and $\boldsymbol{X}(t) = \boldsymbol{X}, \quad t \geq s > r > 0$

If a change in the exogenous variable vector, $\boldsymbol{X}(t)$, takes place at time s, then the vector is invariant subsequently. If there exist u ($\geq s$) and Y^* such that

$$E[Y(t)] = Y^* \quad \text{for any } t,\ t \geq u \geq s \tag{1}$$

then equilibrium is attained at time u.

For notational simplicity, let $Z(t) = E[Y(t)]$. Then, from Equation (1),

$$dZ(t) / dt = 0 \tag{2}$$

if the behaviour is in equilibrium. On the other hand, behaviour is in disequilibrium in the period between s and u. In this period of disequilibrium, $dZ(t)/dt \neq 0$, and $Z(t) \neq Y^*$. This definition of equilibrium allows the notion of random variation in travel behaviour to exist, for example, in daily travel patterns observed over several days. Observed behaviour, when in equilibrium, is assumed to fluctuate randomly around Y^*, which may be assumed to be a function of $\boldsymbol{X}$. This correspondence between $Y(t)$ and Y^* does not exist when the behavioural unit is in disequilibrium. The lack of correspondence implies that biased estimates can result if a contemporaneous model is applied in an attempt to explain $Y(t)$ on the sole basis of $\boldsymbol{X}(t)$, as discussed further below; a dynamic model needs to be used for consistent estimation of behavioural relationships in this case.

Assume now that the behavioural unit adopts an adjustment process when it is disequilibrium, that an equilibrium will eventually be reached, and that the behavioural unit will then remain in equilibrium unless a further change takes place in the contributing factors. The speed of adjustment may be defined in terms of the duration of adjustment, (u - s), or in terms of the rate of change, $dZ(t)/dt$.

Asymmetric adjustment processes may be observed when the speed of adjustment differs by the direction of change in $\boldsymbol{X}(t)$. For example, a disposal of the family car may necessarily lead to an immediate reduction in car use, while an acquisition of a family car may result in an adjustment process before every family member adopts a new lifestyle taking full account of the family car. The speed of adjustment has important implications for the dynamic analysis of travel behaviour.

Suppose there exists a function, G, such that

$$Y(t) = G[\boldsymbol{\beta}(t), \boldsymbol{X}(t), \varepsilon(t)], \quad t \geq 0 \tag{3}$$

where $\boldsymbol{\beta}(t)$ is a vector of model coefficients, and $\varepsilon(t)$ is a random error term. If this relation holds, the behavioural relation between $Y(t)$ and $\boldsymbol{X}(t)$ is said to be contemporaneous. A special case of interest is the following linear model:

$$Y(t) = \boldsymbol{\beta}(t)'\boldsymbol{X}(t) + \varepsilon(t) \tag{4}$$

This has been widely adopted in trip generation analyses, which have, therefore, assumed that travel behaviour is contemporaneous.

When behaviour Y is contemporaneously related to $\boldsymbol{X}$, it implies that a change in $\boldsymbol{X}$ triggers an immediate adjustment in behaviour. Response lags, habit persistence, threshold effects, and behavioural inertia do not exist in this case. Also the behavioural unit is perfectly aware of its environment. Consequently contemporaneous travel behaviour is independent of the history of both the contributing factors and the behaviour itself. Formally, $Y(t)$ may be called history independent when the following holds:

$$E[Y(t)] = E[G(\boldsymbol{\beta}(t),\boldsymbol{X}(t),\varepsilon(t) \mid \boldsymbol{\beta}(s),\boldsymbol{X}(s),\varepsilon(s),\ 0 \leq s < t)] = E[G(\boldsymbol{\beta}(t),\boldsymbol{X}(t),\varepsilon(t))],\ t \geq 0 \tag{5}$$

A contemporaneous relation is also history independent (note that history independence here implies the inter-temporal independence of the error term, ε).

Suppose travel behaviour is contemporaneous. Then observed behaviour does not depend on the trajectory of past behaviour, and the following holds:

$$\text{if } \boldsymbol{X}(s) = \boldsymbol{X}(t), \text{ then } E[Y(s)] = E[Y(t)] \quad t > s \geq 0 \tag{6}$$

irrespective of the trajectory of $\boldsymbol{X}$ and Y prior to s or between s and t. For example, consider an increase in family income from A to B at time t_1, and a subsequent decrease from B to A at time t_2. If behaviour is contemporaneous, this sequence of changes will not leave any trace on behaviour and $E[Y(t_0)] = E[Y(t_3)]$. The behaviour relation in this case, therefore, shall be referred to as reversible or symmetric. Obviously if behaviour is reversible, it is history independent. Cross-sectional analyses of travel behaviour rest on these hypothetical behavioural properties.

The concepts of history independence and reversible or symmetric behaviour can be extended to the case of non-contemporaneous behaviour. The behavioural relationship may be called history independent at equilibrium if the following holds:

Let $X(r) \neq X(s)$ and $X(t) = \mathbf{X}$, $t \geq s > r > 0$. Let u be the time point when an equilibrium is attained, and suppose the following holds:

$$E[Y(t) \mid X, X(s), \varepsilon(s), X(r), \varepsilon(r), s > r > 0] = E[Y(t) \mid X] = Y^* \text{ for any } t, t \geq u > s \quad (7)$$

Then the behavioural process is history independent at equilibrium. The concept of reversibility or symmetry at equilibrium can be defined similarly.

To take account of the pace at which adjustment takes place, consider a simple monotonic model of adjustment in which the speed of adjustment is determined by the difference, $Z(t) - Y^*$, let

$$dZ(t)/dt = r[Y^* - Z(t)] \quad (8)$$

The speed of adjustment is represented by parameter r in this formulation. Solving this with the assumption that a change in the contributing factors takes place at $t = s$ and that the process is in equilibrium prior to the change with $Z(s - 0) = Z_0$, then:

$$Z(t) = Z_0\exp(-rt) + Y^*(1-\exp(-rt)),\ t > s \quad (9)$$

The behaviour of $Z(t)$ toward equilibrium can be seen as a weighted average of the initial value, Z_0 and the new equilibrium, Y^*.

The approach outlined does use the concept of equilibrium in the sense of a target which travel behaviour would reach when the contributing factors assume fixed values for a prolonged period of time, and assumes that, once an equilibrium is reached, the behavioural process will remain in equilibrium unless the contributing factors change their values. However, it does not depend on, or expect, the existence of an achieved equilibrium at any specific moment. Indeed, on the likely condition that behavioural adjustment is not instantaneous, a sample of behavioural units must consist of those that are in equilibrium and those that are in disequilibrium and therefore in the process of adjustment - the proportions depending on what has been happening to the influencing variables. These are likely to include general external factors (e.g., fares, congestion) which will tend to have moved in the same direction for all of the population, and some personal factors (such as income) for which the trend in one direction is likely to be greater than the trend in the other direction. There is therefore no warranty for treating those in disequilibrium as being random compensating errors.

Statistical inferences using such a sample can only be made on the basis of explicit and rigorous treatment of the adjustment process itself.

For this, use of some sort of time series data here is a strictly necessary, but not a sufficient, condition, or in other words analysis of time series data does not guarantee that the results have any dynamic properties.

$$\begin{aligned} D_t &= f(P_t, I_t) \\ D_t &= f(P_t, I_t, D_{t-1}) \end{aligned} \tag{10}$$

This may be illustrated, at the simplest level, by comparing two different general models which are often used in the analysis of time series data. In the first equation, we say that demand in each year *t* is to be explained by the prices and incomes in the same year. Strictly, we are then assuming that in each year the demand is in equilibrium with prices and income, that is, it must have swiftly fully adjusted to changes in their levels. The equation has no memory - it makes no difference to the demand whether prices and incomes have risen or fallen to their current level. In the second equation, the extra term, D_{t-1}, says that demand in each year *t* is, as before, influenced by the prices and demand in that year, but also by the level of demand in the previous year (*t-1*). This may be considered as a sort of inertia - people's behaviour this year is still influenced by what they did last year - but it is also a shorthand way of dealing with a rather complicated longer term process, since demand in year (*t-1*) was itself influenced by the prices and income in that year and the demand in the year before that (*t-2*), and so on. Then demand in any year is the result of the whole history of previous prices and incomes. In this case, the equation has a memory - it *does* make a difference to the demand in one year what prices and incomes have been doing in previous years. The equation allows people to have some delays in responding to changing circumstances[13].

The most successful work in this tradition has mainly been carried out at aggregate level, in a well established literature calculating demand elasticities from time series data analysed with dynamic models incorporating an explicit lag structure. Goodwin (1992) cited 13 studies in which the effect of fuel price on fuel consumption had been calculated (giving a short-term elasticity of around -0.25 to -0.3 and a long-term elasticity of -0.7 to -0.8); 11 studies in which the effect of fuel price on traffic-levels had been calculated (with results of -0.16 for a short-term effect and about-0.3 for a long-term). Sterner et al. (1992) estimated gasoline demand elasticities for 21 countries, separately. The mean result for the preferred model form gave a short-run elasticity of -0.24 and a long-run elasticity of -0.79. A review of Australian evidence

[13] If however people respond instantly, this will be revealed by the estimated coefficients. Thus the static equation is included within the dynamic one, but not vice versa.

by Luk and Hepburn (1993) cited 28 studies, and came to the conclusion that the long term elasticity of traffic levels with respect to fuel costs was -0.1 in the short run and -0.26 in the long run. Such findings are mirrored in similar patterns on public transport fares elasticities, on average about -0.3 in the short run, and -0.6 or more in the longer run.

In such analyses, definition of short and long is not an externally imposed judgement, but is derived empirically from the estimated coefficients: typical empirical results imply that the pace of behavioural adjustment is such that it takes several years to approach an equilibrium, with say 50% of the adjustment in 1-3 years, and 90% in 5-10 years.

It seems reasonable now to say that the key result of such work - that long-term transport demand elasticities are different, and usually bigger, than short term elasticities - is an established research finding, and is not substantially under challenge in principle. While views differ on how much insight about individual behaviour is possible from aggregate methods, one key insight is clearly provided: some behavioural responses *must* be time-dependent, and include adjustments which take years to be completed. While in one sense that observation is so in tune with what we know informally about the human condition as to seem commonplace, even trivial, its implications for the theoretical underpinning of equilibrium analysis, and the interpretation of survey results, is profound.

Work along these lines at the disaggregate level has been longstanding, though until recently in a sort of second division of minority interest.

We draw attention first to an earlier piece of work which, in retrospect, now seems to merit greater attention than it has been given[14]. This is a study by Fried, Havens, and Thall (1977), carried out for the US Transportation Research Board. When referred to at all (which is now rather rare) it is usually in the context of being the first US proposal for an activity-based transport approach, and this was indeed an important element of the work. However, a more important, and less-noticed, aspect was its proposed treatment of adaptation:

> "A process of adaptation, involving changes in activity and travel patterns and adjustments over time, provides a useful theoretical framework for understanding the travel behaviour of individuals and populations: the motivating force behind these behaviours is the effort to reduce imbalances that exist or develop between personal needs and environmental structures..."

[14] Similar comments might be made of very early work by Hägerstrand - not the much-cited 1970 paper usually referred to as a source of the space-time geographic approach to activity analyses, but research in the 1950s on the dynamic diffusion of ideas and techniques.

> "Discrepancies in person-environment fit invoke the adaptional process consisting of informed trial-and-error sequences that continue until the discrepancies are resolved."

Thus while "reducing imbalance" is obviously related to the concept of equilibrium, the research focus is quite different, involving working forward through a process, rather than backwards from the characteristics of the end-state. The approach was based in psychology (or rather "psycho-social studies") and reading it now is remarkable for how closely it reflects current concerns.

An early empirical study using a framework entirely consistent with Fried et al (though apparently unaware of it) was by Tischer and Phillips (1979), using cross-lagged analysis of a two-wave panel spanning about 18 months. They concluded that

> "A strong causal relationship was found to be operating between beliefs and mode use over time. The relationship is mutually causative; beliefs determine behaviour and behaviour reinforces and changes beliefs... In general, support is evidenced for an adaptation or learning process interpretation of the relationship between beliefs and mode choice behaviour."

The role of memory, habit, discontinuous and non-reversible change can be accommodated in process-oriented approaches, as in, inter alia, Pollak (1970) who seeks to identify the characteristics of a long-term equilibrium which includes habit; Wilson (1976) applying catastrophe theory to mode choice; Goodwin (1977) exploring the consequences of habit for response-path to cost changes; Haag and Grutzman (1993) incorporating memory within nested decision processes; Magnusson and Ottosson (1997) on evolutionary economics and path-dependence.

Much of the interesting work in recent years has been analysis on panel data sets, providing information on travel choices made by the same individuals at two or more points in time. Among the more prolific contributors to this area of work are Golob in the US, Hensher in Australia and Kitamura in the US and Japan: Raymond and Hensher (1997) provide a review of studies, including his own influential Australian study of automobile demand (Hensher et al, 1992). Golob, Kitamura and Long (1997) collected many of the most important recent studies. Another focus of panel studies has been the Transport Studies Unit, University of Oxford. The author (Goodwin, 1997) concluded from a selection of ten panel studies analysed there, in the period 1979-1992, that:

> "Even apparently settled aggregate patterns are based on a very high degree of volatility, movement and turnover at the individual level. Statistical relationships established from cross-section data do not provide a 'track' along which responses move over time. Model coefficients estimated from cross section data about the effects of differences in levels of a causal factor are not the same as those established from panel data about the effects of changes in the level of the same factor. Changes in behaviour due to household circumstances are moderated by differential responses to policy variables, such as public transport fares or deregulation."

In this tradition, Cairns et al. (1998) review empirical evidence from around 50 case studies of the effects on traffic and travel behaviour of policies which reallocate road capacity from cars to public transport or pedestrians. They found that the range and scale of responses to changes in capacity were much greater than had usually been assumed when assessing the schemes in advance. The significance of this was explained by reference to research evidence on the variability over time of traffic speeds, and of individual and aggregate travel behaviour. The following key findings are relevant:

> "Experiments show that for a hundred journeys to work, between 10 and 20 will be travelled with an overall door to door time more than 20% faster than the average, and a similar proportion more than 20% slower than the average.
>
> "Survey evidence indicates that the vehicles making up the traffic stream are not, in general, the same set of cars from day to day. Up to 50% of the cars present on a given day will not be present at the same time on the following day, and many will not be there at all. Nearly half of a London sample changed the number of trips they made in a week substantially, and over 20% of employees changed their main mode of journey to work from Spring to Autumn, including 10% of car drivers who chose a different mode half a year later, and nearly 40% of car passengers. In another survey about 20% of the sample increased or reduced the number of trips they made from one year to the next by more than 10 trips per week, compared with a mean recorded trip rate of around 25 trips. Over 10% of the market for public transport in one year was made up of people who, the previous year, were making more than 10 trips per week by car.
>
> "In successive two or three year intervals, net increases in car ownership of 2%-6% can be composed of something like a quarter of the population changing its car ownership level, with around 10% of the population involved in reductions, especially among two car owning households. To a first approximation, about half

of the variation in mileage travelled by cars on a day is not due to stable differences between the travel patterns of the drivers at all, but due to each of them behaving differently from one day to the next, one week to the next, one year to the next. This is consistent with an extensive research literature which establishes that the long term sensitivity of travel demand to changes in the price of travel is different, and usually bigger, than the short term sensitivity, typically being twice or more. The pace of behavioural adjustment is such that it takes several years for the responses to settle down, with 50% of the adjustment in the first 1-3 years, and 90% in 5-10 years.

"The case studies are interpreted in the light of these findings. We conclude that (a)there is no reason in terms of behavioural plausibility to reject the wide range of behavioural responses cited in personal surveys, or the scale and suggested structure of 'disappearing traffic' found in the traffic counts; (b) he process of adaptation to a new policy starts on day one, but takes between five and ten years before it is near enough to completion to get lost in other even longer term processes; (c) when there are circumstances in which an accurate assessment of 'disappearing traffic' would make a difference to the decision whether to implement a capacity-reallocation or not, then it will be necessary to accept the existence of longer term and wider behavioural responses than is usually done".

While the centre of gravity of panel studies, which took off in Europe, now seems to have moved to the US, there are some signs that Netherlands and Scandinavian work now represents the leading edge in approaches combining psychological disciplines within an activity framework - see, for example, the collection edited by Ettema and Timmermans (1997), including the chapter by Gärling et al. (1997) providing a bridge into the current volume.

An example of the difference in insight that is provided by comparing cross-section relationships with changes observed over time is shown in Table 5.1.

These surveys, carried out in the rather special circumstances of South Yorkshire at a time of very low public transport fares show - as we always find - that higher-income people have higher car ownership levels. People who are one income group higher (lower) than each other have, on average, car ownership which is about 0.12 cars per head more (less). This is a strong, static, result. But the dynamic picture is different: when we look at people whose income has *changed*, the change in their car ownership is much less than expected. Those people who changed from one income group to the next over a three year period only changed

Table 5.1
Static and Dynamic Data on Car Ownership and Income.

Effect of Income on Car Ownership Single Cross Section Survey, 1981		Effect of Income on Car Ownership Panel Surveys 1981 & 1984	
Income Group	Cars per head	Change in Income	Change in Cars
1	0.22	-3	-0.08
2	0.29	-2	-0.06
3	0.48	-1	+0.01
4	0.62	0	+0.07
5	0.59	+1	+0.06
6	0.79	+2	-0.06
7	0.75	+3	+0.11
8	1.08		
Strong symmetrical effect of income on car ownership: shift of one income class linked to car ownership higher or lower by about 0.12 cars per head.		Weak, possibly non-symmetrical effect of changes in income: shift of one income class linked to shift of about 0.02 cars per head.	

their car ownership by an average of 0.02 cars per head, a sixth of the "expected" change. Here, as elsewhere, there is also an indication that the effects of increases in income may not be the same as the effects of decreases. Car ownership is rather sticky, with a strong inertia factor.

Interpretation of a number of theoretical and practical policy studies results in a set of (somewhat impressionistic) suggestions for the likely time scale over which various different sorts of transport policies are likely to have an impact (Table 5.2).

Thus, while not discounting the importance of some behavioural responses, by a proportion of the population, that may be very rapid indeed (in the order of seconds, minutes and hours), the range of choices of interest in this book is likely to require at least several years to adjust to changes in influencing factors.

Table 5.2
Suggested Time Scales for Responses to Transport Policies

POLICY	TIME FOR RESPONSE
Pedestrianisation	Immediate effect on traffic; build-up of effect on number of pedestrians, retailing etc over 2-5 years. Possible longer term effects not yet understood.
Traffic calming	Immediate effect on traffic. Specific schemes unlikely to have long-term effect (indeed, effect may be eroded by traffic growth), but comprehensive area-wide traffic calming likely to have cumulative effects as schemes interact.
Public transport priority, operational improvements and fares policy	Impact on demand within the first year, but longer term elasticities about twice as great after approx. 5-10 years.
New public transport infrastructure	Time scale for response probably as above, but time scale for planning and implementation much longer - perhaps doubling entire period.
Road pricing, changes to petrol costs, etc.	Small response within first year; build-up of effects on car use over 5 years, and on car ownership over 10 years or more; long-term elasticities 2-3 times as big as short ones.
Land-use planning to reduce travel distances	Very small short-term impacts, with cumulative impacts over 20 years, perhaps up to 50 years or more.

The proposition that long-term responses are different from short-term responses, once explicitly stated, is so obviously true that it perhaps hardly needs to be argued, but a problem of internal logical consistency then arises if an equilibrium model's parameters have been estimated from a cross-section data base. For any cross-section analysis there are two possibilities.

Case 1. The relationship or system was genuinely in equilibrium at the time it was observed. In this case, the model (if well-specified from other points of view) may have accurately estimated the strengths of the equilibrium relationship. Forecasts using the model can then accurately describe the nature of the equilibrium outcome of changes of interest. The model can have nothing at all to say about how long it will take to reach that equilibrium outcome, nor the path followed in reaching it, since that information could not (by definition) have been contained in the data used which had already reached equilibrium. However, this implies that a forecast for the year 2005, say, (i.e. using input variables expected to apply in 2005) is describing an equilibrium state that can only apply at some unspecified year *later* than 2005 - how much later is indeterminate. Therefore discounting from the *nominal* year to calculate a net present value for evaluation is invalid. This is discussed further below. This is the more favourable case, that is, a static model which succeeds in describing a valid equilibrium.

Case 2. The relationship or system was *not* in equilibrium at the time it was observed. If variables of interest had changed systematically in the period before they were observed, and other variables had not yet completed their process of adjustment, then no matter how carefully or fully or rigorously the model was specified, the estimation *cannot* give parameters which would accurately describe the equilibrium result of changing any of those variables. At best, it will give a forecast of a state something like the same distance along the road to equilibrium as was observed, but we cannot say how far that is, or therefore how long the forecast state will last, since that information could not (by definition) have been contained in the data used which had not already reached equilibrium. This is a very difficult situation, since it means that we have a forecast that may be inaccurate, but even if it is accurate we do not know what it means - a situation closer to augury than forecasting, and putting far too great a premium on the skill of interpretation. This is the less favourable case, that is, a static model which describes an uncompleted stage in an unspecified dynamic process.

In order to tell whether a cross-section analysis is the more favourable or less favourable case, the necessary condition that we need to consider is whether an equilibrium had already been achieved when the data were observed or collected. This requires two pieces of information: (a) the speed of adjustment, and (b) changes in the influencing variables in the period before the observation.

In summary, the necessary condition for cross-section analyses to deliver equilibrium relationships is that the variables of interest should not have changed significantly in the period before the observation, for a time long enough to have allowed the effects to settle down. If as suggested the adaptation period is in the order of years, this condition is rarely, if ever, likely to apply in conditions observable in modern economies in the real world. Thus we conclude that the parameters estimated from cross section observed data will not, in general, describe equilibrium relationships.

All the discussion above tends to suggest that the theoretical basis of equilibrium assumptions and appoaches as an accurate description of the real world is weak to the point of infirmity.

However, that of itself does not tackle Ortuzar and Willumsen (1994) central defense, cited above, that the only way to use models consistently for advice on transport decisions *requires* a general equilibrium framework. (This is similar, perhaps, to the case in which it is difficult now to defend the concept of cardinal utility theoretically, but quite impossible to carry out a tractable social cost-benefit study without it).

Therefore we need to consider whether the equilibrium approach, applied to policy evaluation, carries bias arising from its behavioural inadequacy.

Obviously, if the forecasts of behavioural response are simply wrong, then so will be any evaluation built on them. However, even apart from this there is a problem of bias. This has been demonstrated in work by Dargay, reported in Dargay and Goodwin (1995), of which a summary now follows. (See also Appendix for a demonstration.).

Suppose that choices are made in such a way as to seek an improvement for the individual, thus retaining economic constructs such as utility and consumer surplus, but that this is a process of successive adjustment. The benefits to travellers in period t from a discrete change in travel costs at time 0 can be measured by the change in consumer surplus at time t, which is the value of travel to consumers. This is measured by the area under the demand curve at time t, $D(t)$, where $D(t) = f_t(c)$. Assuming D is integrable, the change in consumer surplus at time t resulting from a change in costs from c to c' is given by

$$B(t) = \int_{c'}^{c} D(t)\, dc \qquad (11)$$

If D is a linear function of costs, this integral is evaluated as:

$$\int_{c'}^{c} D(t)\, dc = -\tfrac{1}{2}[f_t(c') + f_t(c)](c' - c) \qquad (12)$$

where $f_t(c')$ is the level of demand in period t given a change in costs from c to c' in time 0 and $f_t(c) = D(0)$, the level of demand at the initial price, c, which is the same for all t, given that $D(0)$ is in equilibrium. Although equation (12) is exact only for linear functions, this same formula, known as the "rule of a half" is a reasonably good approximation to the actual integral for marginal changes in c even if D is a non-linear function of c.

Since demand does not respond instantaneously to changes in costs, we have a series of demand curves, $D(t)$, representing the level of demand in different time horizons from $t = 0$ to the time of complete adjustment, or equilibrium, t^*, following a change in costs in time 0. From the definition in equation (11), this implies that the consumer surplus $B(t)$ resulting from the change in costs is also dynamic and changes over time. The intertemporal profile of demand, and hence, benefits, $B(t)$, is generally also affected by non-price factors, such as income growth, demographic changes etc., as well as by the initial change in costs, so that $D(t) = f_t(z,c)$. However, in order to simplify the exposition, we will assume that all other factors remain constant, so that $D(t+I) = D(t)\ \forall\ I$ if c is constant. The demand functions $D(t)$ can then be defined purely in terms of dynamic cost elasticities, which specify how demand changes over time in response to a change in costs.

$$\varepsilon(t) \;=\; \frac{(D(t) - D(0))}{\Delta C(0)}\,\frac{c}{D(0)} \;=\; \frac{\partial D(t)}{\partial C(0)}\,\frac{c}{D(0)} \qquad (13)$$

Here $\Delta C(0)$ is a change in costs from c to c' occurring at time 0, $D(0)$ is the level of demand at time 0 given the initial price c, and $D(t)$ - $D(0)$ is the change in quantity demanded after a time interval t has elapsed. In most normal cases, we would expect this elasticity to grow monotonically with increasing time and approach some asymptotic value ε^*, the long-run elasticity, as $D(t)$ approaches its equilibrium value, D^*. It is possible, however, to have other adjustment patterns. For example, the elasticity as well as demand can oscillate over time as they approach their long-run values. In most economic situations, however, $|\varepsilon(t)| < |\varepsilon(t+1)|$ for all $t < t^*$, and $\varepsilon < 0$, so that $D(t) < D(t+1)$ for price falls and $D(t) > D(t+1)$ for price rises, assuming that $D(0)$ is in equilibrium. Given the relationship between the demand curve and consumer surplus, this implies that $B(t) < B(t+1)$ for price falls and $B(t) > B(t+1)$ for price rises. In order to assess the total costs or benefits to travellers of a given change in travel costs, it is normal practice to calculate the present value of the future stream of costs/benefits. Assuming a discount factor, r, and a time horizon, N, the present value can be written, in discrete time as

$$PV(B) \;=\; \sum_{t=0}^{N} \frac{\int_{c'}^{c} D(t)\, dc}{(1+r)^t} \qquad (14)$$

If demand does not respond instantaneously to the change in costs, the present value will thus depend on the dynamic profile of this response. Although it is not uncommon for present value calculations to be made by summing and discounting a stream of future benefits in a way that bears a superficial resemblance to the above formula - including changing demand levels over the years - there is one fundamental difference between normal practice and these equations. Such demand changes as occur are deemed to be the result of other extraneous factors (such as economic growth), *not* delayed responses to the initial cost changes.

The question that arises is, what happens to the calculation of present value when we erroneously use a static elasticity? In particular, can we assume that this may be an acceptable approximation given the advantages of simplicity, or is there the danger of some sort of systematic bias, and if so in what direction?

It emerges that the nature of the error caused depends on the relationship of the static elasticity to the dynamic elasticity, the speed of adjustment to equilibrium, the discount rate r and the time horizon, N. In the first example, the long run elasticity, ε^*, is used instead of the dynamic elasticity. In this case $|\varepsilon^*| > |\varepsilon(t)| \; \forall \; t < t^*$. This implies that $D^*>D(t)$ and $B^*>B(t)$ for price falls and $D^* < D(t)$ and $B^* < B(t)$ for price rises. Thus the benefit to consumers of price decreases is overestimated and the (negative) benefit or loss entailed by price increases is underestimated, as compared with the true situation where initial small responses build up over a period into bigger responses. In both cases, the costs of being in disequilibrium, or the adjustment costs are ignored. In the assessment of intertemporal benefits, each term in the summation in the present value calculation is over- or underestimated as shown above, so that the sum of the discounted benefits will include the sum of these errors, and thus be over- or underestimated in the same fashion. The slower the speed of adjustment the greater will be the discrepancy. However, as the time horizon increases, N approaches or exceeds t^*, so the error becomes smaller. Also, the higher the discount rate, the less weight to future benefits, which are more nearly accurate, so the error increases.

In the second case, suppose we erroneously use a short-run, first-period elasticity, $\varepsilon(1)$, where $|\varepsilon(1)| < |\varepsilon(t)| \; \forall \; t > 1$, instead of the dynamic elasticities $\varepsilon(t)$. Then for $t > 1$, $D(1) < D(t)$ and $B(1) < B(t)$ for price falls and $D(1) > D(t)$ and $B(1) > B(t)$ for price rises, so the benefit of price decreases is underestimated and the loss of price increases is overestimated. Again each term in (6) with the exception of $B(0)$ and $B(1)$ is over- or underestimated, as shown above, as is the discounted stream of future benefits. In this case, however, the effects of differences in the speed of adjustment, the discount rate, and the time horizon are the opposite as in the case above.

It is useful to emphasise that the biases described are not due essentially to the mistake of using an elasticity which is erroneously higher or lower than the true value (although this is, of course, known to be important). They arise from the difference between using a single value, whatever it is, instead of a time-dependent function. However, given that the two limiting cases discussed have opposite patterns of bias, it is interesting to suggest that there is likely to be some notional value of an elasticity - somewhere between the short-term and long-term value - whose use as a static elasticity would give the same value for consumer surplus as a full calculation of the path of response over time. In certain conditions, Dargay and

Goodwin (1995) show that this non-biassed position might be fulfilled in demand conditions when the adjustment process is half completed, which makes intuitive sense, and more generally that it will depend on the relative size of the speed of adjustment and the discount rate.

Such a relationship offers the possibility of a compromise position between biased equilibrium and intractable fully dynamic evaluations. A better understanding of the nature of behavioural adjustments in the order of 5 years or thereabouts might enable some of the short cuts of equilibrium analysis to be used. This does not seem an unreasonable research objective.

CONCLUSION

It is the author's view that we were not in equilibrium when the data we use were collected, we are not in such a state now, there is no guarantee that the system is currently moving towards it, we will never arrive there, and even if we did we would not stay there for long. For all its importance in the history of economic thought, the concept of equilibrium now acts as a barrier to understanding how things are changing, and how they might change.

This is not to say that improved understanding is impossible. But it will depend on treating travel behaviour as a process, not a state. In this way, we will be better able to comprehend the process by which travel habits are formed, or broken, the process by which cultural values and patterns of travel behaviour are transmitted from person to person, between producer and consumer, or from generation to generation, and how the constraints acting as barriers to change themselves get stronger, or relax, in the course of individual, household and social growth.

It may be that in understanding these processes, a mental construct of conditions under which things would be stable, is helpful in defining the rules of motion under which they actually operate. To this extent Fried et al's (1977) "effort to reduce imbalances" may give a role to equilibrium as the counter-factual against which individuals compare their current situation. Such a reduced, residual role is consistent with much of the argument in the chapter. But we cannot sustain two bigger uses of the concept - the assumption that equilibrium is observed in real data, and the assumption that a determinate date can be associated with any future described equilibrium.

Abandoning these, it is true, substantially weakens the scope of cross-section analysis, and the credibility of current methods of long-term project evaluation relying on discounting flows of costs and benefits. On the other hand, it will strengthen the ability to assess issues of great importance for policy about which equilibrium models of travel behaviour have little or nothing to say, notably how long it takes for policies to have an impact, the consequences of different sequences of implementation, and the special problems that may be involved in seeking to reverse well-established trends. Above all, it offers the possibility of policy advice that is consistent with the best available understandings of the effects on behaviour.

ACKNOWLEDGEMENT

The author acknowledges the support of the UK Economic and Social Research Council, and collaborators in earlier joint publications, especially Joyce Dargay and Ryuichi Kitamura. I am grateful for comments by both of them, by referees, and by John Bates, Dieter Helm, David Hensher, Ben Heydecker, Juan de Dios Ortuzar, Ken Small, John Swanson and Pilo Willumsen. Not all of them agreed with the line of argument in the chapter.

REFERENCES

Bell, M. G. H. and Y. Iida (1997). *Transportation Network Analysis*. Wiley, New York.

Cairns, S., C. Hass-Klau and P. B. Goodwin (1998). *Evidence on the impact of highway capacity reduction*. Technical Report, ESRC Transport Studies Unit, University College, London.

Dargay, J. M. and P. B. Goodwin (1995). Evaluation of consumer surplus with dynamic demand. *Journal of Transport Economics and Policy*, **29**, 179-193.

Ettema, D. F. and H. J. P. Timmermans (1997). *Activity-Based Approaches to Travel Analysis*. Pergamon, Oxford.

Fried, M., J. Havens and M. Thall (1977). *Travel behaviour - a synthesised theory*. Technical Report for National Cooperative Highway Research Program, Transportation Research Board, Laboratory of Psychosocial Studies, Boston College, MA.

Gärling, T., R. Gillholm, J. Romanus and M. Selart (1997). Independant activity and travel choices: behavioural principles of integration of choice outcomes. In: *Activity-Based Approaches to Travel Analysis* (D. F. Ettema and H. J. P. Timmermans, eds.), pp. 135-150. Pergamon, Oxford.

Golob, T., R. Kitamura and L. Long (eds.) (1997). *Panels for Transportation Planning*, Kluwer, Dordrecht, The Netherlands.

Goodwin, P. B. (1977). Habit and hysteresis in mode choice. *Urban Studies,* **14**, 95-98.

Goodwin, P. B. (1992). A review of demand elasticities with special reference to the short and long run effects of price changes. *Journal of Transport Economics and Policy,* **26**, 155-170.

Goodwin, P. B. (1997). Have panel surveys told us anything new? In: *Panels for Transportation Planning* (T. Golob, R Kitamura and L Long eds.). Kluwer, Dordrecht, The Netherlands.

Goodwin, P. B., R. Kitamura and H. Meurs (1990). Some principles of dynamic analysis of travel behaviour. In: *Developments in Dynamic and Activity-Based Approaches to Travel Analysis* (P. Jones ed.), 56-72. Avebury, Aldershot, UK.

Haag, G, and K. Grutzman (1993). A new theory of nested decision processes with memory. *Papers in Regional Science,* **72,** 313-335.

Hensher, D. A., N. C. Smith, F. M. Milthorpe and P. O. Barnard (1992). *Dimensions of Automobile Demand: A Longitudinal Study of Automobile Ownership and Use.* North-Holland, Amsterdam.

Luk, J. and S. Hepburn (1993). *New reveiw of Australian travel demand elasticities.* Technical Report, Australian Road Research Board, Victoria, Australia.

Ma, J. and K. G. Goulias (1997). An analysis of activity and travel patterns in the Puget Sound Transportation Panel. In *Activity-Based Approaches to Travel Analysis* (D. F. Ettema and H. J. P. Timmermans, eds.), pp. 189-208. Pergamon, Oxford..

Magnusson, L. and J. Ottosson (1997). *Evolutionary Economics and Path Dependence.* Edward Elgar, Cheltenham.

Mogridge, M. H. M. (1997). The self-defeating nature of urban road capacity policy: a review of theories, disputes and available evidence. *Transport Policy,* **4**, 15-23.

Ortuzar J. de D. and L. G. Willumsen (1994). *Modelling Transport* (2nd ed), Wiley, Chichester, UK.

Pollak, R. A. (1970). Habit formation and dynamic demand functions. *Journal of Political Economy,* **78,** 745-763.

Raymond, T. and D. A. Hensher (1997). Panel surveys and other longitudinal techniques: an annotated bibliographic review. In: *Panels for Transportation Planning* (T. Golob, R. Kitamura and L. Long, eds.). Kluwer , Dordrecht, The Netherlands.

Small, K. A. and X. Chu (1997). *Hypercongestion.* Technical Report, Irvine Economics Paper 96-97-11, Department of Economics, University of California, Irvine, CA.

Sterner, T., C. Dahl and M. Franzen (1992). Gasoline tax policy, carbon emissions and the global environment. *Journal of Transport Economics and Policy,* **26,** 109-120.

Tischer, M. L. and R. V. Phillips (1979). The relationship between transportation perceptions and behaviour over time. *Transportation,* **8**, 21-36.

Walters, A. A. (1961). The theory and measurement of private and social cost of highway congestion. *Econometrica*, **29**, 676-699.

Wardrop, J. G. (1952). Some theoretical aspects of road traffic research. In: *Proceedings of the Institution of Civil Engineers*, pp. 325-378.

Wilson, A. G. (1976). Catastrophe theory and urban modelling: an application to modal choice. *Environment and Planning A*, **8**, 351-356.

APPENDIX

If it is the case that behaviour takes some time to adjust to changes in travel conditions, then there are implications in using observed data to make inferences about consumers' choices and preferences. We now pursue this question in relation to a common approach for estimating the value of travel time.

$$C_k = M_k + v\,T_k$$
$$v = \tau / \mu$$

Consider the case of an individual choosing between modes of transport *k*, on the basis of the modes' generalised costs *C*, the sum of the money cost *M* and the time cost *T* converted to money units by the value of time *v*, that is, where τ is the marginal utility of time and μ is the marginal utility of money.

Let *G* be the difference between the modes' generalised costs, that is, $G=C_{1\text{-}2}$. There exists some value of *G* below which mode 1 will be close and above which mode 2 will be close. We will call this the individual's economic threshold E_i. In the simple case $E=0$, but there will be some specific other attributes favouring one mode or the other, different for each individual. If we assume that *E* is distributed in accordance with some bell-shaped distribution, then the proportion of the population choosing mode 2 is the proportion *P*(*G*) whose threshold value *E* is less than this, that is,

$$P(G) = \int_{\infty}^{G} Z(E)\, dE$$

In this case, we can collect data on choices between the two modes and their relative costs and times, and use some appropriate statistical procedure for estimating the coefficients on money and time, interpreted as their respective disutilities, which adequately explain the choices observed. The ratio of the marginal utilities can be used as a measure of the value of time.

However, suppose that at the time of the survey of observed choices there had recently been a change in the relative times or costs of the two modes, but because of some form of delay, inertia or resistance to change some people had not yet made a change to the now improved mode, or from the now worse one. Suppose we define a habit threshold, *h*, with the same dimensions as *G*. This may be considered as a bias not towards a specific mode, but towards the mode which the traveller has become accustomed to using. Its magnitude will be measured by the amount of further change in *G*, after the economic threshold *E* has been passed, which is required before the traveller changes modes. The value of *G* at which a change of mode is made is now $(E_i + h)$ if *G* is increasing, but $(E_i - h)$ if *G* is decreasing. If *G* is in the range

$(E_i - h) \leq G \leq (E_i + h)$ the individual may be using either mode, and we cannot say which without considering the previous values of G.

We can assume that h is not a fixed habit threshold for all individuals, but is a variable with some probability distribution. When G changes direction, only a small proportion of the population will quickly be persuaded to change modes (less than that which recently changed to that mode when its relative advantage was improving), but the proportion of those transferring will increase as G changes further in the new direction. In other words, at first those travellers with low habit thresholds and close economic thresholds are won, subsequently those with greater habit thresholds or further economic thresholds, lastly those with distant economic threshold and high habit thresholds.

The question now is, what effect would ignoring this possibility have on estimation of the value of time? We report a small simulation exercise designed to assess the answer to this problem. A sample of individuals is invented, each of whom has 5 unique characteristics, namely

ΔT_i	travel time difference between two modes
ΔM_i	travel cost difference
τ_i	marginal utility of time
μ_i	marginal utility of money
h_i	habit threshold.

ΔT and ΔM are random variables, normally distributed, with mean zero. τ, μ, and h are random variables drawn from the positive half of a normal distribution. Each individual is assumed initially to choose deterministically that mode with the least disutility, that is,

$$\tau_i \, \Delta T_i + \mu_i \, \Delta M_i \begin{cases} > 0 \text{ choose mode 2} \\ < 0 \text{ choose mode 1} \end{cases}$$

This provides a notional base situation, at which it is assumed that there have been no recent changes in costs or times and everybody is using that mode which is justified in terms of a simple definition of utility. However, it is then assumed that there is some alteration in the costs or times changing the relative advantage of the two modes, but an actual switch is made only if the change is big enough to offset the inertia factor, h_i, for each individual. In each simulation, a base situation is defined, and then run again with a *trend* in costs or times, that is, by adding or subtracting the same fixed amount of money or time to ΔT or ΔM for each individual. In the first pass, all individuals will be using the minium cost mode (according to their own unique marginal utilities of time and money). In subsequent passes, some individuals will appear to be using the more expensive modes, even by their own marginal utilities, because the extra expense is less than the trouble of switching. This provides a population of individuals for whom we know their relative costs, relative times, and choices. We now "survey" these individuals, and use their observed behaviour to estimate the following model

$$\tau^* \Delta T_i + \mu^* \Delta M_i \begin{cases} > 0 \text{ choose mode 2} \\ < 0 \text{ choose mode 1} \end{cases}$$

Here the object is to estimate not the separate values of the marginal utilities for each individual, but average or population values which best express the relative marginal utilities of the population as a whole. (In the simulation, the estimation procedure used is to select values by iteration which minimise wrong classification. As a statistical technique this would be considered inferior to more rigorous methods, but the bias introduced is immaterial to the current application). Then we estimate a revealed population value of time v^* by

$$v^* = \tau^* / \mu^* .$$

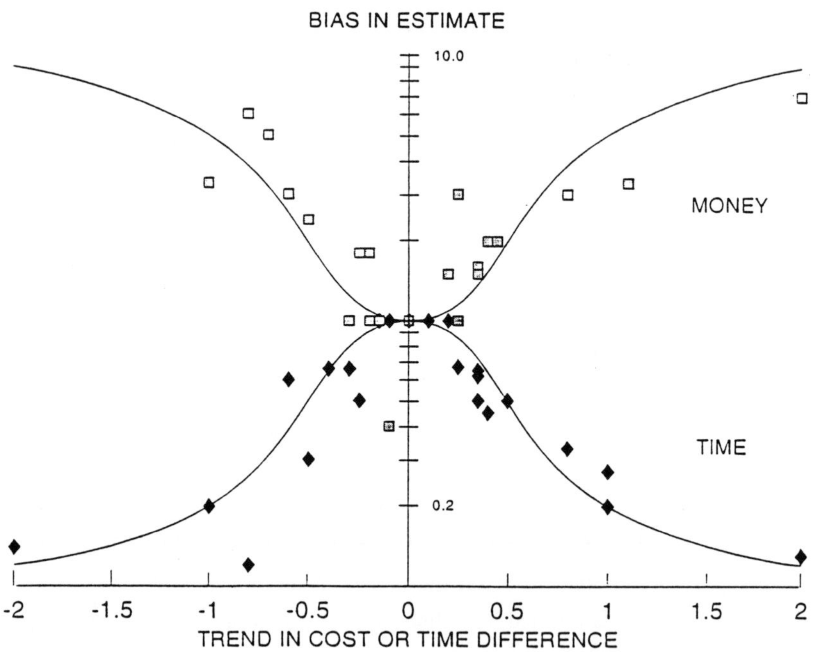

Figure 5.1. Bias in value of time when habit exists, but is ignored

What we have done then is to generate a population of individuals whose behaviour does have this dynamic attribute, but use their observed choices to estimate a value of time assuming that they do not have this dynamic attribute. The results are shown in Figure 5.1.

The bias in the estimate of the value of time is calculated as (estimated value assuming equilibrium) / (true average of values used as input). If there were no bias the ratio should be 1. In the base case, with no cost or time trend and no habit, the estimated single value of time is very close to the true average of individuals' values of time. This reassures us that we are not introducing material bias by the rather simplified procedure adopted. However, if there have been systematic cost or time differences in the period before the survey, there will be a bias in the value of time estimated. In this particular case the bias does not depend on the *direction* of the pre-survey trend (although this will be important in introducing a biased forecast of the size of the behavioural response) but on whether it is times or money costs which have been changing: (a) If money costs have changed, the value of time will be overestimated (because people's behaviour is interpreted as indicating that money differences are not very important to them); (b) If journey times have changed, the value of time will be underestimated (because people's behaviour is interpreted as indicating that time differences are not very important to them).

The above discussion has been conducted entirely on the assumption that the bias is due to a misinterpretation of accurately observed behaviour. However, in recent years it has increasingly been normal to base value of time estimation (and similar exercises) on stated preference (SP) experiments, in which respondents are asked to state their choice or preference in response to a range of experimentally controlled hypothetical levels of travel time, journey cost, and other attributes. There has been little or no consideration of the implied dynamics of this experimental context, but the following three suggestions are offered.

1. It is often argued that SP experiments give the same results as revealed behaviour analysis, and indeed the demonstration that this was the case (albeit subject to some uncertainty in interpretation) was one of the important reasons for giving confidence in SP methods in their early stages. If this is valid, then it proposes that the behaviour implied by SPs corresponds fairly closely with that implied by revealed preferences (RP) - and the danger of bias due to ignoring habit effects would be the same for both. The problem then is fitting an inappropriate model, not an inherent problem in one methodology or the other.
2. However, it might be argued that there is a stronger case for SP methods than is usually claimed. Suppose people are (consciously or unconsciously) able to make reasonably accurate assessments of the strength of habit effects in resisting change in their own behaviour. In that case, SP experiments do have a form of inherent

dynamic unachievable in any cross-section revealed preference survey, in that people are (albeit only hypothetically) faced with *changes* to the base variables. This suggests that there is potential information content in a SP survey which could be revealed with appropriate analysis: models could be tried which are not possible in RPs except over a period of time. In order to obtain the benefit of this, it would be necessary to abandon the comforting assumption that the same models are appropriate for both forms of data. There is a potentially useful line of research here.

3. However, that line of research itself raises new problems. It would be odd to use a model with explicit dynamic attributes without any consideration of the time scale of those attributes. We have to ask "what is the implied time scale of response which respondents in a SP experiment are assuming?" Intuitively, it seems likely that respondents will be assuming that the questions relate to their current situation in terms of home, job, family circumstances, patterns of social life, preference structures for activities, shops, and so on. This implies in turn that the behavioural responses are those in the relatively short run - say the first year or so after a stimulus. (This logic sits uncomfortably with claims that SP results are similar to those gained from models claiming to give equilibrium behaviour).

In summary, these considerations suggest that ignoring dynamic factors could give misleading results not only for behavioural forecasts, but also for interpretation of the values and utilities underpinning them. It is not, of course, self evident that errors of this type are necessarily larger than those arising from the many other known (and unknown) imperfections in forecasting practice, but their particular importance derives from the likelihood of systematic bias, which has the danger of making some policies seem worse than they really are, and others better.

6

MICROECONOMIC THEORY: A COMMENTATORY

Staffan Algers, Royal Institute of Technology, Stockholm, Sweden

INTRODUCTION

Modelling travel choices embeds the complex task of modeling human behavior. The multidisciplinary character of this task is also obvious. A human being may be seen as an agent driven by psychological factors in a system of economic relations and restrictions in a spatial context. It is natural that many disciplines - psychology, economics, geography, sociology, engineering, and others try to deal with this phenomenon, each from its own perspective. Microeconomic theory focuses on the individual related to the consumption of scarce resources induced by travel.

As is well known, models are simplifications of the phenomenon. Therefore, models can always be extended to become a more realistic description. Over time, models are also challenged by new issues. In the case of modelling travel choice, congestion, the introduction of tolls, and, more recently, the introduction of information technology in transportation (e.g., Advanced Travel Information Systems or ATIS) have presented such challenges. Thus, the need to theoretically address new issues evolves over time.

Microeconomic theory has been questioned as a basis for travel choice modelling, in my view to a large extent because it does not address the issue of how to model the choice process itself. The introduction of information technology makes however the need to analyse the choice process urgent. Thereby, problem are raised which microeconomic theory has not solved. It may, on the other hand, be to expect too much from a single discipline to be able to produce a theory that would claim to be the sole basis for travel choice modelling. As seemed to be agreed in many of the chapters of this volume, there is a need for different disciplines to learn from each other and to cooperate in building theories of travel choice.

There is of course still a need for intradisciplinary developments, also in microeconomics, and the four preceding chapters show that the discipline is conscious about the limitations as well as capable of advising new approaches and refinements of existing methods. The chapters thus discuss a broad range of aspects on microeconomic theory. The contents of the chapters will first be briefly described. Then follows a discussion of some of the issues raised in each. The commentatory ends with a few conclusions.

OVERVIEW AND COMMENTS

In Chapter 2 de Palma discusses basic assumptions of microeconomic theory, such as the assumption of rationality. Goodwin in Chapter 5 questions the assumption of equilibrium and recommends a dynamic approach. In Chapter 3 Jara-Díaz extends the theory of allocation of time to a theory of activity/travel choice. Finally, Hensher in Chapter 4 argues for relaxations of restrictions associated with the application of discrete choice theory derived from microeconomic theory.

De Palma addresses basic methodological problems of a general character, not only related to the field of transportation. Many issues are touched upon, each of which could be discussed at length. As he alludes to in the concluding remarks, such an overview may leave the reader with a bitter taste since many of the issues remain unsolved. In my view, it is useful to be reminded of these issues in order not to become too presumptuous, and it is important that research on these matters continues. I think it also stresses the need to validate the models that we use, and for the specific purpose that they are used.

It might be useful to use simulation methods to test the importance of different assumptions made in the analysis, not only to validate models by normal procedures such as comparing the model results with other base case data (like comparing modelled demand to traffic counts not used in the modelling procedure) or, even better, with data before and after some intervention. This would perhaps tell us more about the usefulness of different approaches, and give some guidance concerning where to focus research. It is often more fruitful to discuss to what degree a certain assumption does not hold rather than whether it holds exactly.

De Palma also discusses general equilibrium effects, and to what extent location theory is a separate field related to transportation science. This is an issue that actually receives increased attention, and in my view this question is not as overlooked (among transportation researchers) as the more general questions raised in de Palmas chapter may be. As examples of this, one may take the US regulation that land use analysis is mandatory in connection with larger infrastructure investments.

Political constraints on the use of economic theory in practice is also discussed by de Palma as a reason for the gap between theory and practice. It is certainly true that projects are not chosen only on their merits in cost-benefit terms. On the contrary, studies in Sweden and Norway show that the ranking of projects in politically decided investment plans are only vaguely related to the cost-benefit ratio (Nilsson, 1991, Odeck, 1996). This is however not to say that it does not have a value - on the contrary. As an example, the Swedish National Parliament committee on transportation has recently endorsed cost-benefit analysis as an important tool in the planning process There is however also the insight that current practice does not incorporate all effects in a satisfactory way, and that it may never do so. Therefore, I do not think that trying to incorporate voters' reactions into the model will, as suggested, circumvent the problems of acceptability, but instead, I am convinced, put an end to practice.

Goodwin takes in his chapter a well-argued farewell to equilibrium analysis. In the other chapters, it seem to be some consensus that a choice between process and structural modelling would depend on the purpose of the modelling effort, implying, for instance, that long-term forecasting would not require process-oriented modelling. Goodwin argues, however, that also long-term phenomena need to be analysed in a dynamic framework. He illustrates this with examples of car ownership and changes in fuel price. Goodwin states the dynamic problem as a function of history: demand in period t depends on prices and income in period t-1. Such a relationship may well be a way of capturing effects of habit.

One reason of the delay of responses is that there may be substantial transaction costs associated with a change from one behavioural state to another. The decision to take a transaction cost for transferring to another state is likely to be related to the somehow discounted future effects of that change.. This means that also the expectations of the future will influence the decision and not only the experience of the past. This is not to say that dynamic models built on historical data will not work - there will probably be some general correlation between history and expectations. There may however still be some bias in such a model due to the fact that the model does not incorporate expectations of the future.

Thus, it seems essential to bear the role of expectations in mind when designing panel studies. Maybe experience from macroeconomic analysis is useful, where data on household expectations is regularly collected as one of several important indicators on the economic development.

Goodwin′s chapter also contains an appendix which casts doubt on values of time inferred from cross-sectional revealed preference studies, and also over the current use of the Stated Preference technique in value-of-time studies. The concern is related to the inertia associated with changes, and the argument is illustrated with a simulation exercise. It is also suggested that SP studies may better cope with this phenomenon since they are directly related to changes in base variables. I think this also has an important bearing on the use of combined SP and RP studies.

As in other cases, it may be easy to identify assumptions that are not fulfilled, and again I would like to stress the importance to have an idea of to what extent the differences between the model assumptions and the real world influence the decisions that are based on the models. We will always have to work with approximations. To be able to focus on the most relevant research areas, we need an understanding of which ones are most crucial, and in what context.

Jara-Díaz discusses the role of time and income in travel demand, and sets out a framework for activity allocation. The latter is achieved by a technical conversion of goods into activities, and results in a number of properties, one of which is that travel and activities allocation pertain to the same class. Jara-Díaz points out that the role of labour supply (exogenous or endogenous) is an important factor, as it decides the role of the wage rate.

Jara-Díaz′ theory is formulated in the context of an individual. Most individuals live in households with more than one person, and some attention should perhaps be given to this fact. It may be interesting to note in this context that the time allocation problem within a household - labour supply as well as the intra-family division of work - has been studied as a household bargaining problem using game theory (Ott, 1992).

Although value of time is only marginally mentioned, it may be of interest to note that traditionally household income has been used as the income measure in practical applications of time allocation theories such as value of time studies (Bates, 1987; Hague Consulting Group, 1990)). The general finding is that the value of time is much less than proportionally related to income. It is therefore interesting to note, that in a study of welfare distribution within the household, it is found that only the individuals own share of income has an effect on his or her utility level for private goods (Antonides and Hagenaars, 1990). It might therefore be more relevant to relate the value of time to individual income.

Hensher presents a study demonstrating the difference between the more restrictive concept of a uniform value of time based on the standard logit model on one hand and extensions to value of time distributions based on the heteroscedastic extreme value (HEV) logit model on the other hand. The distribution of the value of time is achieved by specifying the time coefficient in an indirect utility function of a linear additive form as a function of cost and time. A more general form would have been obtained by expanding the general indirect utility function resulting from the general time allocation theory presented in Hultcrantz (1997). Such a formulation results in quadratic terms for both time and cost as well as interaction terms between time and cost and between time, cost, and socio-economic variables (including income).

The relaxation of the constant variance assumption reveals results showing that the relaxation of one constraint does not necessarily produce a more credible result in terms of value of time. However, the task of identifying different relaxation effects seems to be a difficult one, since the least restricted model only contains parameters (except for constants) that are not significantly different from zero (at normal risk levels).

The constant variance linear parameter estimates reflect not only behavioural linearities, but contain also the average effects of not allowing variable variance and nonlinear terms. Therefore, these linear parameter may reflect net effects of restrictions rather than structural behavioural relationships. If the true linear relationships are weak, we may observe lower t-values for these parameters.

As it turns out, we also observe low t-values for all other parameters. This fact may be due to correlation between random effects and nonlinearities. In principle, more data is needed to be able to identify the different effects.

Also, the pseudo rhosquare measure for the models with the relaxed variance assumption is lower than for the models without this relaxation. This fact motivates further discussion.

Based on my own research (e.g., Algers et al., 1995), I do not think that the precision in the estimates would allow one to say that the presented values of time are significantly different between models.

Subsequent studies will tell us more about the relevance of using less restricted but more complex models. As with other more general forms, there will always be a risk of getting peculiarities into the model - validation becomes even more important.

CONCLUSION

Taking as a whole, I think that while it may be possible to say that microeconomic theory does currently not give a satisfactory description of travel choice behaviour, it is not possible to say that we do not need such a theory. We do need a theory for allocation of scarce resources, and therefore also a consistent theory of travel choice behaviour to reflect demand. The contributions given in the four chapters on microeconomic theory highlight the need for improving the current state of art, show how improvements can be achieved and also show that improvements are achieved.

Travel demand modelling is often said to be a multidisciplinary research area. I my view, this should not mean an area from which suitable applications can be taken to many disciplines, but an area into which several disciplines are simultaneously used to solve the research problems. Of this we have so far seen too little.

REFERENCES

Algers, S., J. Lindqvist-Dillén and S. Widlert (1995). *The National Swedish Value of Time Study*. Paper presented at the PTRC European Transportation Forum, Warwick, UK.

Antonides, G. and A. J. M. Hagenaars (1990). *Distribution of welfare in the household*. Papers on Economic Psychology, **81**. Erasmus University, Rotterdam.

Bates, J. (1987). *The Value of Travel Time Savings: A Report on Research Undertaken for the Department of Transport*. Leeds University, England Institute of Transport Studies, Old Vicarage, UK.

Hague Consulting Group. (1990). The Netherlands "Value of time" Study: Final Report. Hague, The Netherlands.

Hultkrantz, L and R. Mortazavi (1997). The value of Travel Time Changes in a random Nonlinear Utility Model. Working paper, The Centre for Research on Transportation and Society CTS, Dalarna University, Borlänge, Sweden.

Nilsson, J-E. (1991). Investment decisions in a public bureaucracy: A case study of the Swedish road planning practices. *Journal of Transport Economics and Policy, **25***, 163-175.

Odeck, J. (1996). Ranking of regional road investment in Norway: Does socioeconomic analysis matter? *Transportation,* **23**, 123-140.

Ott, N. (1992), *Intrafamily Bargaining and Household Decisions*. Microeconomic studies, Springer-Verlag, New York.

PART 2. BEHAVIORAL DECISION THEORY

7

THE PERSPECTIVE FROM BEHAVIORAL DECISION THEORY ON MODELING TRAVEL CHOICE

Ola Svenson, Stockholm University, Sweden

INTRODUCTION

While normative decision theory specifies how decisions should be made, descriptive decision theory focuses on how decisions are actually made. However, in the short history of descriptive behavioral decision theory, normative theory has played and still plays a very important role.

Early authors like Edwards (1954) used the Maximization of Expected Value (EV) theory as a foundation when elaborating the Subjective Expected Utility (SEU) model as a tool describing human decision making. In the former the (long-run) expected value is computed for each alternative through multiplying the value of an outcome with its probability of occurrence. The alternative with the maximum expected value should be the preferred alternative from a normative viewpoint. Equation (1) describes the expected value (EV) of an alternative with two possible outcomes with probabilities p_1 and $(1 - p_1)$. The associated values are denoted V_1 and V_2,

$$EV = p_1 \cdot V_1 + (1 - p_1)V_2 \qquad (1)$$

If subjective values or attractiveness values (utilities, U) are inserted in equation (1) and subjective belief values Ψ_1 and Ψ_2 are used to replace the subjective values equation (2) follows:

$$SEU = \Psi_1 U_1 + \Psi_2 U_2 \qquad (2)$$

Thus, the subjective expected utility (SEU) is computed in the same way as EV. Later research has shown, for example, that Ψ_1 and Ψ_2 do not necessarily add to 1, that the relation between value and utility is not linear, and that the relation changes from one function to another when it passes from losses to gains (cf. Kahneman and Tversky, 1979; Tversky and Kahneman, 1992). Camerer (1988) has presented an overview of different approaches and deviations from classical utility theory.

Both EU and SEU approaches to human decision making model a decision without any particular description of the psychological processes leading to the decision. These approaches also assume that a single decision should be made according to what is the best long-run decision. Such an approach to behavioral decision theory has been called *structural* (Abelson and Levi, 1985; Svenson, 1979). *Process* approaches aim at modeling not only the final decision but also the processes leading up to it (cf. Payne, 1976; Russo and Rosen, 1975; Svenson, 1979).

In the evolution of descriptive behavioral decision research, it is possible to distinguish four different stages (Svenson, 1996). In the first stage, nothing is said about the psychological processes leading to a decision (Edwards, 1954). Decisions were primarily modeled as if they followed an SEU rule and as if they also followed the basic assumptions of normative decision theory such as transitivity and consistency. This first stage of human decision making research represents a purely structural approach.

The second stage of the development of decision research demonstrated behavior inconsistent with the postulates and predictions of the normatively modeled descriptive theory (e.g., Coombs, 1958; May, 1954; Papandreou, 1953). Even here the mediating processes leading to a decision were largely ignored. For a long time the second-stage perspective dominated decision research and a massive research literature now shows us the limits of the subjective expected utility model in describing and predicting individual human decision making (Dawes, 1996; Einhorn and Hogarth, 1981; Slovic et al., 1977). Despite this, most of contemporary research still uses the fundamental utility model elaborated in different ways as a theoretical and empirical research reference (cf. Payne et al., 1992; Yates, 1990).

In what may be called a third stage, one could find researchers who were precursors to later developments and who saw the human decision maker as an information processor with limited capabilities which, in turn, could explain the deviations from theoretical expectations (e.g., Miller, 1956; Simon, 1955). These researchers argued that these limitations were essential in describing and predicting human decisions. One of the limitations of the human decision maker considered by those researchers is the restricted number of entities that can be handled by working memory at one moment in time. This necessitates sequential information processing in more complex decisions. This insight encouraged process-tracing studies and models of the psychological processes leading to a final decision. Almost everything that was said by Simon (1955) can be found in later developments of decision research.

This third stage explored the decision-making process and found evidence for a sequence of other rules different from the subjective expected utility principle. For example, it was recognized that often alternatives were eliminated early in the process by using some cut-off criterion on one or several attributes and that compensatory rules were used more towards the end of the process. The representations of the decision alternatives were always assumed to be stable within uncertainty limits. A similar development occurred within basically structural approaches in models in which alternatives could be eliminated in successive tests (cf. Tversky, 1972).

In addition to assumptions made in the third stage, the fourth stage also included assumptions of changes of representation to solve decision conflicts. This was not completely new in the decision research literature. As early as in 1964, Shepard suggested that the solution of a decision problem can be related to a change of the representation of the decision alternatives in the decision space (Shepard, 1964). He proposed that subjects actively construct a possible solution to a decision problem through changing the perspective which in turn changes the values or projections on the attractiveness attributes. This theme was followed up a decade later by researchers in the structural approach as editing (Kahneman and Tversky, 1979), and in the process approach as a continuous influence of decision rules on changes of the representation system to meet rule requirements (Svenson, 1979). Payne (1982) named the process of adapting decision rules to the structure of a problem contingent decision behavior and Montgomery (1983) selected the rule of dominance to be the guiding principle for the changes of the representation system. Beach (1990) has presented a process theory of decision making called Image Theory and Payne et al. (1993) have further developed the process perspective on decision making. Overviews of the development (and blind alleys) in mainstream decision theory have been presented by Lopes (1988, 1995).

The division of decision research into structural and process approaches is not absolute because there are many process elements in primarily structural approaches as, for example, the editing phase in Prospect Theory (Kahneman and Tversky, 1979) and the elimination by aspects process described by Tversky (1972). Mellers and Biagini (1994) and Mellers and Coke (1994) used a structural perspective to illustrate process phenomena. The term contextual effects (e.g., Mellers and Birnbaum, 1983) can be used to describe the effects of a decision process on a decision. Correspondingly, there are many structural elements in process approaches as, for example, decision rules in the Payne et al. (1993) approach. Both approaches are needed and eventually they will merge in models describing individual decision processes and how they can be influenced by the task, the context, and the person. Through modeling the psychological processes behind decisions it will be possible both to predict and better understand choices people make.

In the following, different kinds of decisions will be described in a system of levels of decision making. Then, different, structural approaches to human decision making will be presented. After this, process approaches are described followed by a presentation of Differentiation and Consolidation (Diff Con Theory) (Svenson, 1992). Lastly, links between Diff Con Theory and travel choices will be identified and discussed.

LEVELS OF DECISIONS

There are at least two different ways of classifying decision making in levels. The first way of doing so is to consider the organizational aggregate level. Then, higher level decisions refer to decisions made at a higher level in a hierarchy of, for example, societal authorities. Higher-level authority decisions can be contrasted with lower-level decisions made by, for example, individual travelers. This way of contrasting perspectives of different parties was used by Svenson and Fischhoff (1985) when modeling environmental decision making to illustrate that rational decision making on the individual level does not automatically correspond to rational decision making on the aggregated regulatory level. To illustrate, perfectly rational behavior at the individual level (e.g., not to use a seat belt on a *single* car trip) is highly irrational on the higher level (e.g., no driver uses a seat belt).

Second, it is also possible to distinguish levels of decision making through reference to the psychological processes used for different decisions made by an individual decision maker. To specify, it is possible to classify decisions according to the extent to which the decision maker uses his or her own value system in a decision. A decision maker's value system is a term used here to summarize all fundamental values, attitudes, motivations, drives, etc. which are of potential importance for a decision. When the value system is mapped onto a decision problem, this creates an attractiveness (utility) representation of the problem. Svenson (1990) suggested four different levels of decisions.

In the following, a decision problem will be represented by *aspects* (e.g., travel time 10 minutes) which are ordered on *attributes* (e.g., the attribute of travel time). In most cases there is an attractiveness representation linked to an aspect when a decision is made. For example, 10 minutes travel time is more attractive than 40 minutes travel time. However, the following

Level 1-decisions illustrate one exception to that rule in that no attractiveness is necessary for making the decision.

Level-1 decisions thus do not need any attractiveness representations at the moment when they are made. Such decisions are often made in quick habitual processes. The decision maker recognizes that a situation is similar to an earlier one, and he chooses the same alternative as last time. This kind of decisions include many quick, largely automatic and unconscious decisions (Shiffrin and Schneider, 1977). Klein (1989) calls such decisions recognition primed and Smith (1992) uses the exemplar model to describe them. Gärling (1992) identifies "routines of performance of everyday activities". Although level-1 decisions do have some relationship to values in the past, the attractiveness representation is not likely to be activated and to influence these decisions every time they are made. Instead, other perceptual or cognitive principles (e.g., deciding through similarity by matching or through imitating some respected other's choice) guide level-1 decisions. However, when the habits or categories were formed in the past, attractiveness and decision rules were guiding principles. In travel decisions, level-1 decisions are likely to predict many daily travel routines. They may represent pre-determined strategies (e.g. always go by subway, always put on seat belt, or always drive a little faster than the speed limit). Level-1 decisions can also function as subprocesses in higher level decisions.

In *level-2* decisions, the decision maker uses attractiveness on one or a few attributes. However, these decisions represent stereotypical and static mappings of the alternatives (e.g., biggest is best). For example, no trade-offs between conflicting attractiveness values may be performed. Decisions which are made through direct holistic references to emotions or affects also belong to this category (cf. Wright, 1975; Zajonc, 1980). Very quickly, some decision alternatives are associated with positive or negative emotions which may be sufficient for a decision. In contrast to level-1, level-2 decisions always refer to an attractiveness (emotional, value, etc.) attribute. But there are no explicit trade-offs. This type of decision processes, as well as level-1 processing, appear in the initial stages of most decision processes of a more elaborate nature.

Level-3 decisions do not presuppose an automatic correspondence between a decision maker's value system and the attractiveness of the alternatives. Decisions at this level may use trade-offs between the attractiveness of aspects on different attributes and transform attributes into new ones, etc. Decisions at this level are fundamental prerequisites for most decisions at levels 1 and 2, which at one time were level-3 (or level-4) decisions later transformed into routine decisions.

In *level-4* decisions the decision maker encounters a new and unfamiliar problem in which decision alternatives have to be elicited or created. Parts of the decision-making processes at this level include problem solving. In for instance travel-mode choice the creation of decision alternatives includes composite alternatives. That is, an alternative is constructed which contains some part of each of a number of different travel modes. For example, the constructive level-4 process may generate one (train + taxi) alternative competing with an (air + rental car) alternative. Langer (1994) calls this active decision making and points out that in comparison with the other levels of decision making, it should lead to greater self-esteem, enhanced perceived control, and less post-decision regret. This kind of decision processes have not been modeled in any great detail in behavioral decision theory. We shall return to this later.

MODELING THE INDIVIDUAL DECISION MAKER

Most modeling efforts have focused on level-3 decisions. There are usually conflicts between attributes for a given set of alternatives in the theoretical approaches at this level. However, recently, for example, Klein (1989) has pointed out how important it is that level-1 decisions are also taken into account.

In the following, decision alternatives are represented by aspects as mentioned above (e.g., the time of a particular trip, a fare of SEK 200) ordered on attributes or dimensions (e.g., travel time, cost). The mapping of goals, values, and other driving forces on decision alternatives results in attractiveness values of the aspects which are essential for all level-3 and level-4 decisions as well as for some level-2 decisions. In the present context, attractiveness and utility differ only in their metric precision. While most utility measures are regarded as commensurable across attributes and related to ratio scales, attractiveness measures can be noncommensurable across attributes and defined on only rank order or interval scales. Thus, utility is a more refined representation of attractiveness than in many decision situations (Svenson, 1979). It is interesting to note that the study decision maker's goals has not attracted much interest from decision researchers. Exceptions from this rule are, for example, Lopes (1987), Jungermann (1983), Gärling et al. (1989) and Lindberg et al. (1987).

The models which will be presented briefly in the following, all share a common representation of decision alternatives as a basis for their respective theoretical approaches. Several approaches have subjective probability (degree of belief) for different events to occur as an important attribute in their theories. This was illustrated by the SEU approach presented above. The theoretical approaches are relevant for human decision making in general including choices of homes, jobs and travel modes. The main focus will be on decisions or choices (excluding a vast literature on judgment) and on process theories for describing those decisions.

Structural Approaches

As mentioned above, Subjective Expected Utility (SEU) Theory has played the major role in behavioral decision research for a long time. Although it has been repeatedly refuted as a behavioral theory, it still serves as a bench mark for most studies of decision making and in particular for structural approaches. The fact that SEU may gain in descriptive strength when data for many decisions are accumulated over people has probably contributed to its longevity also in the descriptive process domain. However, if the basic processes behind behavior approximated by SEU theory are not understood, changes in some underlying conditions may cause severe errors in the theory's ability to predict human behavior (Lopes, 1995).

A multidimensional relative to SEU is Multiattribute Utility Theory (MAUT). Here, the alternatives are characterized by aspects on different attributes, whose utilities are weighed by attribute importance for goal fulfilment (von Winterfeldt and Edwards, 1986). Subjective probabilities or beliefs are estimated if the decision is made under risk (objective probabilities exist) or uncertainty (objective probabilities unknown). However, many decisions are made under certainty, and then an alternative is valued according to its weighed sum of utility values. Note that in the following theoretical approaches, it is always possible to find elements of a process approach (e.g., sequential processing of information at some stage).

Criterion Dependent Choice (CDC) models postulate a sequential evaluation process resulting in an intention to choose one of two alternatives when enough evidence to meet a criterion has been accumulated to make sure that the alternative is the best one (cf. Aschenbrenner et al. 1984; Audley 1960; Busemeyer 1985; Dietrich et al., 1989; Estes 1959; Feller 1957; Wald 1950; Wallsten 1980). Later versions of these models (cf. Albert et al., 1989) have some features in common with psychological process models such as the systematic variability of the criterion value during the sequential search of information. Typically, the attractiveness of an alternative is represented by one attractiveness or criterion dimension only in these models.

Satisficing was introduced by Simon (1955) in a seminal paper which covers the active manipulations of aspiration and acceptance level in response to changes in the environment which poses the decision problem. Later in this paper, Differentiation and Consolidation theory (Svenson, 1992) will be developed to accommodate this perspective. Satisficing may be regarded as partly a process approach to decision making.

Elimination by Aspects Theory (Tversky, 1972) is an early structural theory with very strong implications for process approaches. It describes decision making in terms of two stages. First, the most important attribute is investigated, and all alternatives that do not exceed the conjunctive elimination criterion on this attribute are eliminated. This procedure is then repeated with new attributes successively lower in the lexicographic order or importance order of the attributes.

March's (1978) Posterior Preference Approach relaxes the requirement of priority of preferences and allows preferences to be rationalized after the fact. This perspective focuses on post-decision processes and suggests that we find meaning in our decisions and actions afterwards when the consequences are observed and interpreted.

Regret Theory uses the fact that a decision maker may imagine that if he or she had chosen a certain other alternative, the alternative he or she has now chosen is inferior in some respect. As presented by Loomes and Sugden (1982), the theory assumes that this leads to modified utilities in the theory. Psychological explanations of the phenomenon of regret have been reviewed and elaborated by Landman (1987) who covered cognitive and emotional aspects behind regret.

In a travel choice context regret implies that decision alternatives change in attractiveness depending on available alternatives. An empirical illustration of this is the following. If two alternatives are equal in attractiveness and a third alternative is introduced which is dominated by only one of the first two alternatives, then the alternative which dominates the newcomer will gain in attractiveness (Payne, 1982; Tyszka, 1983).

Norm Theory asserts that judgment standards are constructed in direct response to specific outcomes. Different alternatives evoke different norms and an outcome is compared to the evoked norm for that alternative in the decision process (Kahneman and Miller, 1986). To exemplify, receiving a rejection of a manuscript from a journal may activate a norm (of rejection or acceptance) against which the outcome is evaluated.

Prospect Theory (Kahneman and Tversky, 1979; Tversky and Kahneman, 1992) is the most important and influential theory in contemporary behavioral decision research. The theory was developed as an evolution of SEU theory taking account of important behavioral deviations from this theory. It has a value function which is concave for gains and convex for losses. The convex part for losses is steeper. Probabilities are transformed to weights greater than objective probability for small probabilities, otherwise underestimating the probabilities. Cumulative prospect theory (Tversky and Kahneman, 1992) uses a rank-dependent utility model which transforms a cumulative distribution of probabilities separately for gain and losses. Prospect theory uses probability (or related weights) as an essential component all through its development. In prospect theory, the decision process is started in an editing phase in which information is simplified and structured.

In rank-dependent utility models, the attractiveness of each alternative is evaluated in a cumulative way starting with the least favourable outcome (Lopes, 1995; Tversky and Kahneman, 1992). To illustrate this, assume that one alternative in a decision problem has three possible outcomes 1, 2, and 3. The decision alternative is formulated as a gamble in which 1, 2 and 3 are p=0.05 to win SEK 10; p=0.30 to win SEK 200; and p=0.65 to win SEK 100. To evaluate this gamble the decision maker starts determining the rank order of the utilities of the outcomes and thus orders them 1, 3, and 2. In all cases this alternative gives at least SEK 10. Except for p=0.05 this alternative gives at least SEK 100, that is with p=0.95. In p=0.30 of all cases the alternative gives SEK 200. A rank dependent utility edition of this problem would be (1.00 x 10; 0.95 x 100; 0.30 x 200). A subjective integration of this information would mean that the increase in outcome value plays the most important role instead of SEU processing of the same problem (cf. equation 2).

Naturalistic decision making represents a discontent with contemporary main-stream decision research because of its limited application to real-world decision making. The group of researchers who contributed to the first strong manifestation (cf. Klein et al., 1993) represent a wide field from researchers interested in military applications to everyday decisions. There is no coherent theory uniting this research. Klein´s (1989) recognition primed decision making approach stresses the importance of considering quick decisions which have to or are made without long deliberation (cf. level-1 decisions).

From the above it is clear that leading researchers in the structural approach are gradually incorporating process elements such as editing, elimination, and rank order dependent utility sorting. The dominating role played in contemporary decision research by the family of expected utility approaches for modeling human decision making has lead to an emphasis on results about deviations, errors, and irrational behavior. In fact, Dawes (1996) notes that the most commonly cited articles (in 1992 and 1993) concern anomalies of human decision making in relation to normative decision rules. This is excellent if the purpose is to criticize and adjust, for example, assumptions about human behavior made in economic theory. In travel-mode choices, economic models such as those for stated preferences may benefit from such knowledge and strive for proper adjustments to human anomalies. However, to the present author it seems quite unfortunate for the field of psychology to get stuck into deviations from a framework which is basically at least 50 years old and which has been found insufficient 30 years ago (cf. Lopes, 1995). Furthermore, the development in decision research seems to have proceeded without many serious attempts of connecting the area with progress in other fields of psychology. There are some exceptions as, for example, research in social psychology such as dissonance theory and attitude research which have had some impact on mainstream decision research (Festinger, 1964; Janis and Mann, 1977; Josephs et al., 1992; Roese and Olson, 1995). A more constructive approach of proposing new theories and models for how humans make decisions than what is common today is very much needed. For example, decision theory could be connected to some extent to attitude research as illustrated later in this paper.

Process Approaches

As mentioned earlier, process approaches focus on decision making as a process over time and the psychological characteristics of this process. Here, it is not primarily the result of the process which is the main focus as in the structural approaches. There is a long tradition of process-oriented research starting in the 50´s with Brehm (1956) and related to Festinger's (1957) theory of cognitive dissonance. In contrast with early structural approaches, limited to the relation between input and the decision, the process approaches also studied post-decision processes (e.g., Brehm and Cohen, 1959; Brock, 1963; Zajonc, 1968).

Festinger´s Conflict, Decision, and Dissonance Theory is the first systematic exploration of both pre- and post-decision processes. However, Festinger (1964) did not assume systematic problem restructuring in the pre-decision phase. In the post-decision phase, Festinger´s theory predicts active restructuring, information selection and dissonance reduction, which results from the spontaneously occurring dissonance after a decision (cf. Frey, 1981, 1986). The attributewise representation of decision problems was not used in the Festinger approach which has attracted many followers (cf. ; Gerard, 1994; Greenwald and Ronis, 1978; Josephs et al., 1992). However, recently, a multi-attribute theoretic perspective has been introduced (Josephs et al., 1992).

Janis and Mann´s Conflict, Choice, and Commitment Theory contrasts with Festinger´s in that it considers pre-decisions to involve restructuring more or less consciously until a decision is reached. Regarding post-decision processes, Janis and Mann (1977) postulate that neither regret nor bolstering will occur unless some conditions prevail which make for persisting decisional conflict. Among those necessary conditions are factors inducing hypervigilance, for example, time pressure (Svenson and Maule, 1993) or a decision situation with seemingly unavoidable large negative consequences. Janis and Mann (1977) did not use an attributewise representation of decision alternatives.

In a paper summing up the early research in process-tracing studies of decision making, Svenson (1979) asserted that a number of different decision rules were applied during a decision process and that the representation system and the sequence of applied rules can influence each other. Montgomery (1983) published a theoretical paper on decision rules and the search for a dominance structure, Dominance Structuring Theory. Here, it was asserted that all decision-making processes are made with the purpose of arriving at dominance for one alternative over the other alternatives. Attractiveness restructuring is a way of changing a decision situation into dominance or at least closer to it. Different decision rules can be applied to achieve dominance, which means that an alternative is equal to or better on all attributes in comparison with its competitors. When dominance has been achieved there is no further need of restructuring. Dominance Structuring Theory was not particularly concerned with post decision processes.

Decision Field Theory (Busemeyer and Townsend, 1993) develops the approach-avoidance models presented by Lewin (1935) and Miller (1959). It uses a stochastic process to describe the deliberation process leading to a decision in which avoidance and approach subsystems affect the process. The theory is used to explain among other things violations of stochastic dominance, violations of strong stochastic transitivity, and slower decision times for avoidance as compared with approach conflicts. This theory is presented at some length elsewhere in this volume (Chapter 8).

Contingent Processing and Constructive Decision Making Theory was presented by Payne and co-workers. This work (Bettman et al., 1991; Payne, 1982;) was developed over the years into a Theory of Constructive Decision Making (Payne et al., 1992, 1993). Payne (1976) pointed out that the choice of decision rules was dependent on the decision problem and in Bettman et al. (1991) the contingent strategy selection was structured around three types of influential factors: problem, person, and social context. The theoretical approach of Payne and colleagues (Payne et al., 1993) largely focuses on the choice of decision rules and the relation of this choice to task variables (including, e.g., time pressure and complexity).

Image Theory proposed by Beach (1990) differs from the other theories in that it does not primarily refer to the decision maker's evaluations of attributewise representations of alternatives. Instead, it uses the term image for scenario-like representations governing decision processes. It uses adoption decisions of a preliminary candidate alternative. Image Theory is a process theory which models compatibility tests (testing for adoption of a candidate) followed by profitability tests of the candidate(s) selected in the first phase. In the process there are (1) a value image representing goals and values, (2) a trajectory image representing the mapping of the value image on a particular situation, and (3) the strategic image of plans, tactics, policies, and forecasts used in realizing the decision maker's goals and values. Facts and attractiveness representations are closely related to scenarios, scripts or images for their creation. However, many of the empirical studies in this tradition still use attributewise representations instead of scenarios.

Sequential and Dynamic Decision Making

Sequential and dynamic decision making approaches link different decision to each other in either an environment that only changes as a consequence of the decision makers decisions (a sequence of lotteries), or in a dynamic environment (driving a car). These approaches link structural decision models to sequential decisions but other theories, such as, engineering control theories also inspire the research in this field. Relatively little interest is paid to the decision maker's psychological processes.

Sequential decisions can be linked to each other in different ways. For example, successive related decision problems may (1) form part of a strategy routine or habit, (2) be a similar decision problem which appear soon after a prior decision, and (3) be an earlier decision that changes the conditions for a later decision and partly determine the later decision through, for instance, limiting the number of later available decision alternatives. Each decision may be modeled in a structural or process approach. Most often they are conceptualized in a structural model.

Differentiation and Consolidation Theory (Diff Con) Theory (Svenson, 1992) describes how choices are strengthened in post-decision processes. This may be seen as a fist step towards increasing the chance of a similar choice the next time. If the outcomes are favorable enough, a routine or habit may be formed (Eagly and Chaiken, 1993; Gärling, 1992).

Successive decisions with similar problems is the simplest form of sequential-decision making. There are many terms for describing the fact that a prior decision or the status quo alternative tends to be preferred to competitors such as the sunk cost phenomenon, the status quo effect, the endowment effect, and the omission bias (Ritov and Baron, 1982; Samuelson and Zeckhauser, 1988; Staw, 1976; Thaler, 1980;). Kleiter (1975) reported a decision planning horizon of about 2 steps forward in a sequential decision task which is a typical result in the area. Gärling and co-workers (Gärling et al., 1997; Romanus et al., 1996) have studied the effects of a prior gain or loss on a subsequent decision. They found that prior outcomes were only integrated with potential losses in a later decision.

For travel-mode choices, the third type of linked decisions (when a prior decision delimits the number of alternatives in a later choice) is relevant. Such choices can be made so that the decision concerns a fixed sequence of decisions or a plan (cf. Gärling et al., 1994). Or they can be made one by one at the other extreme. An example of linked decisions is a prior decision concerning where to live and successive travel-mode decisions. If there is no public transportation such travel-mode alternatives are then excluded by the earlier decision. A somewhat similar prior decision may be the purchase of a ticket valid for a month on public transportation which may affect travel mode choices during that month.

Thus, in many contexts such as in decisions of travel modes, the decision process is a level-4 type process. Here, decision alternatives have to be created in which, for example, one alternative contains a travel (e.g, to an island) by car + boat and another by bus + air. In economic contexts this seems somewhat parallel to portfolio selection. Problems with a structure like these examples are interesting in travel-mode choice and other applied contexts. To exemplify further, a great part of the traffic in a city may depend on the decision where to park (following the prior decision to take the car into the city). If this is the situation decision aids concerning parking alternatives can eliminate some city traffic (Petterson, personal communication).

Only a minority of travel-mode choices seem to be level-2 and level-3 decisions which are the most thoroughly researched areas in descriptive behavioral decision research. To exemplify further, if there is a plan activated when a traveler intends to do something (shop, go to the theater), then there is often recognition of alternatives fitting the plan (level-1) which matters (cf. Gärling et al., 1997). The decision process can also be moved up to constructive level-4 where decision alternatives are created according to plans and the decision maker's own goals.

In general, behavioral decision theory has very little to say about level-4 decisions but traditions within operational research have been concerned with normative problems of this kind. Scheduling Theory was developed within operational research and has been introduced to behavioral science by Moray and his collaborators (cf. Dessouky, Moray, and Kijowski, 1995). In particular, the composition of actions in time and resource expenditure have been addressed in this approach which so far is predominantly normative in character. Scheduling Theory attempts to answer questions about strategic decisions in which a sequence of actions should be performed (see, e.g., Gärling et al., in press). Dessouky et al. (1995) give a detailed account of different elements in these kinds of normative theories.

Dynamic decision making models acknowledge that decisions are made in a dynamic environment and view decisions as attempts to control the environment. Brehmer (1992) states that it is hard to find normative models for dynamic decision making within psychology. Instead models from control theory seem applicable. To exemplify, Diehl and Sterman (1995) used simple functional relations and simulated an inventory that should be managed in face of stochastic sales. Dynamic decision making could be applied to the control of a car in a traffic stream. Descriptive results from studies in the dynamic decision making tradition focus on general aspects such as time delay and possible side effects, both of which have been shown to affect success in managing a dynamic system. Empirical research has also shown that in sequential (e.g., Kleiter, 1975) and dynamic decision making (e.g., Brehmer, 1992) decision makers typically search only the present and one step ahead in the decision tree. For optimal decision making across linked decisions it is necessary to use decision aids and/or to adopt an efficient strategy which is satisfactory for the decision maker and does not lead to post-decision regret.

DIFFERENTIATION AND CONSOLIDATION THEORY

In an effort to both integrate earlier research in behavioral decision theory and to develop the theoretical foundations for this kind of research, Differentiation and Consolidation (Diff Con) Theory was developed (Svenson, 1992). Diff Con Theory deviates from other decision theories in that the goal of a decision process is to create an alternative that is *sufficiently* superior in comparison to its competitor(s) through restructuring and application of one or several decision rules. The structuring principles are not derived from one rule only (e.g., SEU or dominance) but from a number of different rules contingent on the situation and the person. This process is named *differentiation*. The corresponding postdecision process in support of the chosen alternative is called *consolidation*. The general framework provided by Diff Con can accommodate different existing decision models as subprocesses or special cases (cf. Svenson, 1992). Note that most contemporary structural or static decision theories do not at all consider degree of differentiation or preference strength (Busemeyer and Townsend, 1993).

Predecision Differentiation

Identifying Decision Alternatives, Goal Elicitation and Editing. The initial processing of a decision problem is partly governed by markers which tell the decision maker where to start. Markers can be perceptual or cognitive and relate to the goal structure and the representation of the decision problem. The nomination of one attribute as most important can be governed by a marker so that the aspects on that attribute are compared first. A marker can be, for example, the label of an attribute (e.g., costs) or the aspects on an attribute (e.g., a very fast travel mode). It is assumed that markers are very important and may explain important parts of otherwise not understood behavior. A marker may not be integrated in the process otherwise (e.g., Svenson, 1983, 1985), or it may be weighed heavily as an important attribute.

Editing or rearranging and grouping the information is also initiated by markers and essential to reduce the demand of energetic resources (cf. Abelson and Levi, 1985; Coupey, 1994; Kahneman and Tversky, 1979; Ranyard, 1989; Tversky and Kahneman, 1992). Identification of alternatives may concern selection among several given ones as in most laboratory studies. In many real-life situations only one decision alternative appears at a time. This single alternative has to be accepted or rejected (c.f. Fischhoff, 1996). In other applied settings a situation may require the creation of new alternatives. To exemplify, one job offer is given and the decision concerns accepting this or to search for more or stay with the status quo.

Selection of a Reference and/or Preliminary Choice Alternative. A reference decision alternative can be selected by a level-1 quick decision involving a classification without any evaluative component. This alternative may serve in determining aspiration levels (Dembo, 1931) or as an anchor in other respects. It may also be derived in more elaborate and deliberate processes. A selection of a preliminary (cf. Montgomery, 1983) alternative can also involve a quick holistic association. In applied settings there may often be just one alternative that is considered and the decision therefore concerns a choice between the status quo alternative and one other alternative (cf. Kahneman et al., 1990; Samuelson and Zeckhauser 1988; Schweitzer, 1994; Svenson, 1990). In travel-mode choice, the habitual mode is the status quo alternative.

Differentiation Leading to a Decision. The differentiation of the preliminarily chosen alternative may be successful or it may fail. However, according to the theory, the probability of an alternative being chosen is rather high than low if it is first selected as the preliminary choice. This is implied by the fact that differentiation attempts to support the first preliminary choice in a biased way favoring that candidate. Empirical evidence for this can be found in studies by, for example, Russo et al., (1998) Shafir and Tversky (1992), Shafir (1993), and in which additional neutral information increases differentiation. If the preliminary alternative cannot be sufficiently differentiated, it will be replaced by another candidate which then becomes the next focus of the differentiation process. There are three kinds of differentiation processes, holistic, process, and structural differentiation.

Holistic Differentiation. Holistic differentiation is quick and may be experienced by the decision maker as something like a classification process (not readily available for conscious control or awareness) of the preferred alternative (Estes, 1994; Klein, 1989). The process may be both totally separated from any attractiveness as for example in well learned habitual decision making. Emotional factors can play an important role in other cases. To exemplify, some emotional reaction may have been elicited instantaneously triggering a holistic decision (cf. Zajonc, 1980). Holistic differentiation often leads to a degree of differentiation which is sufficient for a final choice. However, it may also be part of more complex differentiation processes (e.g., in selecting a preliminary alternative).

Process Differentiation. Process differentiation includes the use of one or more decision rules (e.g., the conjunctive rule, the additive difference rule, cf. Beach, 1990; Svenson, 1979, 1992;). Which rule is elicited depends on both the individual, the context, and the structure of the decision problem. The frequently reported initial use of the conjunctive rule in the screening of the alternatives to reduce the number of options belongs to this category of differentiation. Changing the criterion limits on different attributes, in rules using such criteria, is another and related kind of process differentiation. To exemplify, a conjunctive rule can be applied with successively stricter criteria of rejection in the early stages of a decision process to reduce the set of possible alternatives. In later stages of the differentiation process, the preliminary choice can be tested against other alternatives in different decision rules. Diff Con Theory postulates that a sufficiently superior alternative is chosen. Like additional arguments in a discussion, different decision rules may add differentiation in support of an alternative.

Structural Differentiation. Structural differentiation refers to changes in psychological representations of the decision alternatives. All such changes are made in support of one or several decision rules. Structural differentiation processes are of different kinds depending on which parts of the structure is the target. Different kinds of processes are assumed to take place concurrently and contingent on each other and can be divided into the following categories (cf. Svenson, 1992):

1. Attractiveness restructuring concerns the attractiveness of aspects and is governed by decision rules setting the goals for the process. Restructuring is made possible through uncertainty of attractiveness representations, uncertainty in goal elicitation and the relative importance of different goals. These uncertainties can be kept fuzzy by the decision maker and used in negotiating the structure with her or himself before the decision and thereafter.
2. Attribute importance restructuring refers to changes in the importance of an attribute. For example, differentiation according to a weighed additive difference rule can be performed through changing the relative importance among the attributes so that attributes favoring the hypothetical choice attain more weight.
3. Facts restructuring refers to changes in the perceived reality to support a preliminary choice to a greater extent. A decision maker citing facts which are biased towards supporting the preliminary choice reflects facts restructuring. This is also the case when memory retrievals of earlier facts statements are more supportive than they were before the decision or judgment (cf. hindsight effects,

Fischhoff, 1975; cognitive dissonance, Festinger, 1964, to cite early references). Facts restructuring parallels attractiveness restructuring and the two are sometimes substitutable.

4. Problem restructuring implies using available factors and/or searching for new facts for an alternative way of representing the decision problem. This may involve new attributes, for example, supporting a preliminary choice and/or the creation of new alternatives. Problem restructuring includes level-4 decision processes in which new alternatives can also be created and old ones thoroughly modified. This kind of restructuring may give further support to a preliminary choice alternative but it may also erase it completely from the set of possible alternatives.

Post-Decision Consolidation

Post-decision consolidation is performed by processes similar to those in the pre-decision phase. After a decision the decision maker has lost the opportunity of choice, the unique good aspects of the rejected alternative(s) are gone, and he or she is stuck with the bad aspects of the chosen alternative. This can create post-decision dissonance (Festinger, 1964) or regret (Loomes and Sugden, 1982) which has to be managed psychologically and/or in the implementation phase. In response to these challenging threats pre-decision differentiation can be continued in what is now called post-decision consolidation. In some cases, in particular when decisions have been quick and perhaps immature, post-decision consolidation supporting the decision may be a substitute for pre-decision processes (cf. Svenson, 1992).

Some decision rules such as the conjunctive rule may be less frequent after the decision than before because of its pre-decision screening character. Note that it is also possible to split the post-decision phase into (1) post-choice, (2) implementation, and (3) post-outcome stages. Attractiveness restructuring is likely to depend on the pattern of aspects in a decision problem. If, for example, the preliminarily chosen alternative is initially far better on the most important attribute (e.g., costs) there is less need to consolidate on that attribute than if it is just barely superior. If the second most important attribute (e.g., comfort) also supports the chosen alternative, then there is less need to consolidate through this attribute than if it conflicts (the preliminary choice is less comfortable than its competitor) with the chosen option.

Post-decision consolidation is important in strengthening a decision, forming it into a habit, and developing a general strategy. Post-decision consolidation may explain conservatism in people's behavior such as travel-mode choices and other repeated decisions. Diff Con theory has not so far treated decision maker's strategic plans and how goals are created and mapped onto alternatives and facts representations. However, some discussions about this can be found in Svenson (1996).

Hitherto the dynamics in Diff Con Theory have concerned changes driven by the decision maker himself or herself. However, in many applied contexts decisions are made and implemented in a dynamic environment. Thus, the decision problem changes both in itself and as a result of the implementation of a decision. For example, the choice between bus and taxi may be postponed while the decision maker waits for the next bus. If the bus does not appear in time, the taxi alternative will become sufficiently differentiated by the dynamics of the environment (through, for example, the attributes of arrival in time and comfort becoming sufficiently superior for the taxi alternative to balance out costs).

Decision makers may avoid the responsibility and pain of differentiating alternatives which are rather similar in attractiveness through just postponing the decision. In many cases there is a chance that facts appear and the situation changes so that differentiation will be easier. In addition, waiting means that psychological processes may work towards further differentiation more or less by themselves, or through thinking about the postponed decision. Therefore time gives an opportunity of both external and internal differentiation.

Thus, the particular characteristics of a decision problem plays an important role in determining the time it takes to reach a decision. For example, if waiting or delaying the decision is expected to generate more information concerning the most important attributes or if the decision problem in itself will change so that differentiation is facilitated, time will be longer until a decision has been reached. If the attractiveness of waiting is progressively more negative, this cost works towards a decrease of the time to reach a decision.

Different persons may have different styles in coping with different decision problems in different contexts and time constraints. There is not much research on decision style. Different emotional states (e.g., depression) may lead to different average decision times (e.g., longer for depressed people). Stress and time pressure seem to delay and/or change the differentiation process (Benson, 1993; Malmsten, 1996).

The Role of Uncertainty in Diff Con Theory

Uncertainty or subjective probability concerning events and outcomes is modeled as an independent attribute in Diff Con Theory. However, it has not been given such a prominent role as in expected value theories of decision making. Differentiation of this attribute in Diff Con Theory can be related to wishful thinking (Cyert et al., 1958; Slovic, 1966) and consolidation of the subjective probability attribute to the hindsight phenomenon (Fischhoff, 1975).

Uncertainty concerning future goals and attractiveness mappings are modeled in Diff Con Theory as part of the reasons for differentiation and consolidation, beyond the point where one alternative is just better that its closest competitor. Fischhoff et al., (1980), Jungermann (1983), and Kahneman and Snell (1992) have treated the uncertainty of goals, present and future utilities in a decision theoretic framework, which relates to attractiveness uncertainty in Diff Con Theory.

What Drives Differentiation and Consolidation Processes?

Differentiation processes are primarily driven by the need of reaching a decision. But why should this decision process produce one alternative that is sufficiently superior through application of the different ways of differentiation presented above? Generally speaking, differentiation and consolidation can be seen as driven by at least two component forces. One of these is related to organization of psychological representations so that they preserve and/or adhere to simple structures. The other component force is related to a safety or stability motive.

The first of these two components can be related to arousal driving dissonance reduction (Festinger, 1957). It can also be related to cognitive explanations (Bem, 1965, 1967) and to attribution theory (Kelley, 1967) and self-serving biases (Greenwald, 1980). Empirical studies starting with Zanna and Cooper (1974) have shown that dissonant cognitions give rise to negative emotional arousal (cf. Eagly and Chaiken, 1993). One way of reducing this negative emotional arousal is to restructure a decision problem so that the cognitive dissonance is eliminated (through applying a decision rule or restructuring the problem). Cooper and Fazio (1984) stated that dissonance was driven by negative consequences instead of inconsistency of cognitions. Harmon-Jones et al. (1996) showed that this is not necessary. According to Diff Con Theory there is no need of negative consequences for differentiation and consolidation to take place. In a personal communication Verplanken suggested that the personality trait of a "need for closure" (e.g., Webster and Kruglanski, 1994) may correlate with differentiation and consolidation processes.

However, the first driving factor cannot explain why also an already chosen and superior alternative has to be further differentiated and consolidated. Therefore, a second driving factor must be introduced. In behavioral term it may be seen as a factor driven by a motive of behavioral stability. Thus, in terms of the differentiation, (Svenson, 1992), this intervening variable has to be sufficiently different from zero to keep action stable even if external (e.g., outcome) or internal (e.g., emotional states) changes appear in the post-decision phase.

Empirical Findings Relating to Diff Con Theory

Studies performed during the past few years within the Diff Con framework have mainly treated post-decision consolidation processes. In these studies patterns of rating of attractiveness on different attributes have served as indicators of a differentiation variable *d*. The results indicate (1) that post-decision structural consolidation occurs regularly (Svenson and Benthorn, 1992; Svenson et al., 1994; Svenson and Malmsten, 1996); (2) that subjects differ in degree of importance ascribed to different attributes and that this difference matters in the data analyses (Benthorn, 1994); (3) that aspects on the one and/or second most important attributes are predominantly used in post-decision consolidation; (4) that uninteresting and non-involving decision problems may not lead to post-decision consolidation while interesting and more involving problems do (Benthorn, 1994; Svenson et al., 1994); (5) that an increase in consolidation has not been found to be a result of a need of justification (Svenson et al., 1994); (6) that aspect attractiveness restructuring was much stronger than attribute imporance restructuring (Malmsten, 1996); (7) that decision outcome feedback speeded up post-decision consolidation processes (Svenson and Malmsten, 1996) and (8) that time pressure seems to be detrimental to and to weaken differentiation and/or immediate consolidation (Benson, 1993).

Other studies in other contexts have illustrated pre-decision differentiation in terms of early rejection of several alternatives and selection of one preliminary choice alternative and the successive differentiation in terms of value statements of that alternative (or the rejection of it and differentiation of another alternative) (e.g., Bettman et al., 1991; Kleinmuntz, 1963; Montgomery and Svenson, 1989; Russo et al., (in press); Svenson, 1974; Svenson, 1979).

A Generic Model

It is possible to view Diff Con Theory as describing how a decision maker applies his or her values and attitudes to decision problems and adjusts to the postdecision situation. Therefore,Diff Con theory can be accommodated to a slightly modified contemporary framework used in research on human attitudes (cf. Eagly and Chaiken, 1993) as illustrated in Figure 7.1. This figure illustrates how habit, general evaluation, outcome involvement, norm involvement, and ego involvement affect the decision process, the intention to implement and the actual implementation of the decision. The model in Figure 7.1 makes it possible to integrate findings in attitude research with decision theory in travel-mode choice applications.

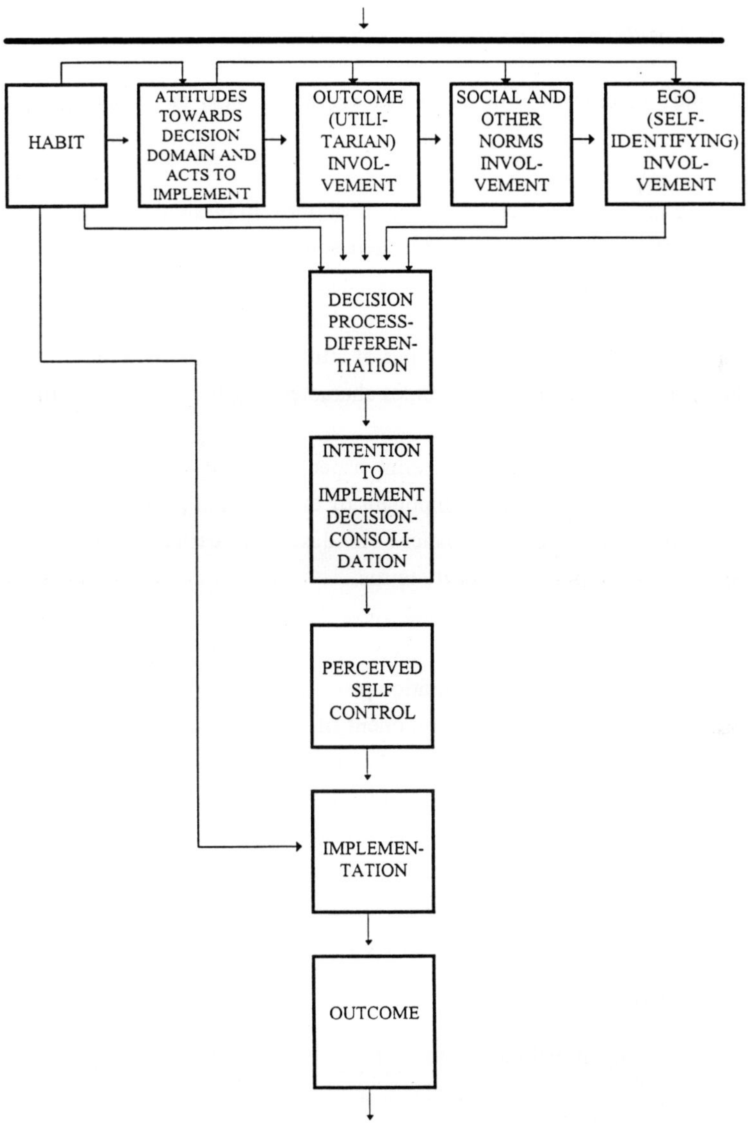

Figure 7.1. Factors affecting the decision process and the implementation of a decision.

The habit component is related to level-1 decisions which in turn are derived from earlier decisions at other levels. A strong car habit may delimit the choice set to car routes and departure times only and thus exclude other travel modes. Exemplifying the strength of the habit component, Verplanken et al. (1994) showed that in the choice of travel mode (car) the attitude towards traveling by car correlated about 0.36 with behavior. For subjects with weak habits (of car use) the correlation between attitude and behavior was 0.65, and for subjects with strong traveling habits the correlation was 0.28. This shows that attitude is a better predictor of the outcome of a decision process when there is no strong habit component involved.

Attitudes can be developed toward travel-mode options, in particular in new situations. At that point level-4 decisions are first used and then some important choices are moved to lower levels. Attitudes toward decision domain and acts of implementation of the decision refer both to general attitudes (toward, for example, a travel mode) and a more specific attitudes (toward using the subway to work). Both these types of attitudes contribute to the process of making a decision through representations on attitudes characterizing the decision alternatives.

Outcome utilitarian involvement concerns aspects related to individual outcomes of a decision. This represents a factual component of the decision process relatively independent of social and personality characteristics. As described earlier, most empirical decision research targets this component in utility terms.

Social norms play an important role in many decision situations. Social norms influence the importance of attributes in the differentiation process. Sometimes a social norm is sufficient to explain behavior. There is a strong component of imitation in human behavior. This means that norms may also develop on the spot. For example, when one pedestrian has started to walk towards a stop signal, the likelihood of the remaining pedestrians to cross increases. The norm of crossing against a stop signal has developed in that situation. Sometimes conflicts appear between outcome and social factors (Gärling, 1998). Biel and Gärling (1998) asserted that people consider (1) individual outcomes, (2) outcomes of the collective consequences, and (3) the individual's outcome of collective consequences when making travel-mode decisions.

Self-identify involvement includes factors which are related to enduring fundamental attitudes which define a person. Self-serving biases may result from self-identity involvement also leading to consolidation effects.

All these different factors feed into a decision process as goals and their mappings onto the current decision problem. In Diff Con theory habits may be processed either in the differentiation process or they can be directly connected to implementation of an action through situation and pattern recognition. In attitude research it has been shown repeatedly that intention (following a decision) to act is a better predictor of behavior than attitudes of a more or less specific character. In addition, it has been shown that the perceived control in a situation is a predictor of behavior.

Verplanken and Svenson (1996) distinguished between (1) value relevant, (2) outcome relevant, and (3) impression relevant involvement determining the goals of a decision maker (cf. Figure 7.1). Value relevant involvement refers to activation of attitudes linked to important values and refers both to the act of decision and the outcome. Outcome relevant involvement is more cognitively colored and relates closer to normative utility approaches. Impression relevant involvement activates goals of self-presentation in a social context.

CONCLUDING COMMENTS

It is important to point out that behavioral decision theory has primarily been developed for describing and understanding human decision making. Thus, the chief purpose has not been to predict behavior. When this has been attempted using, for instance the SEU model, difficulties of measuring the variables independently have appeared. As Dawes (1996) and others have warned us, the prediction of the future based on how we understand the past is a biased-prone activity. However, if it is possible to map habits, and to achieve a sufficient understanding of how a decision maker represents a situation including the attributes and how the goals are activated, then there is hope; knowing the person, the situation, and the habit gives an adequate background for prediction.

Behavioral decision research can help us understand why a certain travel mode or travel route is chosen through the exploration of underlying attributes. Sometimes travelers need information to enable them to make wise decisions. These may concern route choices or decisions related to environmental concerns. Then, psychological theory and empirical findings provide valuable tools when preparing and transmitting the information to the travelers. From a societal point of view it may also be desirable to change travel behavior, a process in which behavioral research can provide valuable insights. For example, marketing and political campaign experts being aware of this have used behavioral science in their trades for some time.

In conclusion, behavioral decision theory and social science are gradually increasing our understanding of how and why people make the decisions they do. At the same time specific models for travel choice have developed so far that it now seems to be a necessity of integrating behavioral and social science into these models. This integration is indeed one of the great challenges in future travel choice modeling.

ACKNOWLEDGEMENTS

This study was supported by grants from the Swedish Council for Research in the Humanities and Social Sciences and from the Swedish Work Environment Fund. The author wishes to thank Anders Johansson, Hans-Erik Petersson, Sanny Shamoun, Nina Zorcec, Bas Verplanken, and Tommy Gärling for valuable discussions, comments, and assistance.

REFERENCES

Abelson, R. P. and A. Levi (1985). Decision making and decision theory. In: *Handbook of Social Psychology* (3rd ed) (G. Lindzey and E. Aronson, eds.), pp. 231-309. Random House, New York.

Albert, D., K. M. Aschenbrenner and F. Schmalhofer (1989). Cognitive Choice Processes and Attitude-Behavior Relation. In: *Attitudes and Behavioral Decisions* (A. Upmeyer, ed.), pp. 61-99. Springer-Verlag, New York.

Aschenbrenner, K .M., D. Albert and F. Schmalhofer (1984). Stochastic choice heuristics. *Acta Psychologica*, **56**, 153-166.

Audley, R. J. (1960). A stochastic model for individual choice behavior. *Psychological Review*, **67**, 1-15.

Beach, L. R. (1990). *Image Theory: Decision Making in Personal and Organizational Contexts*. Wiley, Chichester, UK.

Bem, D. J. (1965). An experimental analysis of self-persuasion. *Journal of Experimental Social Psychology*, **1**, 199-218.

Bem, D. J. (1967). Self perception: An alternative interpretation of cognitive dissonance phenomena. *Psychological Review*, **74**, 183-200.

Benson, III, L. (1993). *Studies in human decision making*. Doctoral dissertation, Department of Psychology, Lund University, Sweden.

Benthorn, L. J. (1994). *On post-decision processes*. Doctoral dissertation, Department of Psychology, Lund University, Sweden.

Bettman, J. R., E. J. Johnson and J. W. Payne (1991). Consumer decision making. In: *Handbook of Consumer Behavior* (T. S. Robertson and H. H. Kassarjian, eds.). pp. 50-84. Prentice Hall, Englewood Cliffs, NJ.

Biel, A. and T. Gärling (1995). The role of uncertainty in resource dilemmas. *Journal of Environmental Psychology,* **15**, 222-233.

Brehm, J. W. (1956). Postdecision changes in the desirability of choice alternatives. *Journal of Abnormal and Social Psychology*, **52**, 384-389.

Brehm, J. W. and A. R. Cohen (1959). Re-evaluation of choice alternatives as a function of their number and qualitative similarity. *Journal of Abnormal and Social Psychology*, **58**, 373-378.

Brehmer, B. (1992). Dynamic decision making: Human control of complex systems. *Acta Psychologica*, **81**, 211-241.

Brock, T. C. (1963). Effects of prior dishonesty on post decision dissonance. *Journal of Abnormal and Social Psychology*, **66**, 325-331.

Busemeyer, J. R. and J. T. Townsend (1993). Decision field theory: A dynamic-cognitive approach to decision making in uncertain environment. *Psychological Review*, **100**, 432-459.

Busemeyer, J. R. (1985). Decision making under uncertainty: A comparison of simple suability, fixed-sample and sequential-sampling models. *Journal of Experimental Psychology: Learning, Memory and Cognition*, **11**, 538-564.

Camerer, C. F. (1988). An experimental test of several generalized utility theories. *Journal of Risk and Uncertainty*, **2**, 61-104.

Coombs, C. H. (1958). On the use of inconsistency of preferences in psychological measurement. *Journal of Experimental Psychology*, **55**, 1-7.

Cooper, J. and R. H. Fazio (1984). A new look at dissonance theory. In: *Advances in Experimental Social Psychology,* (L. Berkowitz, ed.), Vol. 17. pp. 229-264. Academic Press, New York.

Coupey, E. (1994). Restructuring: Constructive processing of information displays in consumer choice. *Journal of Consumer Research*, **21**, 83-99.

Cyert, R. M., W. T. Dill and J. G. March (1958). The role of expectations in business decision making. *Administrative Science Quarterly*, **3**, 307-340.

Dawes, R. M. (1996). Behavioral decision making, judgment and inference. In: *The Handbook of Social Psychology* (D. Gilbert, S. Fiske, and G. Lindgren, eds.). Mc Graw-Hill, Boston, MA.

Dembo, T. (1931). Der Ärger als dynamisches Problem. *Psychologische Forschung,* **15**, 1-144.

Dessouky, M. I., N. Moray and B. Kijowski (1995). Taxonomy of scheduling systems as a basis for the study of strategic behavior. *Human Factors*, **37**, 443-472.

Diehl, E. and J. D. Sterman (1995). Effects of feedback complexity on dynamic decision making. *Organizational Behavior and Human Decision Processes*, **62**, 196-215.

Dietrich, A., K. M. Aschenbrenner and F. Schmalhofer, (1989). Cognitive choice processes and the attitude-behavior relation. In: *Attitudes and Behavioral Decisions* (A. Upmeyer, ed.), pp. 61-99. Springer, New York.

Eagly, A. H. and S. Chaiken (1993). *The Psychology of Attitudes*. Harcourt Brace Jovanovich, New York.

Edwards, W. (1954). The theory of decision making. *Psychological Bulletin*, **41**, 380-417.

Einhorn, H. and R. Hogarth (1981). Behavioral decision theory: Processes of judgment and choice. *Annual Review of Psychology*, **32**, 53-88.

Estes, W. K. (1959). A random walk model for choice behavior. In: *Mathematical Methods in the Social Sciences* (K. J. Arrow, S. Karlin and P. Suppes, eds.), pp. 265-276. Stanford University Press, Stanford, CA.

Estes, W. K. (1994). *Classification and Cognition*. Oxford University Press, New York.

Feller, W. (1957). *An introduction to probability theory and its applications*, Vol. 1 (2nd ed.) Wiley, New York.

Festinger, L. (1957). *A Theory of Cognitive Dissonance*. Stanford University Press, Stanford, CA.

Festinger, L. (1964). *Conflicts, Decision and Dissonance*. Stanford University Press, Stanford CA.

Fischhoff, B. (1975). Hindsight is not equal to foresight: The effect of outcome knowledge on judgment under uncertainty. *Journal of Experimental Psychology: Human Perception and Performance*, **1**, 288-299.

Fischhoff, B. (1996). The real world: What good is it? *Organizational Behavior and Human Decision Processes*, **65**, 232-248.

Fischhoff, B., P. Slovic and S. Lichtenstein (1980). Knowing what you want: Measuring labile values. In: *Cognitive Processes in Choice and Decision Behavior* (T. Wallsten, ed.) pp. 64-85. Erlbaum, Hillsdale, NJ.

Frey, D. (1981). Postdecisional preference for decision-relevant information as a function of its source and the degree of familiarity with its information. *Journal of Experimental Social Psychology*, **17**, 51-67.

Frey, D. (1986). Recent research on selective exposure to information. *Advances in Experimental Social Psychology*, **19**, 41-80.

Gärling, T. (1992). The importance of routines for the performance of everyday activities. *Scandinavian Journal of Psychology*, **33**, 170-177.

Gärling, T. (1998). Behavioral assumptions overlooked in travel-choice modeling. In: *Transport Modeling* (J. Ortuzar, S. R. Jara-Díaz and D. Hensher, eds.). Pergamon, Oxford. In press.

Gärling, T., D.F. Ettema, R. Gillholm and M. Selart (1997). Towards a theory of the intention-behavior relationship with implications for the prediction of travel behavior. In: *World Transport Research: Proceedings of the 7th World Conference on Transport Research:*

Vol. 1. Travel Behaviour. (D. Hensher, J. King, and T. Oum, eds.), pp. 231-240. Pergamon, Oxford.

Gärling, T., J. Romanus and M. Selart (1994). Betting at the race track: Does risk seeking increase when losses accumulate? *Perceptual and Motor Skills*, **78**, 1248-1250.

Gärling. T., R. Gillholm, J. Romanus and M. Selart (1997). Interdependent activity and travel choices: Behavioral principles of integration of choice outcomes. In: *Activity Based Approaches to Travel Analysis.* (D.F. Ettema and H.P.J. Timmermans, eds.), pp. 135-150. Pergamon, Oxford.

Gärling, T., N. Karlsson, J. Romanus and M. Selart (1997). Influences of the past on choices in the future. In: *Decision Making: Cognitive Models and Explanations* (R. Ranyard, R. Crozier and O. Svenson, eds.), pp 167-188, Routledge, London.

Gärling, T., E. Lindberg and H. Montgomery (1989). Beliefs about attainment of life satisfaction as determinants of preferences for everyday activities. In: *Understanding Economic Behavior* (K. Grunert and F. Ölander, eds.), pp. 33-46. Kluwer, Dordrecht, The Netherlands.

Gerard, H. B. (1994). A retrospective theory of Festinger´s A theory of cognitive dissonance. *Contemporary Psychology*, **39**, 1013-1017.

Greenwald, A. G. (1980). The totalitarian ego: Fabrication and revision of personal history. *American Psychologist*, **35**, 603-618.

Greenwald, A. G. and D. L. Ronis (1978). Twenty years of cognitive dissonance: Case study of the evolution of a theory. *Psychological Review*, **85**, 53-57.

Harmon-Jones, E., J. W. Brehm, J. Greenberg, L. Simon and D. E. Nelson (1996). Evidence that the production of aversive consequences is not necessary to create cognitive dissonance. *Journal of Personality and Social Psychology*, **70**, 5-16.

Janis, I. L. and L. Mann (1977). *Decision Making: A Psychological Analysis of Conflict Choice and Commitment.* Free Press, New York.

Josephs, R., R. P. Larrick, C. M. Steele and R. E. Nisbett (1992). Protecting the self from the negative consequences of risky decisions. *Journal of Personality and Social Psychology*, **62**, 26-37.

Jungermann, H. (1983). The two camps on rationality. In: *Decision Making under Uncertainty.* (R. W. Scholz, ed.), pp. 63-86. North Holland, Amsterdam.

Kahneman, D., J. L. Knetsch and R. H. Thaler (1990). Experimental tests of the endowment effect and the Coase theorem. *Journal of Political Economy*, **98**, 1325-1348.

Kahneman, D. and D. T. Miller (1986). Norm theory: Comparing reality to its alternatives. *Psychological Review,* **93**, 136-153.

Kahneman, D. and J. Snell (1992). Predicting a changing taste: Do people know what they will like? *Journal of Behavioral Decision Making*, **5**, 187-200.

Kahneman, D. and A. Tversky (1979). Prospect theory: An analysis of decisions under risk. *Econometrica*, **47**, 263-291.

Kelley, H. H. (1967). Attribution theory in social psychology. In: *Nebraska Symposium on Motivation* (D. Levine, ed.), Vol. 15, pp. 192-238. University of Nebraska Press, Lincoln, NE.

Klein, G. A. (1989). Recognition-primed decisions. In: *Advances in Man-Machine System Research* (W. B. Rouse, ed.), Vol. 5, pp. 47-92. JAI Press, Greenwich, CT.

Klein, G. A., J. Orasanu, R. Calderwood and C. E. Zsambok (1993). *Decision Making in Action: Models and Methods*. Ablex, Norwood, NJ.

Kleinmuntz, B. (1963). Profile analysis revisited: A heuristic approach. *Journal of Counseling Psychology*, **70**, 315-324.

Kleiter, G. D. (1975). Estimating the planning horizon in a multistage decision-task. *Psychological Research*, **38**, 37-64.

Landman, J. (1987). Regret: A theoretical and conceptual analysis. *Journal for the Theory of Social Behavior*, **17**, 135-160.

Langer, E. (1994). The illusion of calculated decisions. In: *Beliefs, Reasoning and Decision Making: Psycho-logic in Honor of Bob Abelson* (R.C. Schank and E. Langer, eds.), pp. 33-53. Erlbaum, Hillsdale, NJ.

Lewin, K. (1935). *A dynamic theory of personality*. Mc Graw-Hill, New York.

Lindberg, E., T. Gärling, H. Montgomery and R. Waara (1987). Peoples evaluations of housing attributes: A study of underlying beliefs and values. *Scandinavian Housing and Planning Research*, **4**, 81-103.

Loomes, G. and R. Sugden (1982). Regret theory: An alternative theory of rational choice under uncertainty. *The Economic Journal*, **92**, 805-824.

Lopes, L. L. (1987). Between hope and fear: The psychology of risk. In: *Advances in Experimental Social Psychology* (L. Berkowitz, ed.), **20**, pp. 255-295. Academic Press, New York

Lopes, L. L. (1988) Economics as psychology: A cognitive essay of French and American Schools of risk theory. In: *Risk, Decision and Rationality* (B. R. Munier, ed.), pp. 405-416. Riedel, Boston, MA.

Lopes, L. L. (1995). On modeling risky choices: Why reasons matter. In: *Contributions to Decision Making Vol. 1* (J. P. Caverni, M. Bar-Hillel and H. Jungermann, eds.), pp. 29-50. Elsevier, Amsterdam.

Malmsten, N. (1996). *Studies of post-decision processes*. Doctoral dissertation, Department of Psychology, Stockholm University, Stockholm.

March, J. G. (1978). Bounded rationality, ambiguity and the engineering of choice. *The Bell Journal of Economics*, **9**, 580-600.

May, K. O. (1954). Intransitivity, utility and the aggregation of preference patterns. *Econometrica*, **22**, 1-13.

Mellers, B. A. and K. Biagini (1994). Similarity and choice. *Psychological Review*, **101**, 505-518.

Mellers, B. A. and M. H. Birnbaum (1983). Contextual effects in social judgment. *Journal of Experimental Social Psychology*, **19**, 157-171.

Mellers, B.A. and A. D. J. Coke (1994). Trade-offs depend on attribute range. *Journal of Experimental Psychology: Human Perception and Performance*, **20**, 1055-1067.

Miller, G. A. (1956). The magical number seven plus or minus two: Some limits of our capacity for the processing of information. *Psychological Review*, **63**, 81-97.

Miller, N. E. (1959). Liberalization of basic S-R concepts: Extensions to conflict behavior, motivation and social learning. In: *Psychology: A Study of a Science* (S. Koch, ed.), pp. 196-292. Mc Graw-Hill, New York.

Montgomery, H. (1983). Decision rules and the search for a dominance structure: Towards a process model of decision making. In: *Analyzing and Aiding Decision Processes* (P. C. Humphreys, O. Svenson, and A. Vasi, eds.), pp. 343-369. North-Holland, Amsterdam.

Montgomery, H. and O. Svenson (1989). A think aloud study of dominance structuring in decision processes. In: *Process and Structure in Human Decision Making* (Montgomery, H. and O. Svenson, eds.), pp. 135-150. Wiley, Chichester, UK.

Papandreou, A. G. (1953). An experimental test of an axiom in the theory of choice. *Econometrica*, **21**, 477.

Payne, J. W. (1976). Task complexity and contingent processing in decision making: An information search and protocol analysis. *Organizational Behavior and Human Performance*, **16**, 366-387.

Payne, J. W. (1982). Contingent decision behavior. *Psychological Bulletin*, **92**, 382-402.

Payne, J. W., J. R. Bettman and E. J. Johnson (1992). Behavioral decision research: A constructive processing perspective. *Annual Review of Psychology*, **43**, 87-131.

Payne, J. W., J. R. Bettman and E. J. Johnson (1993). *The Adaptive Decision Maker*. Cambridge University Press, Cambridge.

Ranyard, R. (1989). Structuring and evaluating simple monetary risks. In: *Process and structure in human decision making* (H. Montgomery and O. Svenson, eds.), pp. 195-207. Wiley, Chichester, UK.

Ritov, I. and J. Baron (1992). Status quo and omission biases. *Journal of Risk and Uncertainty*, **5**, 49-61.

Roese, N. J. and J. M. Olson (Eds.) (1995). *What Might Have Been: The Social Psychology of Counterfactual Thinking*. Erlbaum, Hillsdale, NJ.

Romanus, G., L. Hassing and T. Gärling (1996). A loss-sensitivity explanation of integration of prior outcomes in risky decisions. *Acta Psychologica*, **93**, 173-183.

Russo, J. E. and L. D. Rosen (1975). An eye fixation analysis of multiattribute choice. *Memory and Cognition*, **3**, 267-276.

Russo, J. E., V. H. Medvec and M. G. Meloy (1998). The distortion of information during decisions. *Organizational Behavior and Human Decision Processes*. In press.

Samuelson, W. and R. Zeckhauser, R. (1988). Status quo bias in decision making. *Journal of Risk and Uncertainty*, **1**, 7-59.

Schweitzer, M. (1994). Disentangling status quo and omission effects: An experimental analysis. *Organizational Behavior and Human Decision Processes*, **58**, 457-476.

Shafir, E. (1993). Choosing versus rejecting: Why some options are both better and worse than others. *Memory & Cognition*, **21**, 546-556.

Shafir, E. and A. Tversky (1992). Thinking through uncertainty: Nonconsequential reasoning and choice. *Cognitive Psychology*, **24**, 449-474.

Shepard, R. N. (1964). On subjectively optimum selection among multiattribute alternatives. In: *Human Judgments and Optimality* (M.W. Shelly II and G. L. Bryan, eds.), pp 257-281. Wiley, New York.

Shiffrin, R. M. and W. Schneider (1977). Controlled and automatic human information processing: II. Perceptual learning, automatic attending, and a general theory. *Psychological Review*, **84**, 127-190.

Simon, H. A. (1955). A behavioral model of rational choice. *Quarterly Journal of Economics*, **69**, 99-118.

Slovic, P. (1966). Value as a determiner of subjective probability. *IEEE Transactions on Human Factors in Electronics*, **7**, 22-28.

Slovic, P., B. Fischhoff and S. Lichtenstein (1977). Behavioral decision theory. *Annual Review of Psychology*, **28**, 1-39.

Smith, E. R. (1992). The role of exemplars in social judgment. In: *The Construction of Social Judgments* (L. L. Martin and A. Tesser, eds.), pp. 107-132. Erlbaum, Hillsdale, NJ.

Staw, B. M. (1976). Knee-deep in the big muddy: A study of escalating commitment to a chosen course of action. *Organizational Performance*, **16**, 27-44.

Svenson, O. (1974). *A note on think aloud protocols obtained during the choice of a home.* Report #421 from the Psychological Laboratories. Department of Psychology,. Stockholm University.

Svenson, O. (1979). Process descriptions of decision making. *Organizational Behavior and Human Performance*, **23**, 86-112.

Svenson, O. (1983). Cognitive processes in judging cumulative risk over different periods of time. *Organizational Behavior and Human Performance*, **33**, 22-41.

Svenson, O. (1985). Cognitive strategies in a complex judgment task: Analysis of concurrent verbal reports and judgment of cumulated risk over different exposure times. *Organizational Behavior and Human Performance*, **36**, 1-15.

Svenson, O. (1989). On expert judgment in safety analyses in the process industries. *Reliability Engineering and System Safety*, **25**, 219-256.

Svenson, O. (1990). Some propositions for the classification of decision situations. In: *Contemporary Issues in Decision Making* (K. Borcherding, O. I. Larichev and D.M. Messick, eds.), pp. 17-31. Elsevier, Amsterdam.

Svenson, O. (1992). Differentiation and consolidation theory of human decision making: A frame of reference for the study of pre- and post-decision processes. *Acta Psychologica*, **80**, 143-168.

Svenson, O. (1996). Decision making and the search for fundamental psychological regularities: What can be learned from a process perspective? *Organizational Behavior and Human Decision Processes*, **65**, 252-267.

Svenson, O. and L. J. Benthorn (1992). Consolidation processes in decision making: Post-decision changes in attractiveness of alternatives. *Journal of Economic Psychology*, **13**, 315-327.

Svenson, O. and B. Fischhoff (1985). Levels of environmental decisions: A case study of radiation in Swedish homes. *Journal of Environmental Psychology*, **5**, 55-67.

Svenson, O. and N. Malmsten (1996). Post-decision consolidation overtime as a function of gain or loss of an alternative. *Scandinavian Journal of Psychology*, **37**, 302-311.

Svenson, O. and A. J. Maule (eds.) (1993). *Time Pressure and Stress in Human Judgment and Decision Making*. Plenum, New York.

Svenson, O., A. Ortega Rayo, M. Andersen, A. Sandberg and L. Svahlin (1994). Post-decision consolidation as a function of instruction to the decision maker and decision problem. *Acta Psychologica*, **87**, 181-197.

Thaler, R. (1980). Toward a positive theory of consumer choices. *Journal of Economic Behavior and Organization*, **1**, 76-90.

Tversky, A. (1972). Elimination by aspects: A theory of choice. *Psychological Review*, **79**, 281-299.

Tversky, A. and D. Kahneman (1992). Advances in prospect theory: Cumulative representation of uncertainty. *Journal of Risk and Uncertainty*, **5**, 297-323.

Tyszka, T. (1983). Contextual multiattribute decision rules. In: *Human Decision Making* (L. Sjöberg, T. Tyszka and R. J. A. Wise, eds.), pp. 243-256. Doxa, Lund, Sweden.

Verplanken, B., H. Aarts, A. van Knippenberg and C. van Knippenberg (1994). Attitude versus general habit: Antecedents of travel mode choice. *Journal of Applied Social Psychology*, **24**, 285-300.

Verplanken, B. and O. Svenson (1996). *Personal involvement in human decision making: On conceptualizations and effects on decision processes*. Unpublished manuscript, Department of Psychology, University of Groningen, The Netherlands.

von Winterfeldt, D. and W. Edwards (1986). *Decision Analysis and Behavioral Research*. Cambridge University Press, Cambridge.

Wald, A. (1950). *Sequential Analysis*. Wiley. New York.

Wallsten, T. S. (1980). Processes and models to describe choice and inference behavior. In: *Cognitive Processes in Choice and Decision Behavior* (T. S. Wallsten, ed.), pp. 215-237. Erlbaum, Hillsdale, NJ.

Webster, D. M and A. W. Kruglanski (1994). Individual differences in need for cognitive closure. *Journal of Personality and Social Psychology*, **67**, 1049-1062.

Wright, P. (1975). Consumer choice strategies: Simplifying versus optimizing. *Journal of Marketing Research*, **11**, 60-67.

Yates, F. (1990). *Judgment and Ddecision Making*. Prentice Hall, Englewood Cliffs, NJ.

Zanna, M. P. and J. Cooper (1974). Dissonance and the pill: An attribution approach to studying the arousal properties of dissonance. *Journal of Personality and Social Psychology*, **29**, 703-709.

Zajonc, R. B. (1968). Cognitive theories in social psychology. In: *Handbook of Social Psychology* (E. Aronson and G. Lindzey, eds.), pp. 320-411. Wiley, Chichester, UK.

Zajonc, R. B. (1980). Feelings and thinking: Preferences need no inferences. *American Psychologist*, **35**, 151-175.

8

TRAVEL CHOICE IN CONGESTIONS: MODELING AND RESEARCH NEEDS

Eliahu Stern, Ben-Gurion University, Beer-Sheva, Israel

INTRODUCTION

Many see a potential role of information to relieve traffic congestions (e.g., Schofer et al., 1993; Sinuani-Stern et al., 1997; Stern et al., 1996a), specifically the Advanced Traveller Information Systems (ATIS) currently under various stages of development. It was shown that these systems have a bigger potential in cases of nonrecurrent congestion (Emmerink et al., 1995), which usually account for up to 60% of the congestion delays (Hall, 1993) and may even increase to 70% in the year 2005 (Lindley, 1989). Nonrecurrent congestion is caused by random incidents on the roads which diminish the available road capacity during a certain period of time. Apart from bad weather and incidential road works, non-recurrent congestion may occur due to deviant behavior of drivers (consumers) or authorities (providers). A deviant behavior of drivers includes the so called "human factors" cause of traffic accidents, different traffic violations, and many other, less extreme cases like slowing down due to personal pressure, age, or even presence of policemen on the roadside. A deviant behavior of authorities includes avoidable and unavoidable public events that require a temporary closure of certain roads (e.g., sport races, festivals, ceremonies).

Most of the behavioral choice models in transportation deal only with travel choices. Even the simulation models which aim to estimate the effect of information on congestion relief (e.g., Emmerink et al., 1995, Stern et al., 1996a, 1996b) do not consider choices other than of route, departure time, and mode. Traffic congestion forces people to respond not only by changing their travel behavior but also by changing other types of choices like destination, activity, and driving speed, which affect the transport system. Even if, computation-wise, it is still difficult to incorporate various types of choices in one comprehensive model, one need to have a sound theoretical ground for decision making which could encompass other relevant choices. Although we do not have empirical data to demonstrate the actual distribution of all types of behavioral responses (although partial evidence can be found in Stern et al., 1993) it is clear that modeling choice behavior in congested situations must incorporate other responses than only travel-related choices. Moreover, the effect of travel-related choices on the transport system performance in the presence of information may decline both as market penetration (number of informed drivers) and network patronage increase (Emmerink et al., 1995; Stern et al., 1995, 1996b) and therefore, other responses to congestion could be more effective, thus deserving further research.

The purpose of this chapter is to present, and suggest for further empirical research, a theory aiming to explain the dynamics and the high variability of human preferences and choice. It is believed that such a theory could provide the conceptual mechanism to model a broader range of choices, mainly under conditions of congestion. The chapter first presents the various possible reactions to congestion. Second, it reviews the approaches to travel-choice modeling in the field of transportation and the approaches to behavioral decision making as developed in psychology in order to determine the properties that such a theory should have. The general structure and the incremental stages of the theory are then described and a decision-based framework for congestion responses is elaborated. Finally, modeling and research needs are articulated.

RESPONSES TO CONGESTION

The occurrence of congestion leads to a number of changes in objective travel conditions at the level of the individual trip maker. The main changes are changing travel time on network segments, increasing waiting time at bottlenecks (waiting in queues), increase in total trip duration, decreasing reliability of trip timing (i.e., arrival time reliability), decreasing reliability of trip duration (i.e., travel time reliability), and deteriorating driving conditions (i.e., stop-and-go, speed changes, etc.). In most cases, congestion occurrence as well as congestion delay are highly variable from day to day.

Based on Rapoport's (1975) definition of dynamic decision making, reactions to congestion are assumed to be dynamic. Accordingly, decisions are made sequentially in time, in which the task specification may change over time, either independently or as a result of previous decisions, in which available information may be contingent on the outcome of earlier decisions, and in which implications of any decision may reach into the future.

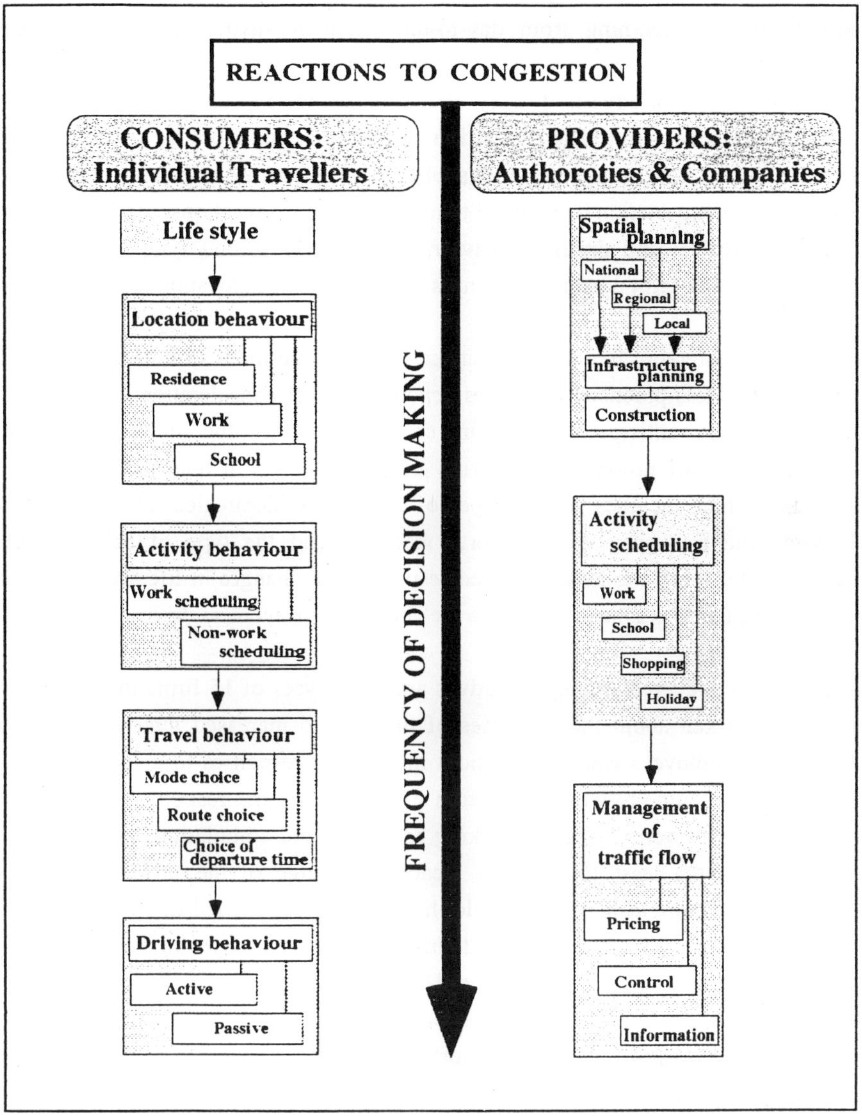

Figure 8.1. Hierarchy of reactions to congestion in the form of frequency of decision making. (After Stern et al. 1996)

Dynamic decision making implies that both choice and knowledge should play a major role in explaining the order of response adoption. Choice depends first on the alternatives the individual traveller subjectively considers to be feasible (Bovy and Stern, 1990). These alternatives are constantly changing with the updating of knowledge. Knowledge, in this case, may be described as learning from day-to-day, firsthand experience (i.e., exposure to congestion), complemented by exogenous information from various media outlets, word-of-mouth, and others. Knowledge is directly connected to two types of expectations that may affect individual reactions to traffic congestion: (1) expectations about the performance of the feasible alternatives, and (2) expectations about the future solutions to the congestion problems. The first expectation is assumed to affect short-term reactions, mainly those related to daily activity travel and driving behavior, whereas the second type of expectation is assumed to affect long-term personal responses such as change in location and/or in life style.

It is assumed that the more frequently practiced reactions are more stimulated by congestion consequences. The less practiced responses depend on other considerations, mainly personal constraints and the anticipated long-run consequences of the individual's response. Accordingly, Figure 8.1 presents an overview of the responses of both individual travellers and authorities which includes several types of choices. A hierarchical order is assumed to exist between the frequency of behavioral reactions and the type of behavior. On the individual traveller level, the least practiced responses are changes in life style and location, and the most frequently practiced responses are changes in travel and driving behavior.

In a recent study conducted among executives and employees of 15 firms in the Netherlands executives were asked about the responses of the firm to increased interurban road traffic congestion. The employees were asked about their likely responses to a further increase of congestion on their home-to-work trip. From a list of potential responses, the most likely reactions of the firms were changing working schedules and stimulating shorter home-work distances, whereas the least likely were locational change of the firm and encouraging employees to work more at home. It should, however, be mentioned that responses differed strongly between types of firms. Industrial firms saw much less potential for adaptation to congestion than the service-sector industries. The responses of individual trip makers was the following in order of likelihood: (1) earlier departure from home; (2) changing working hours; (3) changing route to/from work; (4) working more often at home; (5) using public transportation more often; (6) change job; (7) work less days a week; and (8) change home location. Thus, the least attractive to workers is changing their home, whereas employers, and government as well, see it as a realistic option.

Preventive Responses

Responses can be divided into two types, preventive and reactive responses. Each response has a different theoretical implication (as will be demonstrated later) on modeling choice.

The individual may react in a preventive way when he or she expects or is informed about congestion prior to departure. Thus, preventive response is adopted in cases of recurrent congestion, as well as in the presence of pre-trip information referred to either recurrent or non-recurrent incidents. The best preventive behavior is probably to avoid congested situations in the first place. If possible, this is a costly reaction which may involve residential relocation. Although there is some evidence for relocation responses (Stern and Zarin, 1994), we should not include it yet in a comprehensive travel-choice model because of its low frequency of adoption and the limited market segment to whom this response is applicable.

Other possible behavioral choices include activity and travel. Activity choice has an increasing importance in providing preventive solutions to congestion. Travel patterns, especially during peak hours, are affected by both spatial and temporal dimensions. The spatial dimension reflects the distribution of origins and destinations which are more costly, difficult, and less preferred to change. The temporal dimension reflects the schedule of individuals' work and non-work activities. Changes in work scheduling due to congestion include both daily working hours and working days, and chaining of daily activities. The same may be applied for non-work out-of-home activities where people change their schedules. The main motive for work rescheduling will be decreasing reliability and travel time duration.

Very few studies have investigated the effect of congestion on change of schedules and chaining of daily activities. An example are before and after studies undertaken by Tacken and De Boer (1989, 1991) focused on the way employees use flexible working hours to avoid peak hour traffic. Both studies were undertaken in the Amsterdam area where the later 1991 study was done after an improvement in the urban beltway around Amsterdam resulting in a decrease in congestion levels on the former bottlenecks. Results of the first study showed that people adjust their planned time schedule to the experienced congestion. About 60% of the sample had the option of a major change in working schedule; about 25% were prepared to start more than an hour earlier than usual, and about 40% to start that much later. The second study showed that no preventive behavior was adopted after the congestion decreased where a clear "back-to-normal" reaction was observed with workers returning to more convenient departure times, usually back to the peak.

In contrast to the lack of studies of preventive responses, there is an enormous amount of research on the connection between human activity, travel patterns, and transport planning in general. A recent comprehensive review is Fox (1995). There is also extensive research on travel behavior including studies of travel-mode choice, choice of departure time, and route choice (for a review, see, for example, Bovy and Stern, 1990; Stern et al., 1996b). Modeling in these areas was mainly based on stated preference and laboratory-generated data which widely support the hypothesized preventive reactions to congestion (mainly changes in departure time, mode, and route).

Reactive Responses

Reactive responses are adopted in non-recurrent incidents of congestion, or as a response to enroute information, usually under time pressure. According to Novaco et al. (1990), the concept of impedance, signifies a behavioral constraint on movement and goal attainment. It is claimed that as an aversive, frustrating condition, transportation impedance elevates arousal and elicits negative affect, thereby producing stress reactions that affect both travel and driving. Possible reactive choices include only one travel behavior component - changing route - which is adopted mainly in cases of non recurrent congestion. Ng, Barfield, and Mannering (1995) showed, for example, that most important reasons for choosing an alternative route by both private and commercial drivers, resulting from enroute, in-vehicle information, were accidents and road construction. In contrast, there are several driving behavior reactions. Among the most observed ones are aggressive lane switching, illegal on curb by-passing, illegal turns, in-purpose slowing down (means of "punishment"), and use of prohibited lanes. With the increasing level of congestion, any deviation from a by-law driving will cause further congestion. Moreover, increasing number of traffic violations will result in more cases of non recurrent congestion.

Very few studies have dealt with the effect of congestion on driving behavior. A stated preference study by Bates et al. (1987) showed that the first adjustments considered by commuters in response to 20% increase in current travel durations are, first, "acceptance" reactions, and, second, active driving behavior. Acceptance, or passive reaction, is often compensated by shorter lunch breaks, leave work earlier, and/or bring work home. More active adjustments include attempts to improve journey speeds on later sections of the trip, making illegal "rat runs", and adopt more determined driving styles. Observations on "determined" driving styles were found in Elamatin's (1993) study which looked at the driving behavior of Israeli commuters in a corridor near the city of Beer Sheva. During a week of observations made between 7 am and 9 am, he found that 9.6% of the drivers violated the law in order to overpass the traffic in front of them. Based on personal interviews with both violators and by-law drivers he found that, on average, violators were under a higher time pressure.

Choice modeling in transport mainly aim at predicting travel choices, but there are still some determinants that the existing models fail to consider. These determinants relate to personality attributes which can affect the dynamics of the decision-making process. A recent effort to include personality attributes was made, for example, by Khattak et al. (1993) who showed that drivers who were more inclined toward adventure and discovery were more likely to divert. Higher diversion propensity could imply that those who posses it may be more flexible in their en-route decisions or more impatient in the face of delays.

In order to find a sound theoretical basis for explaining the general mechanism of choice that could be applied to the various responses and reactions to congestion, including personality attributes which are assumed to account for some of the variability in human preferences, in the next section I examine different approaches to travel-choice modeling.

APPROACHES TO TRAVEL-CHOICE MODELING

A recent review of assignment-modeling approaches with emphasis on travel-related choices was presented by Van Berkum and Van der Mede (1994). They distinguished between three dimensions: static versus dynamic, deterministic versus stochastic, and equilibrium versus non-equilibrium. In a transportation environment, the static type of model assumes that network conditions do not vary. The decisions are thus based on invariant conditions, implying that the time element is ignored in the model. In contrast, the dynamic type of model assumes that every process can be seen as an ordered sequence which takes place in time. A dynamic model is one which considers the influence of what has happened at time $t-1$ on what happens at time t, and the influence in turn on what will happen at time $t+1$.

A deterministic model assumes that travellers are perfectly informed and utility maximizers. Travellers are assumed to have a complete knowledge of all routes and other travel-related attributes. A stochastic model, based on perceived knowledge, is usually modelled by a random utility component where a stochastic error term is added to the utility of each identified alternative.

Each of the above models can be based on an equilibrium or a non-equilibrium state. In an equilibrium transport model flows are pushed towards the equilibrium state by route or departure time switching. At equilibrium flows, for example, there is no incentive for route or departure time switching (Sheffi, 1985). Non-equilibrium models are oriented to solve situations that are not at equilibrium such as congested ones. Models of choice in congested situations must be of this type. Although the traveller tends to equilibrium (e.g. the "back-to-the-peak" phenomenon), the system is in permanent disequilibrium.

Most conventional models are based on a utility-maximization principle. Apart from some of their unrealistic assumptions, one of the main weaknesses of the discrete-choice models is their inability to capture many aspects of ambiguity, randomness, and uncertainty through a single random error term. Moreover, the introduction of the choice mechanism in the models is incomplete and could not deal with this weakness. Many of the choice situations are described as a black box (Bovy and Stern, 1990) and does not specify choice rules but only the relationship between input and output variables, not the decision process itself. The more advanced the models, the more they introduce details to explain the interior of the black box. Bovy and Stern (1990) suggested that the content of the black box is a complicated system of filters through which the information is selected and transformed. They determined two types of filters: (1) Perception filter through which the traveller's cognition of alternatives and attributes' values are processed; and (2) Evaluation filter through which the perceptions are transformed into a desirability scale. This approach enables to incorporate changes in the filters under changing realms (or situations) resulting in changes in the mix and relative weight of the attributes (Stern and Krakover, 1993). A somewhat similar black box, referred to as driver's world, was used by Khattak et al. (1993) who revealed its internal linkages with a joint multinomial logit model, without however considering any underlying psychological process.

A relatively new development in modeling travel-related choice behavior was the introduction of artificial neural networks to model driver decisions. This approach was based on the assumption that there is a similarity between the process of driver decision making and the problem solving approach of neural computing (Lyons and Hunt, 1993). In principle, however, the neural network approach remains less representative of real driver decision making than a rule-based model. Rule-based decision models furthermore retain certain advantages over the neural network based approach. The operation of a rule-based model enables adjustment of rules to improve the overall performance of the model. The performance of a neural network model can be adjusted only by re-training using alternative data (e.g., Hunt and Lyons, 1995).

Production rule systems also emerged in the area of artificial intelligence. Basically it is a means of quantifying, codifying, and distributing expertise and opinion, aimed to develop the possibility for a decision maker to navigate in his/her representation of the environment, and to obtain the information necessary to make the relevant decision. A mathematical framework that captures the uncertainties inherent to human perceptions and the linguistic reasoning associated with human cognitive processes by means of decision rules is the fuzzy set theory. The theory was already introduced 1965 by Zadeh as a general approach to express the uncertainty in human systems. Zadeh (1973) proposed that the decision making process does not use the traditional binary valued logic, but rather a logic of fuzzy perceptions and inference rules based on approximate reasoning. The decision-making framework assumes that travellers make their decisions on the basis of simple rules instead of maximizing a utility function. The decisional behavior is, accordingly, described as a series or system of IF, THEN rules. A recent effort to combine the fuzzy set approach with neural network model was presented by Vythoulkas (1994). He utilizes a fuzzy decision framework for modeling and reasoning involved in the decision process, and uses a neural network approach for calibrating the parameters of the fuzzy model. The calibration of a route choice model based on fuzzy sets and approximate reasoning was introduced by Vythoulkas et al. (1994), also through the use of a driving simulator. Similarly, Hunt and Lyons (1995) used a NEURAL DRIVER simulator to collect data for modeling lane and speed change decisions made by a driver.

All the models mentioned above still lack a cognitive explanatory mechanism of the choice process, especially in congested situations where time pressure is a dominant factor. Even the fuzzy set theory neither explains the mechanism nor accounts for the high variability which results from personality and momentary changes in driving styles due to time pressure. Obviously, the use of simulators to collect behavioral data is limited since the effects of time pressure and a road's physical density on the driver are not real. In order to advance research and modeling to overcome these limitations, let me review various approaches to explain decision making processes as emerged from psychology.

THEORETICAL FOUNDATIONS

As argued above, travel choice models, especially in congested situations, should be based on a sound theory that explains the dynamics and the high variability of human preferences. A recent literature review by Maule and Svenson (1993) (see also chapter 7 in this volume) presents some key distinctions between approaches in the area of behavioral decision making. Although there are other distinctions (e.g., Svenson, 1990), those types summarized in Table 8.1 are the principal ones from which we can identify the main properties that such an underlying theory should have.

A structural approach is concerned with the relation between the information about alternatives and the choice between the alternatives. This approach lacks a psychological process to explain the variability of weighting of stimulus attributes. A prospect approach is concerned with a two-phase decision process based only on editing and evaluation of inputs with no time component, thus being a kind of an intermediate type between the structural and

Table 8.1
Approaches to Behavioral Decision Making

Type	APPROACHES		
1	Structural	Prospect	Process
2	Riskless		Risky
3	Normative	Prescriptive	Descriptive
4	Static	Sequential	Dynamic

process approaches. A process approach is focused on the cognitive process which explains how decisions are made and evolve over time. Accordingly, it is aimed to identify the underlying processing strategies described in terms of decision rules used by individuals and to identify the factors determining the adopted rules. A riskless approach is concerned with a decision situation in which the individual knows with certainty the outcome of a particular choice. A risky approach is concerned with a decision situation in which there is uncertainty about the outcome. Since there is no situation in which the outcome does not have at least a minimum level of uncertainty, the distinction between riskless and risky decision making is relative. A normative approach is one which does not consider any effect of human characteristics on the choice process. It always refers to rational behavior. A prescriptive approach is concerned with rational choice behavior but takes into account how people evaluate and integrate information in making decisions. A descriptive approach is aimed at describing how people actually make a choice. It conceptualizes decision making in terms of underlying psychological processes extending over time. The type, order, and form of these processes varies as a function of task, environmental, and personal characteristics (Payne, 1982). A static approach concerns only one decision made on the basis of the information about possible actions, outcomes, and constraints available at the moment of choice. A *Sequential* approach is concerned with a series of decisions, each changing the situation for future decisions. According to Rapoport (1975), such decisions are made sequentially in time, in which the task specification may change over time, either independently or as a result of previous decisions, in which available information may be contingent on the outcome of earlier decisions, and in which implications of any decision may reach into the future. A dynamic approach suites situations in which each decision represents an input to a system that is continuously changing according to its own dynamics even when the decision maker does nothing.

The consequences of congestion and their respective responses, as described earlier, imply that a theory aiming to explain the dynamics and the high variability of human preferences should have the following properties: (1) The underlying choice strategies should be described in terms of decision rules, and therefore it should be of a process nature; (2) Since most reactions to congestion are made under uncertainty, the theory should be of a risky type; (3) As found in the reviews above, travel-choice models do not include underlying psychological processes of decision making and the theory should thus be of a descriptive type; and (4) although travellers tend to equilibrium (e.g., the "back to the peak" phenomenon), the transport system, as already mentioned earlier, is in permanent disequilibrium. Therefore, the theory should suit dynamic situations.

The following chapter presents a newly developed theory called Decision Field Theory which has combined all these properties. Following a summary description of this theory, an expanded version is proposed to be applied to congestion-related choice behavior.

DECISION FIELD THEORY

General Theory Structure

Decision field theory developed by Busemeyer and Townsend (1993) is a synthesis of two prior and independent psychological theories: approach-avoidance theories of motivation, and information-processing theories of choice times. Although the theory is described here in relation to its potential use in the area of modeling travel choice, it is applicable to various choice situations under uncertainty. In order to describe the theory in general, we will first present a prototypic choice situation in which the decision maker must choose between two actions, A_1 and A_2 (Figure 8.2). It should, however, be mentioned that it is possible to develop the theory for decisions with more than two alternative actions (Busemeyer and Townsend, 1989). Later on we will also refer to specific transport-related choices.

For each of the possible actions shown in Figure 8.2, two uncertain events, E_1 and E_2, may occur. Each event has a payoff denoted as Y_{ij} such that if action A_1, for example, is chosen, and event E_1 occurs, a payoff Y_{11} is produced. When decisions are taken under risk, the probability of the occurrence of a certain event can be inferred from external information (e.g., specific behavioral instructions from the media). When decisions are made under uncertainty, the probabilities may be inferred from knowledge and accumulated experience.

Decision field theory is aimed at understanding the motivational and cognitive mechanism that guide a deliberation process involved in decisions under uncertainty. Deliberation is a time-consuming cognitive process that involves information seeking about the consequences

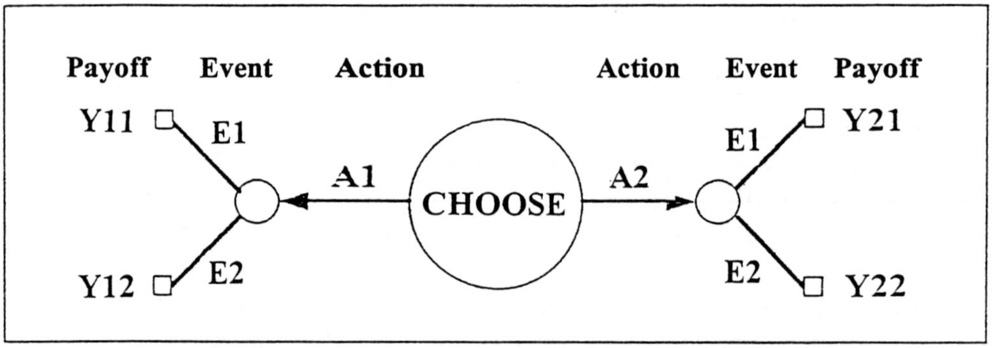

Figure 8.2. A prototypic choice situation in an uncertain environment

(i.e., events E_i) of recognizable actions A_i, weighting of the pay-offs Y_i of the events, and comparing the weighted consequences to establish a preference (choice) state. The deliberation process is manifested by vacillation until a choice is made and an action is taken (Svenson, 1992).

The basic ideas of decision field theory are summarized in a sequence of seven incremental stages mathematically derived from the synthesis of the two prior theories mentioned above. The mathematical formulation, however, will not be presented here but only the conceptual structure and the underlying assumptions.

Beginning with the traditional, well known stages of deterministic and random subjective utility theories, in each subsequent stage, a new and more general theory is formed by including a new processing assumption into the previous theory. Each new processing assumption is needed to represent a fundamental property of the deliberation process involved in decision making. The last stage completes the entire description and introduces the processing time of deliberation. Lower-stage theories are thus special cases of higher-stage theories in which the decision-field stage is the highest. Each of the theory's steps described below will present only the main processing assumption and the limitation of the theory which motivated the development of the next stage.

Incremental Stages of the Theory

Deterministic Subjective Expected Utility (SEU) Theory. Choice is made according to the difference in the subjective expected utilities of two actions. The utility is determined by the distinct payoff and its corresponding subjective probability of occurrence. The probability equals the amount of attention given to each event on each presentation of the choice problem. This amount of attention is represented by a weight assigned to each related payoff. According to this theory, the weight is fixed throughout the choice process and the decision maker uses the same subjective probability on every repetition of a choice problem. In sum, the theory enables to provide explanation only for the direction of choice. As this descriptive SEU theory is of a prospect type, it fails to account for the fundamental variability of human preference. This major limitation is overcomed in the following second-stage theory.

Random SEU Theory. Under this theory the decision maker switches his or her attention from one event to another across choice trials or throughout the deliberation process. The attention weight for a given event is thus a random variable which can change due to attentional fluctuations. Consequently, the SEU for each action is also a random variable, which is called "valence of an action". The difference between the valences of two actions determines the preference state on any choice trial. The theory provides explanation for both the direction and the strength of the preference. However, it is unable to provide a mechanism to explain the relation between choice probability and decision time. This limitation is handled in the following stage.

Sequential SEU Theory. As attention may switch from one event to another within a single choice, the determination of a preference state does not lead to a decision until the preference state exceeds an inhibitory threshold. Until that, vacillation in preference is observed which reflects the fluctuations in valence difference across samples. The total number of samples needed to reach the threshold is a random number which decision time is an increasing function. Decision time increases as the threshold increases and thus the probability of choosing the correct action also increases. Consequently, the threshold controls the speedaccuracy trade-off in decision making. Therefore, shorter decision times lead to less accurate decisions. This theoretical stage fails to explain that choice probability can change from below .50 to above .50 under time pressure which affects the number of samples.

Random Walk SEU Theory. Unlike the previous stage which assumes that the deliberation process always begins at a neutral point, this theory assumes that the initial preference state starts at some anchor point, biased by prior knowledge and experience. This assumption provides explanation for the inverse relation between choice probability and decision time. Thus, more experience may shorten the deliberation time. In addition, the initial anchor point is also used to explain reversals of preference as a function of time pressure. More experience may shorten the deliberation time and therefore more accurate decisions can be made under time pressure. The theory does not provide any mechanism for explaining serial position effects on the final preference state. According to the theory, the preference state is simply the initial state plus the sum of all the valence differences. Thus, the preference state is not affected by the well established effects of primacy and recency in evaluative judgments (e.g., Hogarth and Einhorn, 1992).

Linear System SEU Theory. The theory suggests that the impact of a valence difference may vary depending on whether it occurred early ("primacy effect") or late ("recency" effect) within a sequence. This serial position effect is introduced by a growth-rate parameter to the previous stage of random walk SEU in such a way that the recent samples have greater impact if the parameter is between zero and one, and earlier samples have greater impact if the parameter is less than zero. Thus, primacy, recency, or no serial position effects can occur, depending on the value of the growth-rate parameter. The theory does not account for the fact that the time required to make a decision depends on the approach-avoidance nature of the conflict. Thus, the preference state ignores whether its calculation is based on two rewards or two punishments. (As will be discussed later, in a travel choice situation, reactive and preventive behaviors may differ in this regard). The next theoretical stage rectifies this limitation.

Approach-Avoidance SEU Theory. Approach-avoidance theory allows the average weight given to a pay-off in the deliberation process to be influenced by another dynamic variable called the goal gradient. This variable reflects the relationship between the distance from ones goal and the weight given to the pay-off. The gradient of this relationship is changing whether the consequence of an action is a reward (approach) or a punishment (avoidance). The basic idea is that the attractiveness of a reward or the aversiveness of a punishment is a decreasing function of the distance from the point of commitment to an action. Thus, the consequences of an action become more salient as the preference state for that action approaches its threshold. As the possibility of performing an action increases, the attention (i.e. the weight) assigned to its consequences increases. The theory goes further in proposing that avoidance-avoidance (av-av) decisions produce longer mean deliberation times than do approach-approach (ap-ap) decisions. When av-av decisions are made, the preference state vacillates and slows down the decision process. When ap-ap decisions are made, the preference state is racing toward the threshold and, thus, shortens the decision time. The deliberation time is treated as a discrete time rather than as real time and therefore quantitative predictions of decision time cannot be made. The following stage, the Decision Field Theory, solves it by introducing a time unit into the theory.

Decision Field Theory. Decision field theory is an attempt to formalize the deliberation process. The theory introduces a time element representing the amount of time that it takes to retrieve and process a pair of anticipated consequences of an action before shifting attention to another pair of consequences (of another action). The anticipated consequences are retrieved from an associative memory. The total decision time equals the total time required to retrieve, compare, and integrate all the comparisons of the anticipated consequences. The calculations involved in such a procedure are assumed to be realized by an underlying neural system, and decision field theory is an abstract representation of its essential dynamic properties.

Overall integration. The integration of all the previous stages into one conceptual scheme is illustrated in Figure 8.3. Potential gains and losses are given by corresponding pay-offs in the approach-avoidance subsystems. These pay-offs are inputs to the valence system, connected to the current considered actions by attention weights *W*. The weights change over time as a result of shifts in the decision maker's attention to the various consequences during deliberation. The valence of an action at any moment represents the anticipated value of an action at that moment and it is produced by the weighted sum of the input values connected to that action. The valences produced within this system are fed into the decision system that compares the valences and integrates the comparisons over time to produce a momentary preference state. Finally, the preference state drives a motor system that inhibits responding until a threshold is exceeded and an action is taken.

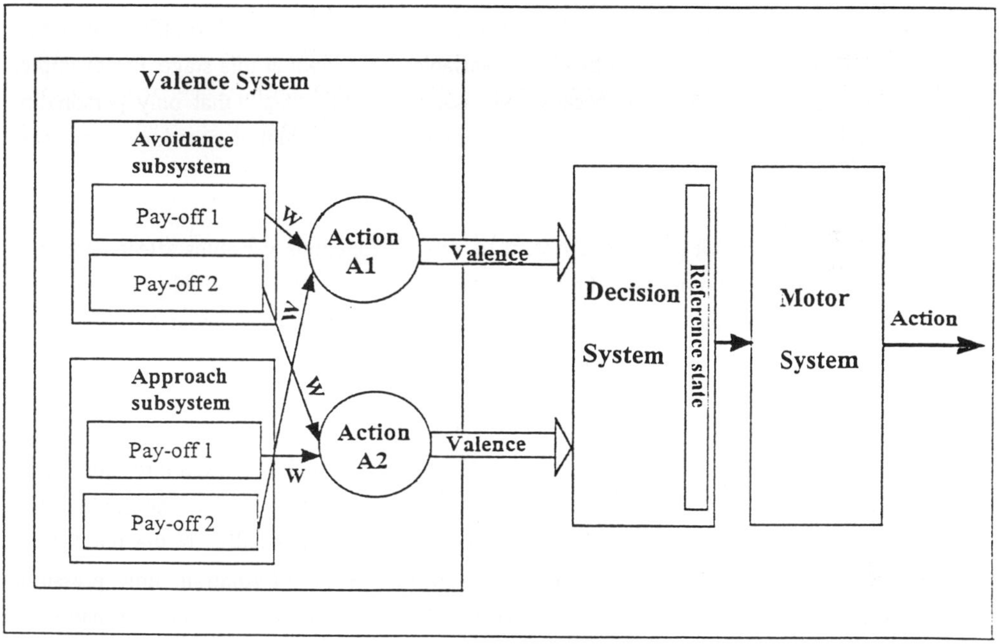

Figure 8.3. A schematic outline of decision field theory. (After Busemeyer and Townsend, 1993.)

The psychological theory is here proposed as a theoretical foundation for modeling choice in congestions, commonly characterized by decision making under changing degrees of uncertainty and time pressure. The following section will present this proposed mechanism within a transportation environment. Furthermore, it will expand the conceptual framework of the decision-making process to include exogenous effects on the deliberation process which are partially supported by empirical evidence.

A DECISION-BASED THEORETICAL FRAMEWORK OF CHOICE BEHAVIOR

Overview

All the transport-related choice models of black-box type lack a choice-process mechanism and use all the variables expected by the researchers to affect the output without determining their effects on the choice process itself. Unlike these models, Figure 8.4 presents an expanded framework of a choice mechanism based on the seven stages of the Decision Field Theory with proposed exogenous variables. These variables are assumed to affect the choice process itself and, therefore, not all black-box variables are shown at this stage. For example, socio-demographics are excluded because we assume, at this stage, that only personality attributes directly affect the cognitive mechanism involved in choice situations. However, they are not substitutes to socio-demographics.

The theoretical framework presents a system in which exogenous variables affect either the deliberation process and/or the threshold. According to Decision Field Theory, the deliberation is a sequential process of calculating and comparing valences of N subjectively identified actions. The time spent on the calculation of a given action depends on the attention given to the events of this action. The attention given to the consequences of the events of alternative action B is not necessarily equal to that of action A. The result of one deliberation cycle is a momentary preference state which is chosen only when it reaches a threshold. Until then, vacillation is observed, thus activating another deliberation cycle. The smaller the distance is between the preference state and the threshold, the smaller is the number of deliberation cycles. The following expands on the effects of information, time pressure, situation, and personality attributes on the two basic components of the choice mechanism - the deliberation process and the threshold.

The decision-making threshold is expressed by an inhibitory criterion presenting a given value of a preference state. Thus, a given action is taken as soon as the momentary preference state exceeds that given value. The theoretical framework shown in Figure 8.4 includes several exogenous variables which are assumed to affect the threshold. We suggest that these effects can vary according to the way by which the threshold is defined. In transportation related situations, the threshold may have several different determinants as discussed below.

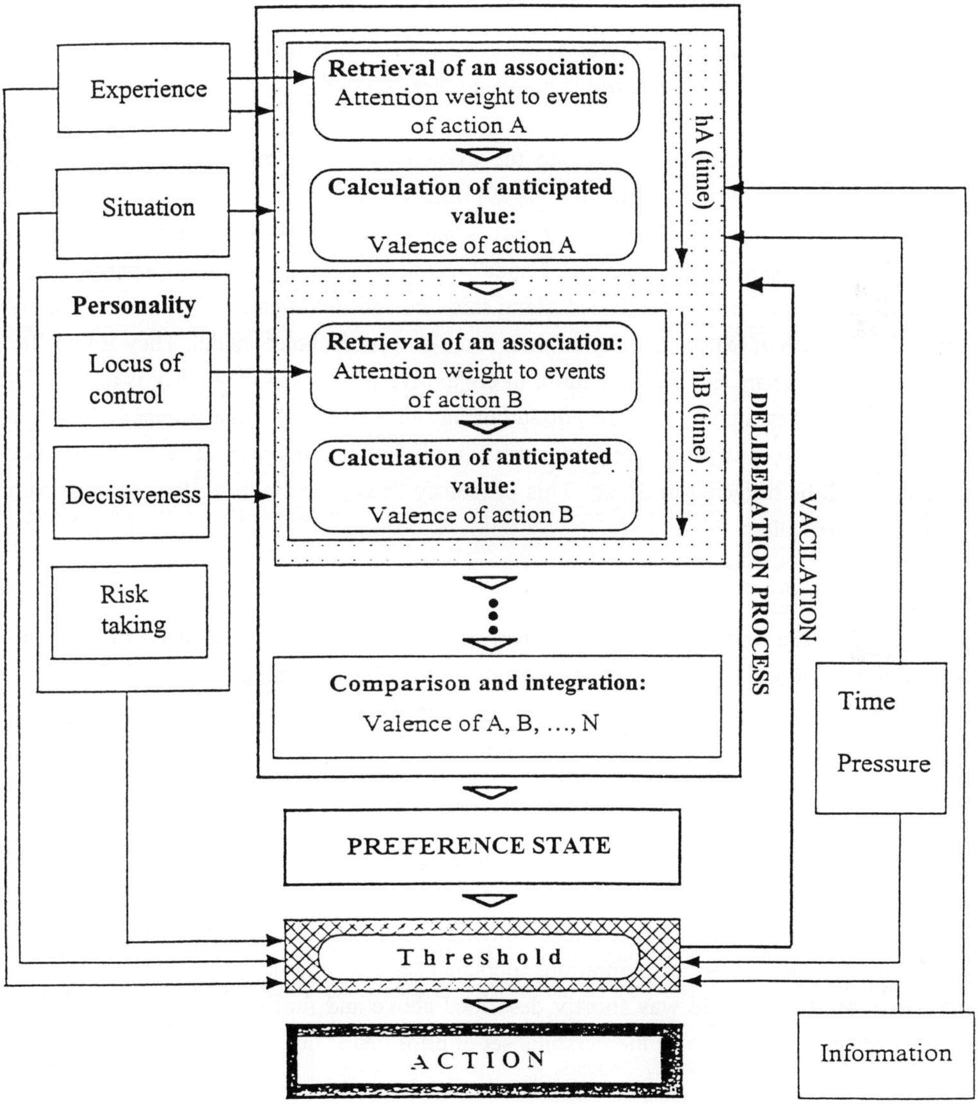

Figure 8.4. The decision-based theoretical framework of choice behavior.

Actual and Expected Delay Time

Actual delay time is assumed to increase the time pressure on the traveller and consequently to decrease the threshold. For example, a driver stuck in a traffic jam may deviate only after a certain period of time. As time pressure increases, this length of time is getting shorter. The length of delay time depends on situation variables and personality (Figure 8.4). Expected delay time, which usually depends on experience, acts in the same way as actual delay time only in cases where reactive behavior is contemplated. When behavior is responsive, the effect of time pressure on the threshold is much smaller.

There is empirical evidence witnessing to the existence of delay-time thresholds in travel decision-making. Huchingson and Dudek (1979) observed driver response to varying lengths of expected delay to be positive although not linear. Moreover, Mahmassani et al. (1990) reports evidence supporting "indifference bands" or thresholds in commuters' departure time choices. Khattak et al. (1993) continued along this line and investigated delay thresholds by incorporating length of expected delay in their route-diversion choice model. They found, for example, that drivers are significantly more likely to divert if the expected delay is at least 20 minutes. Ongoing research by the present author has already revealed that whenever there is an option to deviate, it is most likely that drivers will suffer a delay of at least 10 minutes before they will be triggered to do so. This obviously does not eliminate other factors from affecting the threshold.

Holding time pressure constant, levels of inherent aggression (a personality attribute) can inversely affect the value of the momentary preference state. As the level of inherent aggression increases, the inhibitory value decreases. Under increasing time pressure and/or under certain circumstances, a momentary aggression develops, adding further to the basic inverse effect of inherent aggression on the inhibitory preference value. The threshold could thus be expressed by aggression levels. It should also be mentioned that momentary aggression can be developed independently of socio-demographics and inherent personal levels of aggression. These assumed effects deserve further research.

The threshold can probably be expressed by other determinants like monetary loss or amount of information, depending on situation, personal tastes, and attitudes. The effect of information on the threshold was shortly described above and further discussed below. The use of monetary loss depends on the personal set of priorities as determined by ones tastes and attitudes.

Experience and Information

Knowledge, familiarity, experience, and imply consequences of subjectively identified possible actions. In the theoretical framework a distinction however is made between experience and information. It is only experience from which the consequences of actions are weighted. According to the Random Walk SEU Theory, the initial preference state starts at some anchor point, biased by prior knowledge and experience. More experience may thus shorten the deliberation time.

Information only defines the possible actions and sometimes adds new objective input to the subjective calculation of the valences. Both, however, affect the threshold through expectations. Expected delay time, for example, is affected by experience and real-time information. Availability of real-time traffic information on congestion levels of alternate routes would lower the threshold and trigger actions.

Time Pressure

Only scant research has investigated the effect of time pressure on human judgment and decision making (e.g., Svenson and Maule, 1993). Time pressure is probably the most dominant consequence of congestion affecting not only the deliberation process, as explained by the Random Walk SEU Theory, but also the threshold, either directly or through the mediation of risk taking. It is predicted that time pressure increases risk taking when the expected consequence of an action is negative and decreases risk taking when the expected consequence is positive (Busemeyer, 1985).

Regarding the effect of time pressure on the deliberation process, Payne et al. (1988) concluded that people adapt to time pressure in a progressive way: First they accelerated their speed of processing; if that is not sufficient because there is no possibility to process the information faster, then increased the selectivity of processing; and finally they consider a more attributewise information search and processing. Edland and Svenson (1993) summarized numerous studies and showing that when time pressure does have an effect, as in the case of travelling under congestion, the following are the main changes in deliberation: (a) Increasing selectivity of retrieved information; (b) Salient attributes are given higher weights; (c) Accuracy of decision making is decreasing; and (d) An increased use of non-compensatory decision rules.

In dynamic situations, time pressure is more often connected with the development of the situation itself (e.g., Kerstholt, 1994). As the speed at which the situation evolves towards negative consequences increases, time pressure increases as well and an action is often taken earlier. This is the result of selectivity and acceleration towards the threshold. On the other hand, if the threshold is determined, for example, by delay time, longer evolved delay times increase time pressure and consequently decrease the threshold.

Situation

Although there are many situational factors that have been empirically found to affect both stated and revealed choices (e.g., weather conditions, day/night time, time flexibility, among others), we confine the discussion here to choice situations which affect the deliberation process. Other situational factors may determine different behavior styles rather than directly affect the choice mechanism. If the consequences of a certain situation can be retrieved from memory, it may indirectly affect deliberation through the mediation of experience. A direct effect of a situational factor on the deliberation process can be demonstrated by using the two basic types of response to congestion, preventive and reactive.

Preventive responses can be executed in pre-trip and/or en-route situations, whereas reactive responses can be executed only in en-route situations. By its nature, preventive responses are aimed to minimize risk and will be taken in order to avoid losses of time. It is thus an avoidance-avoidance type of behavior which prolongs deliberation time due to a high number of comparisons, increase vacillation, and consequently slows down the decision process. Reactive responses are motivated by the attractiveness of the reward and it is therefore aimed to gain time. It is obviously an approach-approach type of behavior which speeds up the decision process due to the fact that the preference state is racing toward the threshold. Situations eliciting reactive responses are also characterized by a higher time pressure which further reduces the threshold and consequently shortens vacillation.

To exemplify such situations, let us consider a pre-trip choice where the traveller wants to avoid getting into an expected congestion, and an en-route situation in which the driver wants to gain time by lane switching in a congested road. In the pre-trip choice there are more possible actions and consequences. In the lane-switching situation, the driver is faced only with the choice presented in Figure 8.5. In most cases the driver has no information about the traffic ahead, neither on his or her current lane nor on other lanes, and his or her threshold is constantly affected by the continuos change in road conditions for switching. As a result, vacillation will be shorter in comparison to the pre-trip choice case and actions will be taken more often.

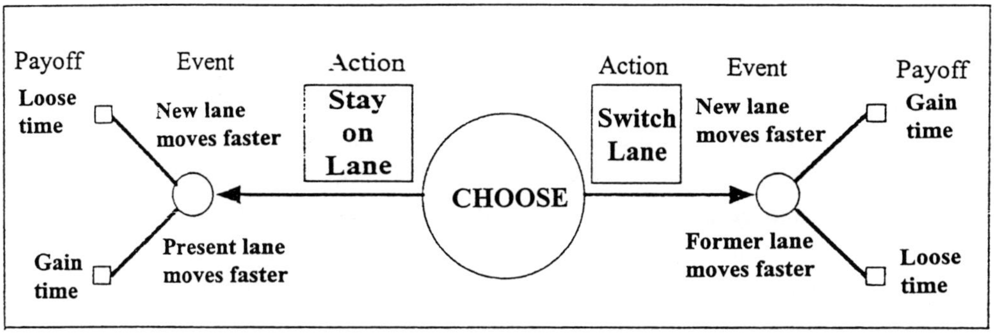

Figure 8.5. A choice situation for lane switching under congestion.

Personality

Among different personality attributes, only three are considered at this stage as having major effects on the deliberation process and the threshold.

Locus of control is a concept that refers to a generalized expectancy about the extent to which reinforcements are under internal or external control (Lefcourt, 1982; O'Brien, 1984). Persons characterized as internals believe that reinforcements are determined largely by personal effort, ability, and initiative, whereas persons classified as externals believe that reinforcements are determined largely by other people, luck, or fate. Accordingly, internals differ cognitively and behaviorally from externals on many tasks and in many life situations (Lefcourt, 1982). There is quite extensive support that persons at the extreme internal end of the continuum attend more actively and selectively to environmental events than do extreme externals (Blankstein, 1984). For example, internals are more ready than externals to search for and use information necessary for the successful handling of their task. Moreover, more extensive information processing by internals includes more comparisons with past experience. Other studies (e.g., Wolk and DuCette, 1974) found that internals have more active attentional processes and more intensive efficient structuring of environmental stimuli. These findings support the assumed linkages between the individual's locus of control and the deliberation process (Figure 8.4). Internals are assumed to have a larger number and variety of past experience associations on which more comparisons will be made, thus affecting the total deliberation time.

Decisiveness is reflected by (a) relatively high speed of decision making, (b) consistency in making and pursuing a decision (stability of choice), and (c) a high degree of self-confidence in the decision. The relatively high speed of decision making is achieved by adopting a heuristic approach and simplification strategy rather than a compensatory strategy which considers the full range of utility values. Accordingly, decisive decision makers focus their attention only on a small number of salient alternatives and situations while ignoring others. A recent route-choice study of two different groups of drivers, professionals (taxi drivers) and non-professionals, showed that the more consistent the driver is, the less criteria he or she uses in the decision making process (Spharadi and Kalmanovitch, 1992). In addition, a non-compensatory, heuristic, approach is cognitively simpler (Einhorn, 1970) and requires therefore shorter time (for empirical evidence, see Payne et al., 1988). With regard to our proposed framework, these conceptions and their empirical evidence suggest that decisiveness speeds up the deliberation process.

Risk taking is, to our best knowledge, the only personality attribute which has been incorporated in travel-choice models (Khattak et al., 1993; Mirchandani and Soroush, 1987; Senna, 1995) without, however, specifying any mediating mechanism. Individuals may behave as risk-averse, risk-prone, or risk-neutral. Risk-neutral travelers look only at average values of the stochastic variables and their choice is influenced by the differences in attributes between alternatives, regardless of the level of the average values (Bovy and Stern, 1990). Risk-averse and risk-prone travellers are sensitive to the probability of extreme outcomes and look at the utility ratios between alternatives. Their choice behavior can be described by an exponential function whereas neutral risk behavior can be expressed by a linear function. Thus, in congested situations we may expect to find that risk-prone drivers tend to have a lower threshold as far as it is determined by delay time. The more positive the attitude is toward risk, vacillation is expected to be shorter and the more often are actions likely to be taken.

MODELING AND RESEARCH NEEDS

The proposed scope of response to congestion and the theoretical framework based on the Decision Field Theory require further data collection, research, and modeling formulation in the following areas.

Since congestion is predicted, in some areas, both to grow absolutely and expand spatially (Stern, 1994), it is important to study other than travel-related reactions to congestion, mainly the least studied effects on choice of activity and driving behavior. The first is a preventive reaction and the later reactive. There is a need to look at: (1) the existing and the potential extent of rescheduling of activities due to congestion, and (2) the variety and the extent of emerging driving styles (a responsive style adopted temporarily in a given situation) under congestion. This latter reaction is important not only for modeling choice but also for road-safety reasons.

Ongoing research by the present author demonstrates the difficulties involved in estimating the revealed threshold for both preventive and reactive reactions. Research is needed to explore the variation in subjective thresholds for each possible action of each response.

The assumed exogenous effects on both the deliberation process and the threshold deserve careful studies of the following topics: (1) The effect of personality attributes on the choice process, mainly (but not only) locus of control, decisiveness, and risk taking; (2) The development and effects of aggression under traffic congestion, especially when momentary aggression is developed; (3) The effects of socio-demographics on the threshold. These effects were not included in the proposed theoretical framework but should nevertheless, be studied. For example, the individual's role at work determines not only personal freedom to choose departure time, but mainly personal limitations. It usually does not allow choice but rather reduces the options.

Two main methodological issues are: (1) The conceptual framework needs to mathematically structured to enable predictions, perhaps taking as starting point the set of equations underlying Decision Field Theory; (2) There is a need to find ways for collecting data on personality and revealed behavior. The inability to use revealed behavior is still the major fault of current modeling since there is no other way to reproduce behavior under real time pressure and physical stress resulting from congested driving conditions. For preventive behavior, in which decisions are made prior to departure, pending on the situation, time pressure and physical stress are usually smaller than in the case of reactive behavior. In these cases one may explore the possibility of in-depth interviews with focus groups. For reactive responses, we may explore the possibility of a two-stage stalking procedure: (a) car stalking along a selected congested section of a road with possible deviating routes while recording plate number and all driving reactions; (b) plate number identification and in-depth interview with drivers. The interviewees should be selected to include a range of driving styles and reactions. Elamatin (1993) has successfully developed this method in Israel.

Aiming to expand the theoretical foundation of travel choice modeling, a behavioral-based theory has been presented and modified. Decision Field Theory provides a mechanism for the inside of the black box. It is not a rule-based model, but rather an explanatory conception of human preferences. It has all the elements of choice theory that are relevant to the dynamics of congested environments. It is general enough to suite all types of congestion response behavior, and it opens ways for further research on threshold determination and change, effects of personality attributes, and new ways for collecting revealed behavior data.

REFERENCES

Bates, J., M. Dix and T. May (1987). Travel time variability and its effect on time of day choice for the journey to work. In: *Proceedings of PTRC Annual Meeting*, pp. 293-311. Education and Research Services, London.

Blankstein, R. K. (1984). Psychophysiology and perceived locus of control. In: *Research with Locus of Control Construct* (M. H. Lefcourt, ed.), pp. 73-208. Academic Press, Orlando.

Bovy, P. and E. Stern (1990). *Route Choice: Wayfinding in Transportation Networks*. Kluwer, Dordrecht, The Netherlands.

Busemeyer, J. R. (1985). Decision making under uncertainty: A comparison of simple scalability, fixed sample and sequential sampling model. *Journal of Experimental Psychology: Learning, Memory and Cognition*, **11**, 38-564.

Busemeyer, J. R. and J. T. Townsend (1989*). Decision field theory: A dynamic-cognitive approach to decision making*. Technical Report No. 89-7, Department of Psychology, Purdue University. West Lafayette, OH.

Busemeyer, J. R. and J. T. Townsend (1993) Decision field theory: A dynamic-cognitive approach to decision making in an uncertain environment. *Psychological Review*, **100**, 432-459.

Edland, A. and O. Svenson (1993). Judgment and decision making under time pressure. In: *Time Pressure and Stress in Human Judgment and Decision Making* (O. Svenson and A. J. Maule, eds.), pp. 27-40. Plenum Press, New York.

Einhorn, H. J. (1970). The use of nonlinear noncompensatory models in decision making. *Psychological Bulletin*, **73**, 221-230.

Elamatin, I. (1993). *Behavioral aspects of israeli drivers in congested Situations*. Unpublished manuscript, Department of Geography & Environmental Development, Ben Gurion University, Beer Sheva, (in Hebrew).

Emmerink, H. M. R., K. W. Axhausen, P. Nijkamp and P. Rietveld (1995). The potential of information provision in a simulated road transport network with non-recurrent congestion. *Transportation Research C*, **3**, 293-309.

Fox, M. (1995). Transport planning and the human activity approach. *Journal of Transport Geography*, **3**, 105-116.

Hall, W. R. (1993). Non-recurrent congestion: How big is the problem? *Transportation Research C*, **1**, 89-103.

Hogarth, R. and H. J. Einhorn (1992). Venture theory: A model of decision weights. *Management Science*, **36**, 780-803.

Hutchinson, R. and C. Dudeck (1979). Delay, time saved, and travel time information for freeway traffic management. *Transportation Research Record, No.* **722**, 36-39.

Hunt G. J. and G. D. Lyons (1995). *Aspects of the application of artificial neural networks to model driver decisions.* Unpublished paper, Cardiff School of Engineering, University of Wales College, Cardiff, UK.

Kerstholt, H. J. (1994). The effect of time pressure on decision-making behavior in a dynamic task environment. *Acta Psychologica*, **86**, 89-104.

Khattak, A. J., L. J. Schofer and S. F. Koppelman (1993). Commuters' enroute diversion and return decisions: analysis and implications for advanced traveler information systems. *Transportation Research A*, **27**, 101-111.

Lefcourt, H. M. (1982). *Locus of Control: Current Trends in Theory and Research*. Elbaum, Hillsdale, NJ.

Lindley, J. A. (1989). Urban freeway congestion problems and solutions: An update. *ITE Journal*, **59**, 21-23.

Lyons, G. D. and J. G. Hunt (1993). Traffic models - A role for neural networks? In: *Proceedings of the Third International Conference on the Application of Artificial Intelligence to Civil and Structural Engineering*, pp. 71-79. Edinburgh, UK.

Mahmassani, H. S., C. Caplice and C. Walton (1990). Characteristics of urban commuter behavior: switching propensity and use of information. *Transportation Research Record, No.* **1285**, 57-69.

Maule, A. J. and O. Svenson (1993). Theoretical and empirical approaches to behavioral decision making and their relation to time constraints. In: *Time Pressure and Stress in Human Judgment and Decision Making* (O. Svenson and A. J. Maule, eds.), pp. 3-25. Plenum Press, New York.

Mirchandani, P. and H. Soroush (1987). Generalized traffic equilibrium with probabilistic travel times and perceptions. *Transportation Science*, *21*, 133-152.

Ng, L., W. Barfield and F. Mannering (1995). A survey-based methodology to determine information requirements for advanced traveller information systems. *Transportation Research C*, **3**, 113-127.

Novaco, R. W., L. Milanesi and D. Stokols (1990). Objective and subjective dimensions of travel impedance as determinants of commuting stress. *American Journal of Community Psychology*, **18**, 231-257.

O'Brien, E. G. (1984). Locus of control, work and retirement. In: *Research With Locus of Control Construct* (M. H. Lefcourt, ed.), pp.7-72. Academic Press, Orlando.

Payne, J. W. (1982). Contingent decision behavior. *Psychological Bulletin*, **92**, 382-402.

Payne, J. W., J. R. Bettman and E. J. Johnson (1988). Adaptive strategy selection in decision-making. *Journal of Experimental Psychology*, **14**, 534-552.

Rapoport, A. (1975). Research paradigms for dynamic decision behavior. In: *Utility, Probability and Human Decision Theory* (D. Wendt and C. Vlek, eds.), pp. 349-369. Reidel, Dordrecht, The Netherlands.

Schofer, L. J., A. Khattak and F. S. Koppelman (1993). Behavioral issues in the design and evaluation of advanced traveler information systems. *Transportation Research C*, **1**, 107-117.

Senna, L. A. (1995). *The values of travel time variability: modeling without support of confronting fairy tales*. Paper presented at the 7th world conference on transport research, Sydney, Australia.

Sheffi, Y. (1985). *Urban Transportation Networks*. Prentice Hall, Englewood Cliffs, N.J.

Sinuani-Stern, Z., E. Stern, Z. Spharadi and E. Holm (1996). The effect of information on commuters' behavior: A comparative micro-simulation approach. *European Journal of Operation Research*, **96**, 455-470.

Spharadi, Z. and I. Kalmanovitch (1992). *Drivers' level of decisiveness and driving decision making process*. Doctoral Dissertation, Department of Industrial Engineering & Management, Ben-Gurion University, Beer Sheva, Israel. (In Hebrew.)

Stern, E. (1994). Traffic congestion and spatial planning in Israel. *Environmental Planning Quarterly*, **48-49**, 19-31.

Stern, E., E. Holm and M. van Maarseveen (1993). Information and commuters behavior: A comparative analysis. In: *Europe on the Move* (P. Nijkamp, ed.), pp. 141-156. Avebury, Aldershot, UK.

Stern, E. and S. Krakover (1993). The formation of a composite urban image. *Geographical Analysis*, **25**, 130-146.

Stern, E. and N. Zarin (1994). *Traffic Congestion and Residential Relocation*. Paper presented at the international geographical union regional conference. Prague.

Stern, E., Z. Sinuani-Stern and Z. Spharadi (1995). *Information effects on commuters behavior in congested situations*. Paper presented at the 7th world conference on transport research, Sydney, Australia.

Stern, E., Z. Sinuani-Stern, Z. Spharadi and E. Holm (1996a). Congestion-related information and road network performance. *Journal of Transportation Geography*, **4**, 169-178.

Stern, E. (1996). Congestion response behavior in Tel Aviv metropolitan area. In: *The Mosaic of Israeli Geography at the Close of the Twentieth Century* (Y. Gradus and G. Lipshitz, eds.). Ben-Gurion University Press, Beer Sheva, Israel.

Stern, E., P. Bovy and M. Tacken (1996b). Traffic congestion and behavioral reactions: policy implications exemplified for the Dutch case. In: *European Transport and Communication: Policies on European Networks* (K. Button, ed.). Kluwer, Dordrecht, The Netherlands. In press.

Svenson, O. (1990). Some propositions for the classification of decision situations. In: *Contemporary Issues in Decision Making* (K. Borcherding, O. I. Larichev and D.M. Messich, eds.), pp. 17-31. Elsevier, Amsterdam.

Svenson, O. (1992). Differentiation and consolidation theory of human decision making: A frame of reference for the study of pre-and-post decision processes. *Acta Psychologica*, **80**, 143-168.

Svenson, O. and J. Maule (eds.) (1993). *Time Pressure and Stress in Human Judgment and Decision Making*. Plenum Press, New York.

Tacken, M. and E. De Boer (1989). *Flexitime and the spread of traffic peak hour. An analysis of conditions and behavior*. Unpublished manuscript, *OSPA*, Delft University of Technology, Delft, The Netherlands. (In Dutch).

Tacken, M. and E. De Boer (1991). *Change in spread of travel and working times due to opening of the Amsterdam Orbital Motorway*. Unpublished manuscript, *OSPA,* Delft University of Technology, Delft, The Netherlands.

Van Berkum, E. and P. van der Mede (1994). *The Impact of Traffic Information-Dynamics in Route and Departure Time Choice*. Delft University Press, Delft, The Netherlands.

Vythoulkas, P. C. (1994). An approach to travel behavior based on the concepts of fuzzy logic and neural networks. In: *Proceedings of Seminar G on Transportation Planning Methods, The 22nd PTRC Transport Forum*, pp. 189-205. Education and Research Services, London.

Wolk, S. and J. DuCette (1974). Intentional performance and incidental learning as a function of personality and task dimension. *Journal of Personality and Social Psychology,* **29**, 90-101.

Zadeh, L. A. (1965). Fuzzy sets. *Information and Control*, **8**, 338-353.

Zadeh, L. A. (1973). Outline of a new approach to the analysis of complex systems and decision processes. *IEEE Transactions on Systems, Man and Cybernetics,* **3**, 28-44.

9

BOUNDED RATIONALITY IN COMMUTER DECISION DYNAMICS: INCORPORATING TRIP CHAINING IN DEPARTURE TIME AND ROUTE SWITCHING DECISIONS

Hani S. Mahmassani, University of Texas at Austin, TX, USA
Rong-Chang Jou, Feng Chia University, Taichung, Taiwan, Republic of China

INTRODUCTION

Investigation of urban commuter behavior dynamics has been of interest to researchers and transportation professionals over the past decade because of its relation to congestion management and the effectiveness of information technologies (Mahmassani, 1997). From a behavior research standpoint, daily commuter decisions are particularly interesting because they constitute an important example of repeated routine decisions that take place in an environment that can vary greatly in terms of predictability and uncertainty. In congested traffic systems that may be in disequilibrium, the consequences of particular decisions cannot generally be known a priori with certainty, as these depend on the decisions made non-cooperatively by all other tripmakers in the system, through nonlinear interactions taking place in the traffic system. Even in systems that appear close to some "equilibrium", considerable fluctuations may be taking place, so trip times and associated schedule delays may be quite variable within the peak period and from day to day. While the consequences of such unpredictability may be high in terms of aggravation, wasted time and fuel, long-term stress and health impacts, and possibly missed opportunities through late arrival, such consequences are not perceived as catastrophic on any given day, and may be changed from day to day, sometimes in real-time, with minimal perceived cost. Daily commuter decisions are therefore characterized by habit, routine, unpredictability, non-catastrophic impacts, many "good" (or equally inconvenient) solutions, and dynamic recourse. As such, these decisions differ markedly from major one-shot investment decision situations, with high initial cost and likely major irreversible impacts, which might warrant formal decision-analytic precepts, accompanied by systematic information acquisition and rational assessment of alternative courses of action. Nonetheless, commuter decision dynamics involve rather complex processes of passive and active information acquisition, judgment, learning, and decision making.

In light of the above, Mahmassani and co-workers have proposed a behavioral process model framework for the day-to-day dynamics of commuter decisions (primarily route selection and trip timing) in traffic systems (Mahmassani, 1990, 1997). A basic premise of this effort has been that commuter daily adjustments is guided by simple heuristic strategies and rules. As such, it has departed from the formal utility maximization paradigm, and viewed the behavior of the individual trip maker engaged in daily commuting as a boundedly-rational search for an acceptable outcome. Trip makers tend to seek to operate within an envelope of predictable outcomes, and will dynamically adjust their decisions and/or their aspirations accordingly. The case for this perspective is supported by evidence in the psychological and behavioral decision theory literature, as articulated in Mahmassani and Chang (1985).

Because of inherent measurement difficulties for the study of complex decision dynamics in nonlinear systems (Mahmassani and Herman, 1990), interactive laboratory-like experiments formed the basis of extensive investigation of departure time and route choices (Chang and Mahmassani, 1989; Mahmassani et al., 1986; Mahmassani and Tong, 1988; Mahmassani, 1997). Such experiments provide a useful observational basis for insights into the underlying behavioral processes, but they may not necessarily correspond to the commuters' actual behavior. The transferability of these insights to commuter behavior in real traffic networks has only been partially established, and remains to be accomplished as the next logical step towards the operational use of such models.

Of course, departure time choice and path selection are not the only choice dimensions available to commuters, nor the only ones that the various demand and congestion management policies seek to influence. However, they constitute the primary response mechanisms available to commuters in the short term, with limited expenditure of additional resources or major shifts in activity patterns. In some contexts, travel-mode choice may be another such dimension; in most US cities, the time frames for mode shifts tend to be of a longer-term nature than the above two dimensions, though one of the premises of Advanced Public Transit Systems (APTS) programs is that pre-trip real-time information can influence mode choice in real-time. On the other hand, the interaction of these decisions with the commuter's activity patterns, individually and at the household and possibly the workplace levels, are integral to meaningful analysis of commuter behavior in congested urban systems. An extensive discussion of activity-based approaches to travel behavior is given in the survey paper by Kitamura (1988); additional comments on the interactions between activity patterns and trip making decisions are also given by Jones et al. (1990).

In this study, we introduce an important choice dimension that was controlled for in the previous laboratory experiments, namely the associated trip-chaining choice behavior of commuters, which has not been investigated and modelled in conjunction with departure time and route choice behavior. The objective is to extend the modeling framework to incorporate trip chaining to capture the day-to-day dynamics of commuter decisions and associated flow patterns. The model development is supported by empirical evidence from a two-week diary survey of commuters in actual networks conducted in two cities in Texas (Jou and Mahmassani, 1996; Mahmassani et al., 1993, 1997). The survey provides a high and heretofore uncommon level of detail, especially with regard to path choice, for which all links used in each reported commute are listed. In addition, trip chaining information in the form of the location, duration, and purpose of stops made along the commute are reported, as are the times of all events (departure, stop arrival, stop departure, arrival). The survey results indicate that the variability of departure time and route choice decisions cannot be adequately explained without considering trip chaining, as trips with stops have a higher likelihood of inducing a departure time or route adjustment (Mahmassani et al., 1997). Previous exploratory analyses of the survey data have set the stage for the development of formal mathematical dynamic models for these phenomena. In addition, analysis of the survey data provides insights on the validity of previously developed models and behavioral insights on the basis of the laboratory experiments.

This chapter is structured as follows. The model development is described in the next section, starting with the classification of activities undertaken as part of the commuting trip chain, followed by the specification of the latent indifference bands for route and departure time switching to reflect such activities. Under general distributional assumptions about the random components of the indifference bands, a multinomial probit (MNP) framework is articulated to perform the model parameter estimation. The following section describes the estimation methodology, followed by estimation results and behavioral interpretation. The estimation of this model structure was considerably facilitated by recent advances in MNP estimation methodology and software (Lam, 1991; Lam and Mahmassani, 1991), allowing the joint estimation of both early and late indifference bands with correlated error structures. The final section provides concluding comments on commuter behavior dynamics incorporating trip chaining, and relates the results to those of previous laboratory experiments.

MODEL DEVELOPMENT

This section discusses the framework for modeling the day-to-day decisions of route and departure time, taken jointly, and incorporating trip chaining. The model extends previous work (Mahmassani, 1990) by incorporating trip chaining. First, we present a classification of activities included along the commuting trip chain for this purpose.

Classification of Activities

Trip chaining, and its daily variability, is influenced by the purpose and nature of activities included along the trip chain. For example, Kawakami and Isobe (1990) recognized that activities can be free or fixed in space and time, and can be discretionary or mandatory. In general, activities fixed in space and time have scheduling priority over those that are free in time/space, and mandatory activities have priority over discretionary ones even if free in time/space. Huff and Hanson (1990) concluded that individual travel-activity patterns are characterized by both repetition (routine) and variability (non-routine). In the present analysis, activities linked as part of a trip chain are grouped into *routine* (R) and *non-routine* (NR) categories. A routine activity is defined as one that a commuter pursues at the same location on his or her daily morning or evening commute. Within each category, *discretionary* activities (e.g., private leisure) and *mandatory* activities (e.g., drop-off children) are

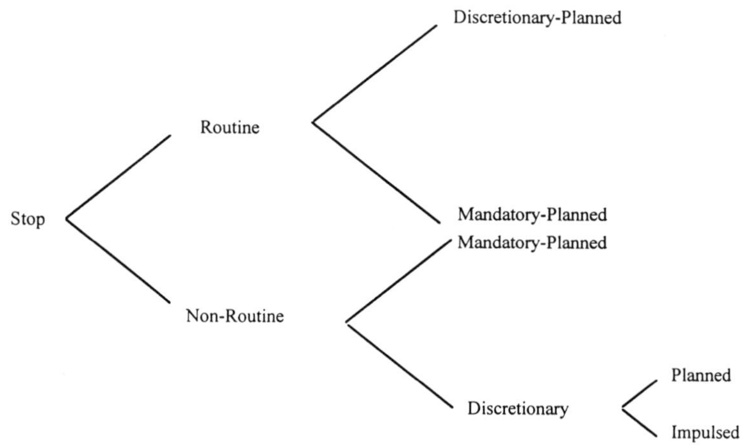

Figure 9.1. Classification of activities pursued en-route to or from work

distinguished. Activities can also be characterized as either *planned* or *impulse*. Commuters take planned activities into account for the selection of departure time and route prior to their commutes. In contrast, impulse activities are pursued without premeditation. Both types of activities occur, of course, during the commute period. All routine activities, discretionary or mandatory, are planned because they are part of daily commutes. For non-routine activities, all mandatory activities (e.g., see a dentist) are planned. However, discretionary activities can be impulse-driven or planned depending on whether a scheduled appointment is needed. In general, non-routine activities exert greater influence on commuter daily behavior than routine activities since the former exhibit greater variability in time and space. In addition, impulse-driven activities have less effect on commuter decisions than planned activities. The classification of activities is summarized in Figure 9.1.

Model Framework

The conceptual framework of the dynamic decision model incorporating the effect of trip chaining is shown in Figure 9.2. The departure time and route decisions of commuter i on day $t+1$ are assumed to be conditioned on the following two events: (1) whether the experienced schedule delay on day t is within the tolerable schedule delay for that day; and (2) whether there is trip chaining on day $t+1$. The effect of the latter is manifested through the tolerable schedules delay, as described hereafter. The principal decision mechanism is rooted in the notion of bounded rationality (Simon, 1955), operationalized for commuter decision dynamics through the concept of indifference bands (or intervals) of schedule delay, which govern the acceptability of the current departure time and/or route (Mahmassani, 1990; Mahmassani et al., 1986). An arrival time AT_{it} is considered acceptable when it falls within the time interval $[PAT_i - EB_{it}, PAT_i + LB_{it}]$. The threshold values, EB_{it} and LB_{it}, represent early-side and late-side indifference bands with respect to the preferred arrival time (PAT). The indifference band, indicating an individual's aspiration level on a given day, is not necessarily constant over time. To incorporate trip chaining, a plausible mechanism comprising seven components (in addition to the unobserved component) is proposed here. The trip chaining effect is separated into routine, non-routine, planned, and impulse components to investigate their relative effect on the indifference bands, and hence on the daily switching of departure time and route. For compactness of notation, the early-side and late-side indifference bands can be specified in one expression, and subsequently jointly estimated. In previous work, which did not include trip-chaining effect, these were estimated separately, primarily due to software limitations (Tong, 1990). The resulting combined (i.e., early and late) indifference bands for departure time and route, denoted by IBD_{it} and IBR_{it} respectively, for user i on day t, can be expressed as follows:

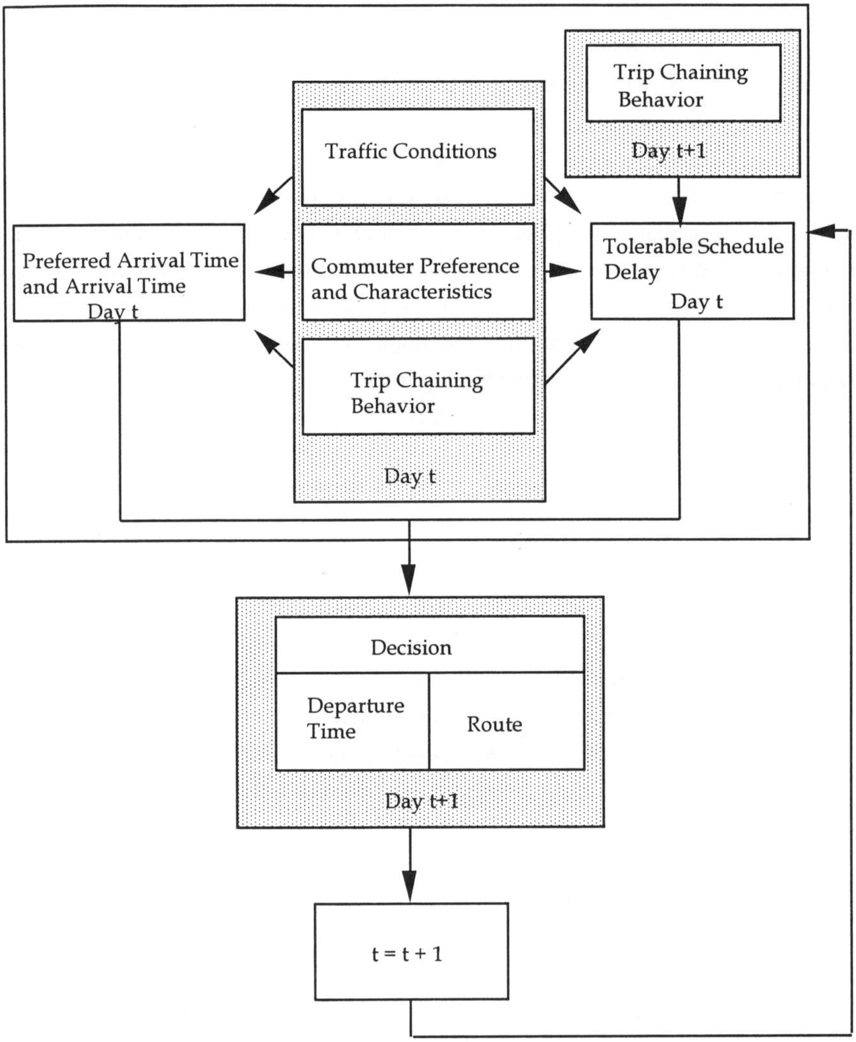

Figure 9.2. Conceptual framework of the dynamic decision model incorporating the effect of trip-chaining

$IBD_{it} = f_1(X_i, Z_{it}, \theta_{it})$	Non Trip Chaining Component
$+R_{it} \cdot f_2(X_i, Z_{it}, \theta_{it})$	Routine Trip Chaining Component (day *t*)
$+NRI_{it} \cdot f_3(X_i, Z_{it}, \theta_{it})$	NR Impulse Trip Chaining Component (day *t*)
$+NRP_{it} \cdot f_4(X_i, Z_{it}, \theta_{it})$	NR Planned Trip Chaining Component (day *t*)
$+R_{i(t+1)} \cdot f_5(X_i, Z_{i(t+1)}, \theta_{i(t+1)})$	Routine Trip Chaining Component (day *t+1*)

$+NRI_{i(t+1)} \cdot f_6(X_i, Z_{i(t+1)}, \theta_{i(t+1)})$	NR Impulse Trip Chaining Component (day *t+1*)
$+NRP_{i(t+1)} \cdot f_7(X_i, Z_{i(t+1)}, \theta_{i(t+1)})$	NR Planned Trip Chaining Component (day *t+1*)
$+ \varepsilon_{it}$	Unobserved Component
$IBR_{it} = h_1(X_i, Z_{it}, \theta_{it})$	Non Trip Chaining Component
$+R_{it} \cdot h_2(X_i, Z_{it}, \theta_{it})$	Routine Trip Chaining Component (day *t*)
$+NRI_{it} \cdot h_3(X_i, Z_{it}, \theta_{it})$	NR Impulse Trip Chaining Component (day *t*)
$+NRP_{it} \cdot h_4(X_i, Z_{it}, \theta_{it})$	NR Planned Trip Chaining Component (day *t*)
$+R_{i(t+1)} \cdot h_5(X_i, Z_{i(t+1)}, \theta_{i(t+1)})$	Routine Trip Chaining Component (day *t+1*)
$+NRI_{i(t+1)} \cdot h_6(X_i, Z_{i(t+1)}, \theta_{i(t+1)})$	NR Impulse Trip Chaining Component (day *t+1*)
$+NRP_{i(t+1)} \cdot h_7(X_i, Z_{i(t+1)}, \theta_{i(t+1)})$	NR Planned Trip Chaining Component (day *t+1*)
$+ \tau_{it}$	Unobserved Component

$$\varepsilon_{it} \sim \text{MVN}(0, \Sigma_\varepsilon)$$
$$\tau_{it} \sim \text{MVN}(0, \Sigma_\tau) \qquad (1)$$

The various terms incorporated in the above expressions are defined in Table 9.1. The systematic components of the above indifference bands are principally functions of (the vector of) user characteristics X_i and performance characteristics, Z_{it}, which capture commuter *i*'s experience up to day *t*; the vector of parameters to be estimated is θ. Zero-one indicator variables (*R*, *NRI*, *NRP*) are introduced to activate the applicable components of the indifference band for the particular combination of trip chaining on consecutive days.

The possible combinations of trip chaining on consecutive days are shown in Figure 9.3. A trip maker experiences one of the eight trip chaining alternatives on each commute. Note that the symbol *NR* is used to denote a commute with both impulse and planned non-routine stops. On two consecutive days, the commuter experiences one of 32 alternatives. There are 32 alternatives (4 x 4 + 4 x 4) instead of 64 (8 x 8) because the definition of routine stops implies that a commuter always experiences a routine stop(s) on day *t+1* if he or she had a routine stop on day *t*. The 32 combinations of trip chaining result in the indifference bands summarized in Table 9.2. For simplicity and conceptual clarity, these 32 combinations of trip chaining can be grouped into three types, illustrated in Figure 9.4: (1) keep routine (habitual) trip chaining, (2) return to routine (habitual) trip chaining, and (3) non-routine (non-habitual) trip chaining behavior. The first type includes no stops on days *t* and *t+1*, as well as routine stop(s) without non-routine activities on days *t* and *t+1* (cases 1 and 5 in Table 9.2). The second type includes non-routine activities (with or without routine stops) on day *t* and no stops or routine stop(s) without non-routine activities on day $t+1$ (cases 9, 13, 17, 21, 25, 29 in Table 9.2). The third type consists of all other cases.

Table 9.1.
Definitions of Indifference Band Elements in Equation 1

Element	Definition
$f_1(\cdot)$ and $h_1(\cdot)$	Tolerable schedule delay without trip chaining effect, for departure time and route, respectively.
$f_2(\cdot)$ and $h_2(\cdot)$	Effects of routine trip chaining experienced on day t.
$f_3(\cdot)$ and $h_3(\cdot)$	Effects of NR impulse trip chaining experienced on day t.
$f_4(\cdot)$ and $h_4(\cdot)$	Effects of NR planned trip chaining experienced on day t.
$f_5(\cdot)$ and $h_5(\cdot)$	Effects of routine trip chaining experienced on day $t+1$.
$f_6(\cdot)$ and $h_6(\cdot)$	Effects of NR impulse trip chaining experienced on day $t+1$.
$f_7(\cdot)$ and $h_7(\cdot)$	Effects of NR planned trip chaining experienced on day $t+1$.
R_{it}	Routine trip chaining index; $R_{it} = 1$ if commuter i has routine stop(s) on day t; otherwise, $R_{it} = 0$.
NRI_{it}	Non-routine impulse trip chaining index; $NRI_{it} = 1$ if commuter i has NR impulse stop(s) on day t; otherwise, $NRI_{it} = 0$.
NRP_{it} (and $RP_{i(t+1)}$)	Non-routine planned trip chaining index; $NRP_{it} = 1$ if commuter i has NR impulse stop(s) on day t; otherwise, $NRP_{it} = 0$.
ε_{it} and τ_{it}	Normally distributed disturbance terms for departure time and route switching indifference bands, respectively, for commuter i on day t.

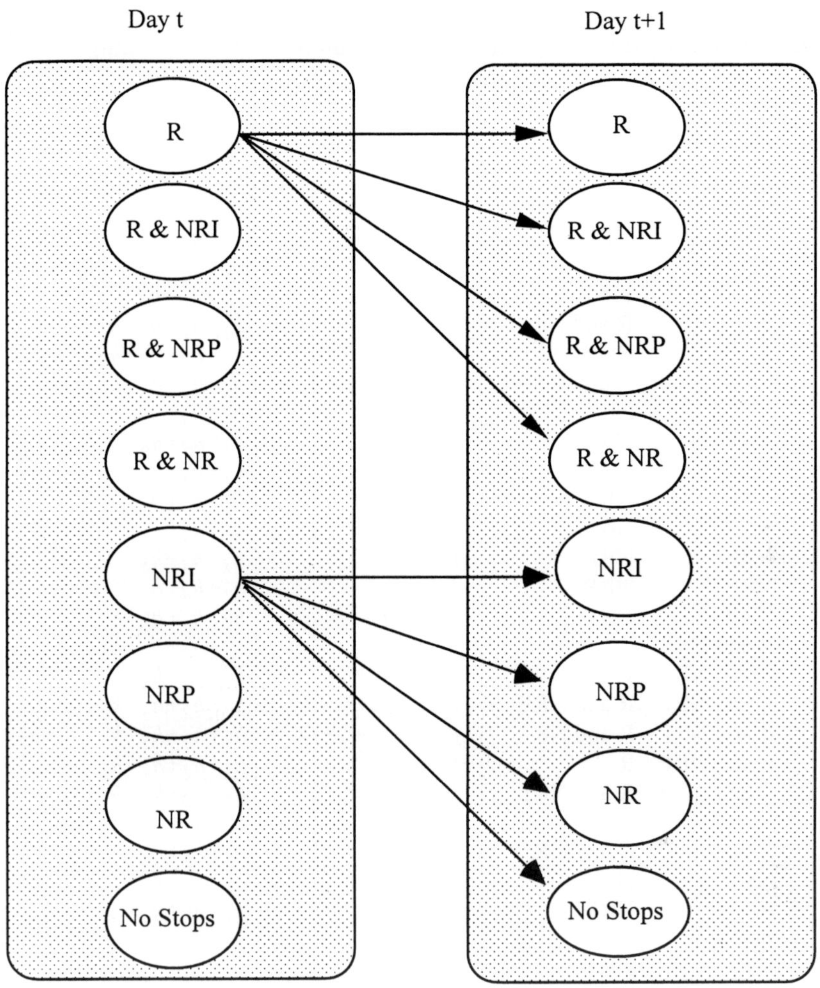

Figure 9.3. The 32 possible combinations of trip chaining on consecutive days (R: Routine activity; NRI: Non-Routine Impulse activity; NRP: Non-Routine Planned activity; NR: both planned and impulse non-routine activities)

The grouping recognizes the multiple patterns of daily trip chaining exhibited by commuters, and also recognizes that a given commuter may exhibit more than one trip chaining pattern. As discussed in a previous section, non-routine activities, both planned and impulse, are more likely to induce changes in commuter trip timing and path selection decisions than routine

Table 9.2
Indifference Band Components for Combinations of Trip Chaining on Two Consecutive Days

Case	Day *t*	Day *t+1*	IBD_{it}	IBR_{it}
1	No	No	f_1	h_1
2	No	NRI	f_1+f_6	h_1+h_6
3	No	NRP	f_1+f_7	h_1+h_7
4	No	NR	$f_1+f_6+f_7$	$h_1+h_6+h_7$
5	R	R	$f_1+f_2+f_5$	$h_1+h_2+h_5$
6	R	R and NRI	$f_1+f_2+f_5+f_6$	$h_1+h_2+h_5+h_6$
7	R	R and NRP	$f_1+f_2+f_5+f_7$	$h_1+h_2+h_5+h_7$
8	R	R and NR	$f_1+f_2+f_5+f_6+f_7$	$h_1+h_2+h_5+h_6+h_7$
9	NRI	No	f_1+f_3	h_1+h_3
10	NRI	NRI	$f_1+f_3+f_6$	$h_1+h_3+h_6$
11	NRI	NRP	$f_1+f_3+f_7$	$h_1+h_3+h_7$
12	NRI	NR	$f_1+f_3+f_6+f_7$	$h_1+h_3+h_6+h_7$
13	NRP	No	f_1+f_4	h_1+h_4
14	NRP	NRI	$f_1+f_4+f_6$	$h_1+h_4+h_6$
15	NRP	NRP	$f_1+f_4+f_7$	$h_1+h_4+h_7$
16	NRP	NR	$f_1+f_4+f_6+f_7$	$h_1+h_4+h_6+h_7$
17	R and NRI	R	$f_1+f_2+f_3+f_5$	$h_1+h_2+h_3+h_5$
18	R and NRI	R and NRI	$f_1+f_2+f_3+f_5+f_6$	$h_1+h_2+h_3+h_5+h_6$
19	R and NRI	R and NRP	$f_1+f_2+f_3+f_5+f_7$	$h_1+h_2+h_3+h_5+h_7$
20	R and NRI	R and NR	$f_1+f_2+f_3+f_5+f_6+f_7$	$h_1+h_2+h_3+h_5+h_6+h_7$
21	R and NRP	R	$f_1+f_2+f_4+f_5$	$h_1+h_2+h_4+h_5$
22	R and NRP	R and NRI	$f_1+f_2+f_4+f_5+f_6$	$h_1+h_2+h_4+h_5+h_6$
23	R and NRP	R and NRP	$f_1+f_2+f_4+f_5+f_7$	$h_1+h_2+h_4+h_5+h_7$
24	R and NRP	R and NR	$f_1+f_2+f_4+f_5+f_6+f_7$	$h_1+h_2+h_4+h_5+h_6+h_7$
25	NR	No	$f_1+f_3+f_4$	$h_1+h_3+h_4$
26	NR	NRI	$f_1+f_3+f_4+f_6$	$h_1+hy+h_4+h_6$
27	NR	NRP	$f_1+f_3+f_4+f_7$	$h_1+h_3+h_4+h_7$
28	NR	NR	$f_1+f_3+f_4+f_6+f_7$	$h_1+h_3+h_4+h_6+h_7$
29	R and NR	R	$f_1+f_2+f_3+f_4+f_5$	$h_1+h_2+h_3+h_4+h_5$
30	R and NR	R and NRI	$f_1+f_2+f_3+f_4+f_5+f_6$	$h_1+h_2+h_3+h_4+h_5+h_6$
31	R and NR	R and NRP	$f_1+f_2+f_3+f_4+f_5+f_7$	$h_1+h_2+h_3+h_4+h_5+h_7$
32	R and NR	R and NR	$f_1+f_2+f_3+f_4+f_5+f_6+f_7$	$h_1+h_2+h_3+h_4+h_5+h_6+h_7$

activities. Also, non-routine planned activities have greater effect than non-routine impulse activities. Thus, in Table 9.2, case 1 (without stops on day *t* and day *t+1*) is expected to have the least influence while case 32 (with routine and non-routine activities on day *t* and day *t+1*) would exert greatest influence on commuter decisions. Generally, type one trip chaining

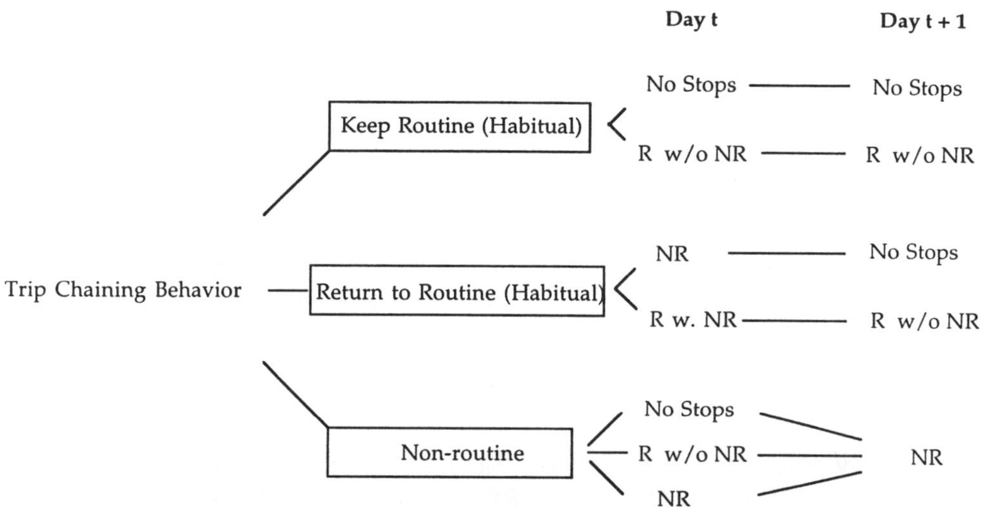

Figure 9.4. Grouping of trip chaining on consecutive days

without non-routine activities on consecutive days (including no stops) exerts the least influence on daily commuting among the three types, while type three (excluding no stops) has the greatest influence. These hypotheses are examined against empirical evidence through model parameter estimation and testing, following description of the model specification.

Model Specification

The specification of the various functions included in the indifference bands' systematic components takes as its starting point the preliminary models described in Jou and Mahmassani (1994) for daily departure time and route switching, which did however not explicitly recognize trip chaining effects. As such, these contribute primarily to the non-trip chaining components. The indifference band is assumed to include the following five components: (1) initial band, (2) commuter characteristics component, (3) dynamic component, (4) myopic component, and (5) trip chaining component. While the long-term dynamic effect is captured by the dynamic component, the myopic component captures the short-term dynamic effect. Prior estimation results have suggested that the myopic component is not significant for the route switching indifference band (Tong, 1990; Jou and Mahmassani, 1994). The trip chaining component is further separated into routine, non-routine, planned, and impulse components to investigate their relative effect on the indifference bands. In addition, as noted previously, the indifference band specifications must reflect different behaviors for late vs. early arrivals (relative to the *PAT*), though these are estimated jointly. This is achieved through the indicator variable $W_{it} = 1$ for early arrival, and $W_{it} = 0$ for late arrival. The detailed departure time model specification can then be expressed as:

$$IBD_{it} = W_{it}\, a_1 + (1-W_{it})\, a_2 + W_{it}\, a_3\, AGE_i + (1-W_{it})\, a_4\, AGE_i + W_t\, a_5\, GENDER_i + (1-W_{it})\, a_6\, GENDER_i + W_{it}\, a_7\, NFAIL_{it} + (1-W_{it})\, a_8\, NFAIL_{it} + W_{it}\, a_9\, \lambda_{it}\, \Delta ETR_{it}/\Delta EDT_{it} + (1-W_{it})\, a_{10}\, \lambda_{it}\, \Delta ETR_{it}/\Delta EDT_{it} + W_{it}\, a_{11}\, STDTR_{it} + (1-W_{it})\, a_{12}\, STDTR_{it} + W_t\, a_{13}\, RS_{it} + (1-W_{it})\, a_{14}\, RS_{it} + W_{it}\, a_{15}\, RS_{i(t+1)} + (1-W_{it})\, a_{16}\, RS_{i(t+1)} + W_{it}\, a_{17}\, NP_{it} + (1-W_{it})\, a_{18}\, NP_{it} + W_{it}\, a_{19}\, NP_{i(t+1)} + (1-W_{it})\, a_{20}\, NP_{i(t+1)} + W_{it}\, a_{21}\, NI_{it} + (1-W_{it})\, a_{22}\, NI_{it} + W_{it}\, a_{23}\, NI_{i(t+1)} + (1-W_{it})\, a_{24}\, NI_{i(t+1)} + \varepsilon_{it} \qquad (2)$$

The route indifference band model specification can be expressed as:

$$IBR_{it} = W_{it}\, a_{25} + (1-W_{it})\, a_{26} + W_t\, a_{27}\, GENDER_i + (1-W_{it})\, a_{28}\, GENDER_i + W_{it}\, a_{29}\, NFAIL_{it} + (1-W_{it})\, a_{30}\, NFAIL_{it} + W_{it}\, a_{31}\, STDTR_{it} + (1-W_{it})\, a_{32}\, STDTR_{it} + W_t\, a_{33}\, RS_{it} + (1-W_{it})\, a_{34}\, RS_{it} + W_{it}\, a_{35}\, RS_{i(t+1)} + (1-W_{it})\, a_{36}\, RS_{i(t+1)} + W_{it}\, a_{37}\, NP_{it} + (1-W_{it})\, a_{38}\, NP_{it} + W_{it}\, a_{39}\, NP_{i(t+1)} + (1-W_{it})\, a_{40}\, NP_{i(t+1)} + W_{it}\, a_{41}\, NI_{it} + (1-W_{it})\, a_{42}\, NI_{it} + W_{it}\, a_{43}\, NI_{i(t+1)} + (1-W_{it})\, a_{44}\, NI_{i(t+1)} + \tau_{it} \qquad (3)$$

The definitions of the variables included in equations (2) and (3) are given in Table 9.3. Several assumptions are made in conjunction with the above model specifications, reflecting the results of prior investigations and exploratory analyses. First, an initial tolerable schedule delay exists for both departure time and route; this initial band is asymmetric for late vs. early arrivals. Second, a commuter's age may affect departure-time switching behavior, with older commuters likely to tolerate greater schedule delay than younger ones. Third, a commuter's gender may influence departure-time and route-switching decisions, with female commuters generally exhibiting wider indifference band than males. Fourth, the indifference bands may increase in response to more switches and higher fluctuations of travel time experienced over a period of time, reflecting the relaxation of aspiration levels due to the uncertainty of experienced traffic conditions. Fifth, the indifference band governing departure time switching tends to increase as the commuter experiences large changes in travel time resulting from small changes in departure time; this effect can be captured by $\Delta ETR_{it}/\Delta EDT_{it}$. Sixth, the duration of activities pursued at stops along the commute both on day *t* and day *t*+1 may

Table 9.3

Definition of Variables in the Indifference Band Specifications

Element	Definition
AGE_i	Age of commuter i; 1, if age<18; 2, if age $\in$ [18,29]; 3, if age $\in$ [30,44]; 4, if age $\in$ [45,60]; 5, if age>60
$GENDER_i$	Gender of commuter i; =1, if male; =0, otherwise
$NFAIL_{it}$	Number of unacceptable early and late arrivals (number of departure time changes) until day t
ΔETR_{it}	Difference between travel times of commuter i on day t and $t-1$ (minutes)
ΔEDT_{it}	Amount of departure time adjustment by commuter i on day t with respect to day $t-1$ (minutes)
$STDTR_{it}$	Standard deviation of travel time up to day t (minutes)
RS_{it}	Routine stop duration on day t
NI_{it}	NR-impulse stop duration on day t
NP_{it}	NR-planned stop duration on day t
$RS_{i(t+1)}$	Routine stop duration on day $t+1$
$NI_{i(t+1)}$	NR-impulse stop duration on day $t+1$
$NP_{i(t+1)}$	NR-planned stop duration on day $t+1$
ε_{it} and τ_{it}	Error terms for commuter i on day t

affect the indifference bands; the longer the duration, the smaller the indifference bands. Also, routine, non-routine, planned, and impulse stops may influence switching differently. Generally, non-routine activities exert greater influence on the commuters' daily behavior than routine activities since the former exhibit more variability in time and space. Moreover, impulse activities have less effect on commuter decisions than planned activities.

With regard to the specification of the random disturbance terms ε_{it} and τ_{it}, a structure essentially similar to that described in Mahmassani (1990) is adopted. It assumes the following: (1) serial correlation in the random terms of the indifference bands for departure time over the sequence of decisions made by a given commuter, ε_{it}, t=1,...,T; (2) Serial correlation in the random terms of the route indifference bands over the sequence of decisions, τ_{it}, t=1,...,T; and (3) Contemporaneous correlation between the departure time and route indifference bands on a given day (ε_{it} and τ_{it}), for t=1,...,T, reflecting dependence on the same set of experienced traffic conditions and trip chaining behaviors. Note that unlike previous estimations with the laboratory experiments data, the error terms of the indifference bands for early arrival situations may be correlated with those corresponding to late arrivals. The elements in the variance-covariance matrix Σ of the 2xT-dimensional vector of random variates [ε_{it}, t=1,...,T; τ_{it}, t=1,...,T] are specified as:

$$
\begin{aligned}
&E(\varepsilon_{it}^2) = W_{it}\,\sigma_1^2 + (1-W_{it})\,\sigma_3^2 \\
&E(\tau_{it}^2) = W_{it}\,\sigma_2^2 + (1-W_{it})\,\sigma_4^2 \\
&E(\varepsilon_{it},\varepsilon_{ik}) = W_{it}W_{ik}\gamma_1 + (1-W_{it})(1-W_{ik})\gamma_4 \\
&\qquad + W_{it}(1-W_{ik})\,\gamma_{14} + (1-W_{it})\,W_{ik}\,\gamma_{41} \qquad t, k = 1,\ldots\ldots T,\ t \neq k \\
&E(\tau_{it},\tau_{ik}) = W_{it}W_{ik}\gamma_2 + (1-W_{it})(1-W_{ik})\gamma_5 \\
&\qquad + W_{it}(1-W_{ik})\,\gamma_{25} + (1-W_{it})\,W_{ik}\,\gamma_{52} \qquad t, k = 1,\ldots\ldots T,\ t \neq k \\
&\mathrm{E}(\varepsilon_{it},\tau_{ik}) = 0 \qquad t, k = 1,\ldots\ldots T,\ t \neq k \\
&E(\varepsilon_{it},\tau_{it}) = W_{it}\gamma_3 + (1-W_{it})\gamma_6
\end{aligned}
\tag{4}
$$

All parameters to be estimated in the indifference band formulation and variance-covariance matrix are listed along with their interpretation in Table 9.4.

Table 9.4

Parameters in the Indifference Band Specifications

Parameter	Definition
a_1, a_2	Initial early and late-side departure time (DT) bands
a_3, a_4, a_5, a_6	Early and late-side commuter characteristics component coefficients(DT)
a_7, a_8, a_{11}, a_{12}	Early and late-side dynamic component coefficients (DT)
a_9, a_{10}	Early and late-side myopic component coefficients (DT)
$a_{13\text{-}16}$	Early and late-side routine trip chaining component coefficients (DT); day t and day $t+1$
$a_{17\text{-}20}$	Early and late-side NR planned component coefficients (DT); day t and day $t+1$
$a_{21\text{-}24}$	Early and late-side NR impulse component coefficients (DT); day t and day $t+1$
a_{25}, a_{26}	Initial early and late-side route bands
a_{27}, a_{28}	Early and late-side commuter characteristics coefficients; route band
a_{29}, a_{32}	Early and late-side dynamic component coefficients; route band
a_{33}-a_{36}	Early and late-side routine trip chaining component coefficients
a_{37}-a_{40}	Early and late-side NR planned component coefficients
a_{41}-a_{44}	Early and late-side NR impulse component coefficients
σ_1 and σ_3	Variances for early and late-side indifference bands (DT)
σ_2 and σ_4	Variances for early and late-side indifference bands (Rte)
γ_1 and γ_4	Covariances for early-side and late-side indifference bands (DT), respectively (separate early and late serial correlation terms for DT)
γ_2 and γ_5	Covariances for early-side and late-side indifference bands (Rte), respectively (separate early and late serial correlation terms for Route)
γ_{14}, γ_{41}	Covariances for (early,late) indifference bands (DT)
γ_{25}, γ_{52}	Covariances for (early,late) indifference bands (Rte)
γ_3 and γ_6	Covariances for early and late-side indifference bands (DT, Rte), respectively (contemporaneous correlation terms)

MODEL ESTIMATION

Data

A sample of 196 commuters in the North Central area of Dallas, Texas, provided responses to a two-week (10 working days) diary survey of both AM and PM commutes. These diaries constituted the second stage of a two-stage survey procedure; the first stage, sent to a large random sample of households in the area of interest, was intended primarily for screening purposes to help identify suitable participants for the second stage (Mahmassani et al., 1993). The length of the trip diary stage (10 working days) was determined to be sufficient for examining short-term dynamic behavior but not so lengthy as to harm the respondents' goodwill. This diary contained detailed morning and evening trip information for a period of two weeks, including actual trip departure and arrival times, link-by-link route descriptions, and information on the location, purpose, and timing of stops in multi-purpose chains. Commuters were asked to provide official work start times. Questions pertaining to the trip from work were similar to those for the trip to work. A detailed discussion, including survey areas, sample pages, sampling strategies, and survey results, for the first- and second-stage surveys can be found elsewhere (Joseph et al., 1992; Jou and Mahmassani, 1996; Mahmassani et al., 1993).

Before the estimation, the trip chaining revealed in the observations for each decision instance needs to be classified as routine, non-routine, planned, and impulse, as defined in the previous section. As mentioned earlier, all routine stops are planned since they are part of the daily commute, whereas non-routine stops can either be planned or impulse-driven, depending on whether or not a prior appointment is scheduled. Twenty-one different stop purposes are possible for both routine and non-routine stops, and are classified as planned or impulse-driven according to Table 9.5.

Because the survey diary includes observations of commuter behavior over 10 consecutive decision days, the resulting data leads to multinomial models with a high dimension and intricate variance-covariance matrix, as discussed before. Note that estimation is based on a maximum of 8 different decision days, as no switch can be recorded on day 1, and some of the variables included in the specification require at least two prior days for meaningful calculation.

Estimation Process

Two separate models for departure time and route switching decisions were specified in Equations (2) and (3). Each model consists of early and late-side (with respect to *PAT*) indifference bands, jointly estimated to properly capture the correlated error structures described in the previous section. According to equations (2) and (3), 44 parameters are to be estimated in the indifference band formulation, in addition to 14 parameters in the variance-covariance matrix in Equation (4). In order to perform the estimations efficiently and

Table 9.5
Classification of Activities

Activities	Routine	Non-routine
serve passenger	P*	P
food	P	I**
recreation	P	P
post office	P	I
gas	P	I
cleaners	P	I
social	P	P
bank	P	I
shopping & (groc.)	P	I
medical	P	P
church	P	P
car repair	P	P
VCR movie rental	P	I
ATM transaction	P	I
rental equipment	P	P
meeting	P	P
work related	P	P
haircut/salon	P	P
newspaper	P	I
miscellaneous	P	I
other personal business	P	I

*P: Planned
**I: Impulse-driven

systematically, the following procedure is followed: (1) A restricted model in which the error terms are assumed identically and independently distributed (IID) across the population of trip makers, decisions, and over time (days) is estimated; (2) The estimated values of the restricted model from the previous step are used as initial values in the estimation of the more general specification in the next step; (3) The full model with a more general variance-covariance matrix that includes serial correlation, as specified in equation (4), is estimated; (4) Likelihood ratio tests are performed to examine the equality of parameters for the trip-chaining variables. A final model specification is then obtained following this procedure.

The model parameters are estimated on panel data of 8 decision days, following Lam's (1991) reestimation of Tong's (1990) specification for the separate early and late sides bands. The sequence of decisions over T days is translated into "$2T+1$" pseudo-utility functions based on the equations in Mahmassani (1990). One difficulty encountered here and in surveys of this type is that participants may have completed an unequal number of days (while some commuters completed the full 10 diary trips, others completed only 8 or 9 trips). The standard approach to deal with such situations is to specify alternatives with large negative "utilities" for the missing days in the model estimation process. The MNP models are then estimated with the same number of days for each observation.

Estimation Results

Before estimating the complete variance-covariance matrix specified in equation (4), a simple case in which the error terms are identically, independently distributed across the population of commuters and decisions and over time is considered. Note that separate variance terms are specified for departure time and route switching indifference bands, respectively. Thus the error terms are identical within a particular choice dimension, but different between route and departure time. Under the above independence assumption, the equations in (4) become:

$$
\begin{aligned}
&E(\varepsilon_{it}^2) = W_{it}\delta_1^2 + (1-W_{it})\,\delta_3^2 \\
&E(\tau_{it}^2) = W_{it}\delta_2^2 + (1-W_{it})\,\delta_4^2 \\
&E(\varepsilon_{it}, \tau_{it}) = 0 \\
&E(\varepsilon_{it}, \varepsilon_{ik}) = 0 && t, k = 1,\ldots,T,\ t \neq k \\
&E(\tau_{it}, \tau_{ik}) = 0 && t, k = 1,\ldots,T,\ t \neq k \\
&E(\varepsilon_{it}, \tau_{ik}) = 0 && t, k = 1,\ldots,T,\ t \neq k \qquad (5)
\end{aligned}
$$

The estimation results for the above model are not reported; most of the estimated values exhibited meaningful signs and strong statistical significance at the 95% level. Hence, the results provided a good initial point for the estimation of the more elaborate model specifications. The interpretation and behavioral implications of the estimation results are discussed hereafter in conjunction with the full model estimation results.

The estimation results from the iid model suggested that the coefficients associated with the trip chaining component (*RS*, *NP*, *NI*) on day t and day $t+1$ may not significantly differ from each other for departure time and route, early-side and late-side. To statistically compare consecutive days' trip chaining effects on the indifference bands, the coefficient estimates of trip chaining attributes (*RS*, *NP*, and *NI*) on day t and day $t+1$ are tested for equality by using likelihood ratio tests. To perform the tests, models with and without equal (restricted) coefficients for the day t and day $t+1$ components are estimated by maximum likelihood. The models without restriction of equal coefficients for the trip chaining attributes are as given by equations (2) and (3). The estimation results of these models, with the error structure specified by the equation (4), are presented in Table 9.6.

Table 9.6
Estimation Results for the Joint Indifference Band based on Eight Decision Days (without restriction of equal coefficients for the trip chaining component indicators, *RS*, *NP*, and *NI*, on day t and day *t+1*)

Component	Attribute/ Parameter	Estimates	t
DT Initial Band (e)	DT Initial Band (e)/a_1	2.74	3.39
DT Initial Band (l)	DT Initial Band (l)/ a_2	0.96	3.03
DT Social-economic 1 (e)	AGE_i (e)/ a_3	1.16	7.37
DT Social-economic 1 (l)	AGE_i (l)/ a_4	0.89	2.00
DT Social-economic 2 (e)	$GENDER_i$ (e)/ a_5	-1.14	-4.79
DT Social-economic 2 (l)	$GENDER_i$ (l)/ a_6	-1.89	-4.91
DT Dynamic 1 (e)	$NFAIL_{it}$ (e)/ a_7	1.35	2.31
DT Dynamic 1 (l)	$NFAIL_{it}$ (l)/ a_8	1.16	2.90
DT Myopic 1 (e)	λ_{it} $\Delta ETR_{it}/\Delta EDT_{it}$ (e)/ a_9	0.13	17.63
DT Myopic 1 (l)	λ_{it} $\Delta ETR_{it}/\Delta EDT_{it}$ (l)/ a_{10}	0.67	4.36
DT Dynamic 2 (e)	$STDTR_{it}$ (e)/ a_{11}	0.52	4.65
DT Dynamic 2 (l)	$STDTR_{it}$(l)/ a_{12}	0.15	12.34
DT Routine 1 (e)	RS_{it} (e)/ a_{13}	-0.16	-4.27
DT Routine 1 (l)	RS_{it} (l)/ a_{14}	-0.21	-10.06
DT Routine 2 (e)	$RS_{i(t+1)}$ (e)/ a_{15}	-0.13	-5.07
DT Routine 2 (l)	$RS_{i(t+1)}$ (l)/ a_{16}	-0.26	-4.50
DT NR plan 1 (e)	NP_{it} (e)/ a_{17}	-0.38	-2.32
DT NR plan 1 (l)	NP_{it}(l)/ a_{18}	-0.40	-16.36
DT NR plan 2 (e)	$NP_{i(t+1)}$ (e)/ a_{19}	-0.48	-4.12
DT NR plan 2 (l)	$NP_{i(t+1)}$ (l)/ a_{20}	-0.44	-2.91
DT NR impulse 1 (e)	NI_{it} (e)/ a_{21}	-0.71	-3.01
DT NR impulse 1 (l)	NI_{it} (l)/ a_{22}	-0.38	-2.73
DT NR impulse 2 (e)	$NI_{i(t+1)}$ (e)/ a_{23}	-0.67	-3.23
DT NR impulse 2 (1)	$NI_{i(t+1)}$ (l)/ a_{24}	-0.40	-2.35

Table 9.6 (cont.)

Component	Attribute/ Parameter	Estimates	t
R Initial Band (e)	R Initial Band (e)/a_{25}	25.25	10.62
R Initial Band (1)	R Initial Band (l)/ a_{26}	22.78	4.57
R Social-economic 1 (e)	$GENDER_i$ (e)/ a_{27}	-0.98	-4.65
R Social-economic 1 (l)	$GENDER_i$ (l)/ a_{28}	-1.58	-5.04
R Dynamic 1 (e)	$NFAIL_{it}$ (e)/ a_{29}	1.95	3.64
R Dynamic 1 (1)	$NFAIL_{it}$ (l)/ a_{30}	1.35	5.03
R Dynamic 2 (e)	$STDTR_{it}$ (e)/ a_{31}	1.42	4.28
R Dynamic 2 (l)	$STDTR_{it}$ (l)/ a_{32}	.016	2.43
R Routine 1 (e)	RS_{it}(e)/ a_{33}	-0.32	-7.18
R Routine 1 (l)	RS_{it} (l)/ a_{34}	-0.14	-2.79
R Routine 2 (e)	$RS_{i(t+1)}$ (e)/ a_{35}	-0.33	-3.04
R Routine 2 (l)	$RS_{i(t+1)}$ (l)/ a_{36}	-0.37	-2.46
R NR plan 1 (e)	NP_{it} (e)/ a_{37}	-0.68	-4.37
R NR plan 1 (l)	NP_{it} (l)/ a_{38}	-0.90	-2.23
R NR plan 2 (e)	$NP_{i(t+1)}$ (e)/ a_{39}	-0.72	-5.68
R NR plan 2 (l)	$NP_{i(t+1)}$ (l)/ a_{40}	-1.02	-3.28
R NR impulse 1 (e)	NI_{it}(e)/ a_{41}	-0.48	-16.93
R NR impulse 1 (l)	NI_{it} (l)/ a_{42}	-0.69	-18.40
R NR impulse 2 (e)	$NI_{i(t+1)}$ (e)/ a_{43}	-0.52	-2.52
R NR impulse 2 (l)	$NI_{i(t+1)}$ (l)/ a_{44}	-0.56	-2.34
DT Standard Deviation (e)	σ_1	12.07	8.72
DT Standard Deviation (1)	σ_3	11.73	5.46
R Standard Deviation (e)	σ_2	13.67	13.37
R Standard Deviation (1)	σ_4	12.23	15.94
Covariance for the Serial Correlation of DT (e)	γ_1	33.08	7.35
Covariance for the Serial Correlation of DT (1)	γ_4	11.79	3.36
Covariance for the Serial Correlation of DT (e & 1)	γ_{14}	12.97	12.34
Covariance for the Serial Correlation of DT (1 & e)	γ_{41}	13.00	6.93

Table 9.6 (cont.)

Component	Attribute/ Parameter	Estimates	t
Covariance for the Serial Correlation of R (e)	γ_2	100.00	21.03
Covariance for the Serial Correlation of R (l)	γ_5	14.57	14.41
Covariance for the Serial Correlation of R (e & l)	γ_{25}	15.01	15.01
Covariance for the Serial Correlation of R (l & e)	γ_{52}	15.00	2.71
Covariance for the Contemporaneous Correlation of R and DT (e)	γ_3	28.63	5.95
Covariance for the Contemporaneous Correlation of R and DT (l)	γ_6	7.38	2.45
Log-likelihood at convergence	- 1499.08		

The models with equal coefficients for the trip-chaining indicators, *RS*, *NP*, and *NI*, on day *t* and day *t*+1 can be expressed as:

$$IBD_{it} = W_{it} b_1 + (1-W_{it}) b_2 + W_{it} b_3 AGE_i + (1-W_{it}) b_4 AGE_i + W_t b_5 GENDER_i + (1-W_{it}) b_6 GENDER_i + W_{it} b_7 NFAIL_{it} + (1-W_{it}) b_8 NFAIL_{it} + W_{it} b_9 \lambda_{it} \Delta ETR_{it}/\Delta EDT_{it} + (1-W_{it}) b_{10} \lambda_{it} \Delta ETR_{it}/\Delta EDT_{it} + W_{it} b_{11} STDTR_{it} + (1-W_{it}) b_{12} STDTR_{it} + W_t b_{13} (RS_{it} + RS_{i(t+1)}) + (1-W_{it}) b_{14} (RS_{it} + RS_{i(t+1)}) + W_{it} b_{15} (NP_{it} + NP_{i(t+1)}) + (1-W_{it}) b_{16} (NP_{it} + NP_{i(t+1)}) + W_{it} b_{17} (NI_{it} + NI_{i(t+1)}) + (1-W_{it}) b_{18} (NI_{it} + NI_{i(t+1)}) + \varepsilon_{it} \quad (6)$$

$$IBR_{it} = W_{it} b_{19} + (1-W_{it}) b_{20} + W_t b_{21} GENDER_i + (1-W_{it}) b_{22} GENDER_i + W_{it} b_{23} NFAIL_{it} + (1-W_{it}) b_{24} NFAIL_{it} + W_{it} b_{25} STDTR_{it} + (1-W_{it}) b_{26} STDTR_{it} + W_t b_{27} (RS_{it} + RS_{i(t+1)}) + (1-W_{it}) b_{28} (RS_{it} + RS_{i(t+1)}) + W_{it} b_{29} (NP_{it} + NP_{i(t+1)}) + (1-W_{it}) b_{30} (NP_{it} + NP_{i(t+1)}) + W_{it} b_{31} (NI_{it} + NI_{i(t+1)} + (1-W_{it}) b_{32} (NI_{it} + NI_{i(t+1)}) + \tau_{it} \quad (7)$$

The estimation results are presented in Table 9.7. The likelihood ratio test results for consecutive days' trip chaining effects on departure time and route indifference bands are summarized in Table 9.8. The results indicate that one cannot reject the null hypotheses that the coefficients of trip chaining attributes (*RS*, *NP*, and *NI*) on day *t* are respectively equal to the corresponding coefficients on day *t+1*. This conclusion is true for departure time and

Table 9.7

Estimation Results for the Joint Indifference Band based on Eight Decision Days (with restriction of equal coefficients for the trip chaining component indicators, *RS*, *NP*, and *NI*, on day *t* and day *t+1*)

Component	Attribute/ Parameter	Estimates	*t*
DT Initial Band (e)	DT Initial Band (e)/b_1	2.69	2.93
DT Initial Band (l)	DT Initial Band (l)/ b_2	0.90	11.51
DT Social-economic 1 (e)	AGE_i(e)/ b_3	1.17	2.58
DT Social-economic 1 (l)	AGE_i (l)/ b_4	0.90	10.00
DT Social-economic 2 (e)	$GENDER_i$ (e)/ b_5	-1.22	-12.29
DT Social-economic 2 (l)	$GENDER_i$ (l)/ b_6	-1.78	-8.29
DT Dynamic 1 (e)	$NFAIL_{it}$ (e)/ b_7	1.31	7.37
DT Dynamic 1 (l)	$NFAIL_{it}$ (l)/ b_8	0.95	4.76
DT Myopic 1 (e)	λ_{it} $\Delta ETR_{it}/\Delta EDT_{it}$ (e)/ b_9	0.42	2.99
DT Myopic 1 (l)	λ_{it} $\Delta ETR_{it}/\Delta EDT_{it}$ (l)/ b_{10}	0.87	2.45
DT Dynamic 2 (e)	$STDTR_{it}$ (e)/ b_{11}	0.38	6.07
DT Dynamic 2 (l)	$STDTR_{it}$ (l)/ b_{12}	0.13	4.08
DT Routine (e)	RS_{it}, $RS_{i(t+1)}$ (e)/ b_{13}	-0.11	-3.55
DT Routine (l)	RS_{it}, $RS_{i(t+1)}$ (l)/ b_{14}	-0.29	-2.83
DT NR plan (e)	NP_{it},$NP_{i(t+1)}$ (e)/ b_{15}	-0.44	-3.17
DT NR plan (l)	NP_{it},$NP_{i(t+1)}$ (l)/ b_{16}	-0.67	-9.32
DT NR impulse (e)	NI_{it},$NI_{i(t+1)}$ (e)/ b_{17}	0.12	2.25
DT NR impulse (l)	NI_{it},$NI_{i(t+1)}$ (l)/ b_{18}	-0.20	-1.45
R Initial Band (e)	R Initial Band (e)/ b_{19}	25.25	4.51
R Initial Band (1)	R Initial Band (l)/ b_{20}	22.79	11.28
R Social-economic 1 (e)	$GENDER_i$ (e)/ b_{21}	-1.05	-14.34
R Social-economic 1 (1)	$GENDER_i$(l)/ b_{22}	-1.56	-4.02
R Dynamic 1 (e)	$NFAIL_{it}$ (e)/ b_{23}	1.99	6.63
R Dynamic 1 (l)	$NFAIL_{it}$ (l)/ b_{24}	1.32	9.20
R Dynamic 2 (e)	$STDTR_{it}$ (e)/b_{25}	1.38	5.16
R Dynamic 2 (l)	$STDTR_{it}$ (l)/b_{26}	0.38	2.66
R Routine (e)	RS_{it},$RS_{i(t+1)}$ (e)/ b_{27}	-0.19	-0.83
R Routine (l)	RS_{it},$RS_{i(t+1)}$ (l)/ b_{28}	-0.23	-1.78
R NR plan (e)	NP_{it},$NP_{i(t+1)}$ (e)/ b_{29}	-0.43	-14.69
R NR plan (l)	NP_{it},$NP_{i(t+1)}$ (l)/ b_{30}	-0.55	-0.99

Table 9.7 (cont.)

Component	Attribute/ Parameter	Estimates	t
R NR impulse (e)	$NI_{it}, NI_{i(t+1)}$ (e)/ b_{31}	-0.34	-3.89
R NR impulse (l)	$NI_{it}, NI_{i(t+1)}$ (l)/ b_{32}	-0.47	-5.70
DT Standard Deviation (e)	σ_1	12.12	2.60
DT Standard Deviation (l)	σ_3	11.69	14.11
R Standard Deviation (e)	σ_2	13.68	2.09
R Standard Deviation (l)	σ_4	12.25	2.40
Covariance for the Serial Correlation of DT (e)	γ_1	33.08	5.07
Covariance for the Serial Correlation of DT (l)	γ_4	11.78	15.73
Covariance for the Serial Correlation of DT (e & l)	γ_{14}	12.98	2.85
Covariance for the Serial Correlation of DT (l & e)	γ_{41}	12.98	10.69
Covariance for the Serial Correlation of R (e)	γ_2	100.00	3.52
Covariance for the Serial Correlation of R (l)	γ_5	14.57	3.54
Covariance for the Serial Correlation of R (e & l)	γ_{25}	15.00	2.92
Covariance for the Serial Correlation of R (l & e)	γ_{52}	15.00	2.45
Covariance for the Contemporaneous Correlation of R and DT (e)	γ_3	28.62	2.26
Covariance for the Contemporaneous Correlation of R and DT (l)	γ_6	7.37	2.36
Log-likelihood at convergence		- 1503.72	

route, for both early-side and late-side. The results indicate that the coefficients associated with the trip chaining component on day t and day $t+1$ do not significantly differ from each

Table 9.8
Log likelihood Ratio Test for Trip-Chaining Parameters between Day *t* and Day *t+1*

Restricted on	$L(U)$	$L(R)$	Test Statistic	Significant
RS, NP, NI	-1499.08	-1503.72	9.28	ns

other, implying that trip chaining attributes, *RS*, *NP*, and *NI*, have a stable influence on the indifference band over time.

The behavioral interpretation and discussion are based on the results in Table 9.7, as equations (6) and (7) are considered the final models, with equal coefficients for the trip chaining component indicators, *RS*, *NP*, and *NI*, on day t and day $t+1$. The results in Table 9.7 reveal that the initial band for late-side is smaller than that for early-side. This is shown by the respective values of b_1 and b_2 for departure time, and b_{19} and b_{20} for route. These estimates confirm the earlier finding obtained by Mahmassani and Stephan (1988) and Tong (1990) from laboratory experiments data.

The socio-economic variable coefficients, b_3 through b_6 for departure time and b_{21} through b_{22} for route, have the correct signs. The results suggest that (1) older commuters tend to tolerate greater schedule delay than younger ones for departure time switching decisions, but not for route switching since the coefficient of AGE_i is not statistically significant; and (2) male commuters have narrower indifference bands (i.e., are more likely to switch) than females for both departure time and route switching decisions.

The dynamic component parameters are b_7, b_8, b_{11}, and b_{12} for departure time switching decisions and b_{23} through b_{26} for route. These parameters capture the dynamic effect of commuters' learning through previous experience. The results indicate that commuters tend to engage in less departure time or route switching after experiencing higher travel time fluctuation or conditions that require them to switch on the previous t days. Mahmassani and Chang (1986) observed a similar finding in a laboratory experiment where the only choice dimension available was that of departure time. The parameters for the myopic component are b_9 and b_{10} for departure time only. The results imply that if commuters have recently experienced a substantial increase in travel time as a result of a small adjustment in departure time, they will be more likely to tolerate greater schedule delay in subsequent decisions (i.e., are less likely to switch), as a way of absorbing the possibly large fluctuations in trip time associated with small adjustments in departure time.

The parameters for the trip-chaining component are b_{13} through b_{18} and b_{27} through b_{32} for departure time and route switching decisions, respectively. The estimates have negative signs indicating that commuters are likely to narrow the indifference bands (are more likely to switch) if they have experienced stops on day t or if there are stops on day $t+1$. Also, as the duration of a stop increases, commuters are more likely to switch departure time and route. The stops may be routine, non-routine planned, or non-routine impulse. Among the trip chaining attributes, non-routine planned stops exert greater weight on commuter decisions than non-routine impulse stops since commuters may need to change their normal departure time or route to meet the scheduled appointment. Non-routine stops, both planned and impulse, have greater coefficient values than routine stops, reflecting greater influence on commuting behavior.

One of the concerns in the model specification is the presence of both correlation and state dependence. The latter, to a large extent, is captured by specification of the model's systematic components, such as the dynamic and myopic components (b_7 through b_{12} for departure time and b_{23} through b_{26} for route). Serial correlation is reflected in the covariance matrix of the random terms. As expected, the σ's and γ's in Table 10.7 are significant at a reasonable level, which confirms the need to incorporate the respective serial correlation terms for departure time and route for early-side and late-side (γ_1, γ_2, γ_4, and γ_5), the correlations between route and departure time bands (γ_3 and γ_6), and the correlations between early-side and late-side (γ_{14}, γ_{41}, γ_{25}, and γ_{52}). The covariance terms (γ's) are generally smaller than the main variance terms (σ's). The contemporaneous correlations are smaller than the serial correlations of departure time or of route ($\gamma_3 < \gamma_1$ and γ_2, and $\gamma_6 < \gamma_4$ and γ_5) for early-side and late-side. The correlations between early-side and late-side are smaller than the other terms. As expected, all the correlation terms exhibit positive signs indicating positive correlation between the unobserved disturbances.

Comparing the results for early-side with those for late-side, the coefficients of socio-economic variables for early arrival are greater than those for late arrivals for both departure time and route switching, reflecting the different influence of commuter characteristics (*AGE* and *GENDER*) on the early-side and late-side indifference bands. This is also true for the dynamic components, implying that the fluctuation of experienced travel time (*STDTR*) and the number of unacceptable arrivals (*NFAIL*) have different effect on the early-side and late-side bands for both departure time and route switching. The coefficients of trip chaining components (*RS*, *NP*, and *NI*) for early-side are greater than those for late-side for departure time and route switching. The results indicate that trip chaining, including routine, non-routine planned, and non-routine impulse-driven, has different effects on the early and late indifference bands. The late-side parameters are smaller than the early-side, which suggests that commuters are more cautiously expanding the late-side than early-side bands, that is commuters are indeed more sensitive to late arrivals than early ones (Tong, 1990). Comparing the results for route with those for departure time, the estimated values for the departure time switching band are usually smaller than those for route switching, with respect to the same attributes, both for early and late side components. This confirms that a commuter is very likely to switch departure time whenever he or she switches route (Mahmassani and Stephan, 1988; Mahmassani and Herman, 1990).

CONCLUDING COMMENTS

This chapter establishes the model framework for dynamic commuter decisions incorporating trip-chaining. Activities linked as part of a trip chain are classified into routine and non-routine categories for modeling purposes. Activities are also characterized as either planned or impulse-driven depending on when the commuter decides to engage in the activity. Essentially, one might expect that routine and non-routine activities would have different effects on daily commuting, and planned and impulse activities would also have different effects. Commuter departure time and route decisions are assumed to be based on two elements. The first is a function of traffic conditions, trip-chaining, and commuters' preferences and characteristics on day t. The second is a function of trip-chaining behaviour on day $t+1$.

The model specifications for both non-trip-chaining and trip-chaining components are illustrated. The trip-chaining component is separated into routine, non-routine, planned, and impulse components to examine their relative effect on the indifference bands. Two different models including departure time and route are specified. Each model combines early and late-side indifference bands. The model specifications provide the basis for the estimation of a complex dynamic travel behavior model incorporating trip-chaining.

The models are estimated using a MNP framework, taking advantage of new estimation software that allows a relatively large number of parameters, alternatives, and special correlated error structures.

Several substantive conclusions regarding commuter behavior can be obtained from this analysis:

1. In the departure time model, older commuters tend to tolerate greater schedule delay than younger ones. Also, female commuters exhibit a wider mean

indifference band than male commuters for both departure time and route switching decisions.

2. Commuters are inclined to tolerate greater schedule delay (associated with a particular departure time decision) if they have recently experienced a substantial increase in travel time resulting from a small adjustment in departure time.
3. Commuters are reluctant to continue switching route and departure time in response to greater experienced travel time fluctuations.
4. Commuters tend to increase their indifference band (i.e., switch less) in response to more failures; this is true for departure time and route models.
5. Non-routine planned stops exert a greater effect on commuters' decisions than non-routine impulse stops since commuters may need to change their normal departure time or route to meet the scheduled appointment. Non-routine stops, including planned and impulse, have greater estimated coefficients than routine stops.
6. The estimates of the variance and covariance terms, including serial correlations between route or departure time, the correlations between route and departure time bands, and the correlations between early-side and late-side, are all statistically significant which confirms the need to incorporate serial correlations in the specification.
7. The mean indifference band for departure time switching is much smaller than that for route switching, both for early and late-side components, again confirming earlier findings from laboratory experiments.
8. The late-side attribute coefficients are usually smaller than the early-side, which confirms that commuters are more cautiously expanding the late-side than early-side indifference bands.
9. The estimated values for the departure time switching band are usually smaller than those for route switching, with respect to the same attributes, both for early and late side components.

From a more fundamental standpoint, the results of this analysis suggest two important conclusions. First, the modeling framework based on bounded rationality theory, and operationalized through the notion of indifference bands, to represent the day-to-day dynamics of commuter decisions of departure time and route, is sufficiently rich and flexible to incorporate activity trip chaining as a behavioral determinant. Second, it is highly encouraging to find that the behavioral insights as well as model specifications derived from interactive laboratory experiments with simulated traffic systems transfer quite well to the behavior of commuters in actual systems, as revealed in the diary surveys. Of course, considerable additional evidence is necessary to develop a deeper understanding of the complex interactions underlying the dynamics of commuter behavior, and to develop a more rigorous basis for transferring insights from laboratory experiments to the field.

REFERENCES

Chang, G.-L. and H. S. Mahmassani (1989). The dynamics of commuting decision behaviour in urban transportation networks. In: *Travel Behavior Research, Fifth International Conference on Travel Behavior*, pp. 15-26. Avebury, Aldershot, UK.

Huff, J. and S. Hanson, (1990). Measurement of habitual behavior: examining systematic variability in repetitive travel. In: *Developments in Dynamic and Activity-Based Approaches to Travel Analysis* (P. Jones, ed.), pp. 272-298. Avebury, Aldershot, UK.

Jones, P., F. S. Koppelman, and J.-P. Orfeuil (1990). Activity analysis: state-of-the-art and future directions. In: *Developments in Dynamic and Activity-Based Approaches to Travel Analysis,* (P. Jones, ed.), pp. 34-55. Avebury, Aldershot, UK.

Joseph, T., H. S. Mahmassani and R. C. Jou (1992). *Dynamic framework for the analysis of user responses to traffic system disruptions and control actions.* Technical Report 1216-F, Center for Transportation Research, The University of Texas at Austin, TX.

Jou, R. C. and H. S. Mahmassani (1994). Day-to-day dynamics of commuter travel behaviour in an urban environment-departure time and route decisions. In: *Proceedings of the International Association for Travel Behavior Research Conference,* (J. D. Ortuzar, ed.),Vol. I, pp. 210-222. Santiago, Chile.

Jou, R. C. and H. S. Mahmassani (1996). Comparability and transferability of commuter behaviour characteristics between cities: Departure time and route switching decisions. *Transportation Research Record, No.* **1556**, 119-130.

Kawakami, S. and T. Isobe (1990). Development of a one-day travel-activity scheduling model for workers. In: *Developments in Dynamic and Activity-Based Approaches to Travel Analysis* (P. Jones, ed.), pp. 272-298. Avebury, Aldershot, UK.

Kitamura, R. (1988). An evaluation of activity-based travel analysis. *Transportation*, **15**, 9-34.

Lam, S.-H. (1991). *Multinomial probit model estimation: Computational procedures and applications*. Doctoral Dissertation, Department of Civil Engineering, The University of Texas at Austin, TX.

Lam, S.-H. and H. S. Mahmassani, (1991). Multinomial probit model estimation: computational procedures and applications. *Methods for Understanding Travel Behaviour in the 1990's*, IATB, pp. 229- 242.

Mahmassani, H. S. (1990). Dynamic models of commuter behaviour: Experimental investigation and application to the analysis of planned traffic disruptions. *Transportation Research A,* **24**, 465-484.

Mahmassani, H. S. (1997). Dynamics of commuter behaviour: recent research and continuing challenges. In: *Understanding Travel Behaviour in an Era of Change* (M. Lee-Gosselin and P. R. Stopher, eds.), pp. 279-314. Pergamon, Oxford.

Mahmassani, H. S. and G.-L. Chang (1985). Dynamic aspects of departure time choice behaviour in a commuting system: Theoretical framework and experimental analysis. *Transportation Research Record, No.* **1037**, 88-101.

Mahmassani, H. S. and G.-L. Chang (1986). *Specification and estimation of a dynamic departure time acceptability mechanism.* Paper presented at the 65th annual meeting of the Transportation Research Board. Washington, DC.

Mahmassani, H. S., G.-L. Chang, and R. Herman (1986). Individual decisions and collective effects in a simulated traffic system. *Transportation Science*, **20**, 258-271.

Mahmassani, H. S., S. Hatcher and C. Caplice (1997). Daily variation of trip chaining, scheduling, and path selection behaviour of work commuters. In: *Understanding Travel Behaviour in an Era of Change* (M. Lee-Gosselin and P.R. Stopher, eds.), pp. 351-380. Pergamon, Oxford.

Mahmassani, H. S. and R. Herman (1990). Interactive experiments for the study of tripmaker behaviour dynamics in congested commuting. In: *Developments in Dynamic and Activity-Based Approaches to Travel Analysis Systems* (P. Jones, ed.), pp. 272-198. Avebury, Aldershot, UK.

Mahmassani, H. S., T. Joseph and R. C. Jou (1993). A survey approach for the study of urban commuter choice dynamics. *Transportation Research Record, No.* **1412**, 80-89.

Mahmassani, H. S. and D. Stephan (1988). Experimental investigation of route and departure time dynamics of urban commuters. *Transportation Research Record, No.* **1203**, 69-84.

Mahmassani, H. S. and C.-C. Tong (1988). Availability of information and dynamics of departure time choice: Experimental investigation. *Transportation Research Record, No.* **1085**, 33-47.

Simon, H. (1955). A behavioral model of rational choice. *Quarterly Journal of Economics*, **69**, 99-118.

Tong, C.-C. (1990). *A study of dynamic a departure time and route choice behavior of urban commuters.* Doctoral Dissertation, Department of Civil Engineering, The University of Texas at Austin, TX.

10

TIME IN TRAVEL CHOICE MODELING: FROM RELATIVE OBSCURITY TO CENTER STAGE[15]

Eric I. Pas, Duke University, Durham, NC, USA

INTRODUCTION

Considerable interest in improving the state-of-the-art and state-of-the-practice in travel demand analysis and modeling has arisen in recent years, because the models currently used in practice are woefully inadequate to address many of the important contemporary policy analysis and planning questions, primarily in relation to air quality and sustainability. The first generation of travel demand models were developed in the earliest urban landuse/ transportation planning studies conducted in the USA in the mid-1950's, primarily to support planning of highway systems to cater to the expected large increases in automobile travel. Addressing such questions is relatively easy, and does require neither particularly accurate nor precise analyses. Because the number of highway lanes is indivisible, a precise analysis is not needed to be able to decide on the number of lanes to be built for a particular facility. Furthermore, in the case of highways it is generally true that "if you build it, they will come;" therefore, to some extent the forecasts were self-fulfilling prophecies.

The state-of-the-art in travel demand analysis and modeling has progressed considerably since the earliest models were developed and applied, although the state-of-the-practice has increasingly lagged behind the research. One of the most fundamental ways in which the state-of-the-art in travel demand analysis and modeling has changed over the past 40 years is the role that time plays in such models and analyses, although there is a serious imbalance in

[15] Within weeks after delivering his final chapter draft, Eric Pas unexpectedly died. We, the editors, regret the untimely death of an eminent scholar and exceptionally friendly and considerate human being who we are fortunate to have had the chance to learn to know. We have made the final revision of Eric's chapter in a way we hope he would have agreed to.

the way space and time are treated in the travel demand models used in practice, with more attention being paid to space than time. In fact, Goodwin (1991) noted that travel demand modelers typically consider space at the maximum level of detail possible, and he cites the large number of cells in the typical origin-destination matrix as evidence. Goodwin attributes the emphasis on space to the fact that the primary use of travel forecasting models was to address infrastructure issues, and infrastructure is spatially fixed and has a long life.

While the first generation travel demand models treated time very narrowly, primarily in terms of the travel time between each pair of traffic analysis zones by each of the different modes of travel (with some consideration given to the time of day of travel, in terms of the recognition of peak and off-peak periods of travel), the conceptualization of time in travel demand modeling and analysis has broadened considerably over the past 20 years. The purpose of this chapter is to examine the way in which time is conceptualized and employed in travel-demand models, and how this has changed since the first models were developed. The second section describes how, beginning in the early 1970's, the role of time in travel demand analysis and modeling has become increasingly important and broad. The third section provides an overview of the recent and on-going work in modeling time use in the context of travel behavior. This discussion shows clearly how time currently plays a central role in travel behavior modeling, at least from the perspective of the activity-based approach to travel demand analysis and modeling. The final section of the chapter presents some concluding thoughts and suggests that time could be the unifying concept around which an integrated theory of travel choice can be developed.

EMERGING FROM RELATIVE OBSCURITY: THE CHANGING VIEW AND ROLE OF TIME

This section of the chapter examines the development of the field of travel demand analysis and modeling, beginning with the first generation travel forecasting models developed in the mid-1950's in the USA, with an emphasis on the view and role of time. This examination shows that over the past 40 years time has become a much broader concept and has therefore become more important in travel demand analysis and modeling.

Time was treated very narrowly and superficially in the earliest travel-demand models that were developed and applied in the pioneering urban transportation planning studies in the USA, in such cities as Detroit, Chicago, and Pittsburg. Specifically, time was considered in two different ways in the first generation models. First, time was thought of as travel time; that is, as a measure of the spatial separation or impedance between zone centroids. As such, travel time appeared as an explanatory variable in the aggregate trip distribution, modal split, and traffic assignment models of that era (see, for example, Stopher and Meyburg, 1975). Second, the time at which trips took place was recognized in a rather coarse and *ad hoc* manner by dividing the 24-hour day into peak and off-peak periods, although the traffic assignment was typically run for a 24-hour period.

The early to mid-1970's saw the development of disaggregate (or individual) choice models, based on random utility theory. These models, typically discrete-choice, logit-type models, started to become the standard for travel-mode choice modeling in the US during the 1980's, followed by the application of nested logit models in the later 1980's. At one level, the role of time in travel demand analysis and modeling did not change very much with the introduction of the discrete-choice models in the 1970's, in that travel time is still used as one of the important components in the utility function in destination, mode, and route choice models. On the other hand, from the early to mid-1970's, the concept of time started taking on an increasingly important and broader role in relation to travel demand analysis and modeling.

The discrete-choice models developed in the early to mid-1970's have seen many applications in the context of travel-mode choice, although they have also been applied to destination and route choice. The emphasis on mode choice is probably a result of the fact that there was considerable interest at the time in the development of new public transportation systems, primarily rail, and the important question at hand was predicting the ridership on these systems. As a result, the earliest research work in the area of discrete-choice models was done in the context of mode choice, then called modal split by, for example, Lave (1968), Lisco (1967), Quarmby (1967), and Stopher (1967). As Goodwin (1991) notes, one might imagine a different development path for travel-demand models if these early works in discrete-choice modeling had emphasized time-of-day choice rather than mode choice. Goodwin goes on to suggest that had this been the case, the "four step" travel forecasting model might now comprise trip generation, time of day, destination, and route choice, and he points out that the inherent analytical or definitional problems of incorporating trip time as a discrete choice are no more difficult than those which have been solved or ignored in incorporating mode choice.

The emphasis in the 1970s and 1980s on travel-mode choice in travel demand models may have been misplaced (Goodwin, 1991) since a number of studies in the 1980s, including that by Jones et al. (1983), showed that in terms of routine traveler decision-making, time-of-day choice is more relevant than is mode-choice. Furthermore, studies have shown that travel-mode choice is often greatly influenced by scheduling considerations, in that the choice of a mode of travel for say the outbound trip to work is influenced by the traveler's plans later in the day for the return trip home.

In spite of the possibly misplaced emphasis on travel-mode choice modeling, the introduction of the individual choice models in the 1970's forced travel demand modelers to think about travel and related phenomena as decisions made by individuals and households, as opposed to the more mechanistic treatment of travel in the earlier, aggregate travel demand models. As a result, it soon became recognized that the various travel and related decisions are made at different time-scales, or over different time horizons. Specifically, travel modelers began to differentiate between (1) mobility decisions, comprising employment location, residential location, housing type, automobile ownership, and mode of travel to work, and (2) travel decisions for non-work trips, comprising frequency, destination, travel-mode, route, and time of day (Ben-Akiva, 1972).

Ben-Akiva suggested modeling the various decisions in a choice hierarchy, in which the decisions lower in the hierarchy (i.e., travel choices for non-work trips), which are made on a more short-term basis, are modeled conditional on those higher up (i.e., mobility decisions), which are made on a longer-term basis. Subsequently, Salomon (1983) extended the choice hierarchy to include "lifestyle choice," the longest-term decision, including family formation, participation in the labor force, and orientation towards leisure, with mobility and travel choices being conditional on lifestyle choice.[16]

The recognition of different time-scales for various travel and related decisions contributed to the developing awareness of the dynamic nature of travel. Thus analysts became interested in how travel behavior varies over time in response to changes in the factors that affect travel. Time was therefore beginning to be used not only as an explanatory variable, but also as a dimension to structure the conceptualization and development of travel demand analyses and models. The emerging interest in the dynamics of travel behavior in the 1980's was also stimulated by the considerable interest in dynamics in the social sciences that developed at about the same time. The interest in understanding and modeling how travel behavior varies over time was also made possible by the increasing availability of longitudinal data sets, specifically panel data sets.[17] The research in the 1980s and early 1990s into the dynamics of travel behavior represents one of the most important developments. This work built on the

---

[16] Salomon (1983) also suggested revisions to the lowest level in the choice hierarchy. Instead of the frequency, destination, mode, route, and time of day for non-work trips, the lowest level of Salomon's hierarchy deals with the activity type, activity duration, destination, route and travel-mode for non-work activities and related travel choices.

[17] I use the term to refer to any data set in which repeated measurements have been taken on the same observational units. Thus, this term includes data sets in which the observational units have, for example, reported their travel behavior at two or more periods of time, at say 6-month or 1-year intervals, and/or data sets in which the observational units have reported their travel behavior over a period of, say, 5 consecutive days.

foundations laid during a previous period of interest in longitudinal data for travel analysis (Worral, 1967).

One can identify three major areas of emphasis in the research into the dynamics of travel behavior that has taken place over the past approximately 15 years; namely, long-term dynamics (such as the effects of changes in socio-demographics and/or the urban environment on, for example, car ownership and mode choice for the work trip), day-to-day dynamics (such as, for example, variations in trip rates from day-to-day due to day-to-day variations in needs or desires, or changes in departure time and/or route in response to traffic conditions on previous days), and within-day dynamics (including the scheduling and rescheduling of activities and travel in response to time constraints or opportunities as the day progresses).

The research into the long-term dynamics of travel behavior has examined and identified the existence of a number of important dynamic phenomena, including response lags and response leads (where changes in travel behavior are made some time after, or in anticipation of, changes in, for example, employment status), habit (the same behavior is undertaken over time, even though the underlying circumstances may have changed), threshold or cumulative effects (a series of small, cumulative changes may be necessary to bring about a change in behavior), and behavioral asymmetry (the change in behavior in response to an increase in, for example, income, might not be equal and opposite to the change brought about by an equal decrease in income). The identification of the existence of these dynamic phenomena means that the assumptions about travel behavior implicitly made in the typical approach of making predictions from cross-sectional data and models are generally false (Goodwin et al., 1990).

In terms of day-to-day variability in travel behavior, the research has examined, for example, variations in the number of trips made by an individual on consecutive days. This research shows that a substantial fraction of the total variability in trip rates is due to within-person, day-to-day variability in trip rates (Koppelman and Pas, 1984; Pas and Koppelman, 1987). This research also shows that the importance of day-to-day variability varies significantly across sociodemographically defined sub-groups of the population (Pas and Koppelman, 1987). Further, this research shows that because of the existence of day-to-day variability in individuals' travel behavior, it might well be cost-effective to collect more than a single day of travel behavior data from each sample unit (Pas, 1986).

Another aspect of day-to-day variability in travel behavior is the possibility that there are cycles of behavior, such as weekly or monthly patterns. For example, one might go grocery shopping each Wednesday on a regular basis, in which case the day-to-day variability in travel behavior would be systematic. Of course, in order to examine whether such patterns of behavior exist, one needs a long sequence of multiday data. There appears to be only one data set that is suited to such analyses; namely, the 1971 Uppsala, Sweden, household travel survey (Hanson and Hanson, 1980), which obtained information on all out-of-home travel-activity behavior using self-administered travel diaries for a 35-day period. In the 1980's, Hanson and Huff (1988) conducted a series of studies of a representative sample of 149 individuals who completed diaries for the entire 35-day period to examine the day-to-day and week-to-week variability in travel behavior. When examining the systematic (nonrandom) portion of intra-individual variability in travel behavior, Hanson and Huff found that relatively few stops occurred at regular intervals in the individual's longitudinal record. Of the stops that were non-randomly distributed throughout the record, more were distributed in a clustered fashion than were distributed uniformly. They conclude that little of the day-to-day variability present in the individual's travel-activity pattern could be said to be "regularly systematic."

Mahmassani and his collaborators have studied the day-to-day dynamics of departure time choice, trip chaining, and route choice on the journey-to-work, under congested network conditions, through a series of controlled laboratory experiments and a survey of commuters in Austin, Texas (see Chapter 9 in this volume; Mahmassani, 1997; for an overview of this line of research into dynamics). They found, for example, that departure time switching is more frequent than route switching in both the laboratory and field data and they develop models to explain the daily variation in departure time and route choice. They note, however, that their results are dependent on the measures of variability used and the criterion employed to identify a "switch" in behavior. Hanson and Huff (1982) also investigated day-to-day variability in complex activity-travel patterns.

The latter part of the 1970's saw the first introduction of the notion of "time budgets" in the context of travel demand analysis and modeling.[18] Hupkes (1982)[19] first introduced the idea of constant travel time budgets in his paper entitled "The Law of Constant Travel Time and Trip-Rates." Zahavi (1977) extended Hupkes' notion of fixed travel time budgets to include a monetary budget for travel, by assuming that the percentage of income spent on travel remains fixed. Zahavi incorporated the time and cost budget idea into a travel forecasting model known as the "Unified Mechanism of Travel" (UMOT) model, which was designed to provide long-term forecasts of travel distances. The idea of a constant travel time budget was subjected to quite a lot of debate and subsequently this notion (which leads to counter-intuitive predictions with the UMOT model) was relaxed in such a way that time or money

[18] The discussion here of fixed and flexible travel budgets draws on the thorough treatment of the subject by Kraan (1996) in her dissertation.

[19] Hupkes published his work on travel budgets in Dutch in 1977 but it did not appear in English until 1982.

not spent on travel could be spent on non-travel activities or goods (Downes and Emmerson, 1985; Golob et al.; 1981).

In the early 1970's, the Swedish geographer Hägerstrand (1970) and his colleagues developed the notion of time geography which allowed for an integrated view of time and space, especially in regard to the constraints faced by human beings. These ideas were subsequently used by Burns (1979) to extend the notion of accessibility to include the temporal dimension. Burns' work showed that extending opening hours would be far more effective in increasing opportunities than would increases in travel speed. Burnett and Hanson (1979) and Pas (1980) also built on Hägerstrand's time-space framework in describing complex activity-travel patterns.

At about the same time as individual choice models were being developed and introduced to replace the aggregate travel demand models first developed and applied in the mid-1950's, researchers at the Transport Studies Unit at Oxford University began seminal work on developing a completely different framework for modeling travel behavior (Jones et al., 1983). This framework which is known as the activity-based approach to travel-demand analysis and modeling represents a paradigm shift, not simply an incremental change in the way travel is conceptualized and modeled (Pas, 1990).

The fundamental premise underlying the activity-based approach to travel demand analysis and modeling is that travel is a derived demand based on activity participation, in that travel takes place when people need or wish to undertake activities that are at different spatial locations. The idea that travel is a derived demand is not new, of course. However, before only lip service was paid to this important concept. In addition to the derived demand nature of travel, a number of other themes characterize the variety of research in the activity-based approach to travel demand analysis and modeling. These themes include the scheduling of activities, the constraints, both spatio-temporal and interpersonal, in activity and travel choices, interdependencies between the activity and travel decisions over a day or longer time period, as well as between persons, and the structure of the household and the roles played by the household members (Jones et al., 1990; Pas, 1985).

The focus on activity participation in the activity-based approach, with travel truly considered as a derived demand, brought time into a new role, since activity participation can be seen as a question of time allocation. That is, activity participation, from which the demand for travel is generated, concerns decisions about whether and for how long to participate in an activity.[20]

[20] Of course, some activities and their durations are mandatory. However, other activities are discretionary, although the time available for them is affected by the time spent on mandatory activities.

As a result, with the introduction of the activity-based approach to travel demand analysis and modeling, time began to take on the role of the dependent variable, with research into the development of models of time allocation and activity duration by researchers interested in travel behavior dating back to the 1970's (Allaman et al., 1981; Bain, 1976). Many of the other developments in the role of time that are discussed earlier in this section can also be attributed, at least to some extent, to the development of the activity-based approach. However, the central role that time is beginning to play in travel demand analysis and modeling is due to the increased recognition of the activity-based approach.

The activity-based approach to travel demand analysis and modeling has gained increasing acceptance in recent years, in response to the recognition that traditional trip-based travel forecasting methods are inadequate for addressing current issues in transportation planning and policy analysis. The move towards activity-based modeling has brought time into a key role.

Interestingly, during the period when time was assuming a more important role in travel demand analysis and modeling, time was also gaining an important role in micro economic theories of consumer choice. As Juster (1990) points out, over the past 100 years, the classical microeconomic theory of consumer choice has been broadened substantially from one in which utility was seen to be gained only from the consumption of goods and services to a theory in which consumers are seen to make choices about how they spend their time. (See Ettema, 1996; Jara-Díaz, Chapter 3 in this volume; Juster, 1990; for perspectives on these developments in microeconomic theory of consumer choice, and how time devoted to activities has become central to these theories).

Becker (1965) played a key role in the development of consumer choice theory with his formulation of a household production framework, that extended the idea of productive activity to include that households used elements of their own time as inputs to the production process, along with goods purchased in the market place, to produce commodities that were the ultimate objects of "utility." According to Becker, time could be devoted to (1) work, (2) leisure (treated as a commodity by Becker), or (3) production within the household to produce commodities (such as meals, children, and going to plays, etc.), with these commodities being the ultimate sources of utility. Becker's seminal work led to a number of extensions in consumer choice theory in the past 30 years. Specifically, one can note the work of Evans (1972), deSerpa (1971), and Train and McFadden (1978). These works are reviewed by Jara-Díaz (Chapter 3 in this volume) who presents a new model based on an extension of Evans' work.

ARRIVING AT CENTER STAGE: RECENT DEVELOPMENTS IN MODELING TIME USE AND TRAVEL BEHAVIOR

The previous section of this chapter shows clearly that the concept of time emerged from playing a relatively minor role to become increasingly important in the field. The previous section demonstrates that time is now viewed from a variety of standpoints and is incorporated in travel demand analysis and modeling in a number of different ways. The discussion below confirms that time is at the center of the stage in recent and current developments.

This section of the chapter provides an overview of recent and current developments in activity-based travel analysis and modeling with respect to the development of time use or time allocation models, in order to demonstrate the central role that time has assumed. The models discussed here deal with the time devoted to each of a set of activities over some time period, typically a day or week. Most of the models deal with time allocation by individuals, while some deal with the interdependent time allocation of members of the same household. These models are discussed in more detail by Pas (1997).

The most fundamental change in the role of time in travel demand analysis and modeling is that time use, time allocation or activity participation is now being modeled endogenously in models of travel behavior. Earliest efforts in activity-based travel research focused on analysis and descriptive modeling, but many recent efforts have been directed at the development of time allocation, as well as activity duration, models. Further, as discussed below, some of these efforts have attempted to develop joint models of time allocation and travel behavior, but in many cases these models focus on activity participation.

For the purposes of the present overview, the recent efforts to develop daily and weekly time allocation models are categorized in terms of the methodology used for the development of the model: (1) utility maximizing models, in which travelers are seen to make choices concerning the allocation of time, and sometimes money as well, to activities, in such a way as to optimize some posited objective function; (2) structural equation system models, in which systems of equations relating a large number of variables are estimated. The endogenous variables in the latter models include activity participation and often travel as well. These models allow one to examine hypotheses concerning the complex interdependencies among the endogenous variables while taking account of the effects of the exogenous variables.

Utility Maximizing Models of Time Allocation

As noted in the previous section, the classical microeconomic theory of consumer choice has been broadened substantially from one in which utility was seen to be gained only from the consumption of goods and services to a theory in which how consumers spend their time is seen as central. In concern with these developments in consumer choice theory, a number of transportation researchers have developed models of time allocation to activities based on the assumption of utility maximization. Although there have been many critiques of the assumption of utility maximization in regard to modeling travel choice, it does provide an elegant framework for modeling that leads to tractable models under the appropriate assumptions.

A number of different models of time allocation, based on the utility maximization principle, have been developed in the context of travel behavior. While these models have a common underlying framework, they differ greatly according to (1) what is included in the utility function, (2) the form of the utility function, and (3) the constraints to be satisfied. In Chapter 3 in this volume, Jara-Díaz discusses many of the early models in this category. Further reviews of utility-maximizing time-allocation models can be found in Ettema (1996), Ettema and Timmermans (1997), and Kraan (1996).

Kitamura et al. (1996) proposed a discrete-continuous choice model of time allocation to two types of discretionary activities, based on random utility maximization. The model formulation takes into account the fact that on a given day a particular type of activity might not be undertaken. The model deals with in-home and out-of-home time allocated to discretionary activities and extends the earlier work by Kitamura (1984). The model is formulated as a doubly-censored Tobit model, while requiring only the assumption that one of the two activity types is engaged in on the day in question. That is, the model formulation assumes that the person engages in some discretionary activity, either at home, out-of-home, or both. The explanatory variables in the model are work schedules, commute characteristics, as well as residential, household, and personal attributes. A weekly time-use data set from the Netherlands is used in the empirical analysis. An error component is introduced into the model to deal with the heterogeneity in the data set comprising repeated measures of daily time use for each of the respondents. The model is estimated using a non-parametric approach, using mass points.

Kitamura et al. (1996) found that individuals who work on a given day tend not to engage in discretionary out-of-home activities (this is the single most significant variable in the model). However, those who work more hours per week tend to spend a larger fraction of their discretionary time out-of-home. Individuals who spend more time commuting spend more time on in-home discretionary activities. Gender does not, by itself, seem to affect in-home/out-of-home time allocation but child rearing does. Larger households tend to be more in-home oriented. Income and number of vehicles and flexible work hours are not significant explanatory variables.

Yamamoto and Kitamura (1997) extend the Kitamura et al. (1996) discrete-continuous model of time allocation. The extension incorporated in this model deals with interactions between working and non-working days based on the expectation that there are differences in time allocation to discretionary activities on working days and non-working days.

Bhat and Misra (1997) framed the problem of weekly discretionary time allocation to either in-home or out-of-home activities (a coarse representation of the location of activities), on either weekdays or weekend days (a coarse representation of the temporal context of activities), as a utility maximization problem. They assumed that total weekly discretionary time is predetermined, so that the problem becomes one of modeling the fraction of time on each of the four activity types. Bhat and Misra found that the allocation of weekly discretionary time between weekdays/weekends and between in-home/out-of-home was affected by household and personal sociodemographics and work-related characteristics of the individual. The single most important explanatory variable in the model is age, followed by work duration on weekdays. Interestingly, Bhat and Misra found that their results support the finding by Kitamura et al. (1996) that income does not appear to have any impact on the allocation of discretionary time between in-home and out-of-home activities. They also found that income does not affect the allocation of time between weekdays and weekends.

Kraan (1996) developed a theoretical model of the allocation of time and money to travel and activities, building on the foundations of the travel budget research discussed above, and the microeconomic theories of time and cost allocation that began with Becker's work. Kraan's model's objective function takes the form of a Cobb-Douglas function and includes out-of-home activity duration, the distance traveled, the frequency of out-of-home activities, the duration at home, and the amount of money spent on goods and savings. The model comprises two constraints, one for time and one for money. Kraan conducted an extensive investigation of the properties of her model to ensure that the impacts of the various changes in the exogenous variables were logical. The empirical portion of Kraan's work was limited to time expenditures since she was unable to identify a suitable data set that incorporated both time and money expenditures. Her empirical research shows substantial differences in time allocation among segments of the population.

Jara-Díaz (Chapter 3 in this volume) proposes a theoretical model of time allocation to activities and travel, based on the utility maximizing framework. This model takes Evans' (1972) work as the starting point, but also builds broadly on the microeconomic research on travel-mode choice and time allocation modeling. Jara-Díaz's model, which includes activity participation time (including fixed and variable work hours) and travel time in the utility function, incorporates time and money constraints as well as relationships among activity participation and the purchase of goods. Jara-Díaz's model has not been operationalized as yet, but it does yield conditional demands for goods, work, and activities as intermediate results when modeling travel-mode choice.

Structural Equation System Models of Time Allocation

In spite of the fact that it is a powerful framework for modeling interrelationships among a large number of variables, the structural equation modeling methodology has seen relatively little application in travel-demand modeling until recently. The early applications of structural equation models to travel-demand modeling were conducted by Golob and his collaborators (see, for example, Golob and Meurs 1987; Golob 1990a, 1990b). Recently, other researchers have started using the same methodology to develop activity-based travel demand models, specifically models dealing with time allocation (Fujii and Kitamura, 1997), while Golob has extended the range of applications of this methodology to include time allocation modeling (Chapter 16 in this volume; Golob, Bradley, and Polak, 1996; Golob and McNally, 1997).

Fujii and Kitamura (1997) used the methodology to model commuters' time use and travel after work hours using data collected in the Osaka-Kobe metropolitan area. Interestingly, the model predicts that of a 10-minute time "savings" for the commute trip, slightly more than 7 minutes will be used for in-home activities, with only a small increase in travel. This result brings into question the idea of a constant travel time budget.

Golob and McNally (1997) developed a joint model of the amount of time spent on work, maintenance and discretionary activities out-of-the-home, and the resulting travel of couples. The research aimed at identifying the interactions between activity participation and travel and between the two individuals being modeled. This research demonstrates the existence, and provides quantitative estimates, of the effects of out-of-home activity participation on travel behavior. The results of this work also demonstrate the interdependencies between the male and female household heads in their activity participation and travel. Possible interdependencies in the activity and travel behavior of members of the same household is one of the basic concepts of the activity-based approach to travel demand modeling, but few models have been estimated that capture this effect.

Golob (Chapter 16 in this volume) used the structural equation modeling approach to model demand for activity participation and mobility and included one category of in-home activity (work) in the model. The model is formulated to allow for a number of hypothesized behavioral phenomena including: travel demand derived from activity participation, time budget effects, mobility demand (activity participation affects vehicle ownership), and accessibility (vehicle ownership affects activity participation).

In his most recent application of the structural equation methodology in the context of time allocation modeling, Golob (1997) developed and estimated a model in which trip generation (of trip chains), activity participation, and travel time are modeled endogenously and simultaneously. This model which comprises 10 endogenous and 13 exogenous variables captures the effects of in-home work on both trip chaining and out-of-home activity participation. The model was extended to include another exogenous variable, an index of accessibility, showing that accessibility is positively related to participation in out-of-home non-work activities, the generation of simple home-based trip chains for non-work purposes and more complex trip chaining. Golob concluded that joint models of trip generation and time use can reveal how the generation of simple and complex chains is interrelated with the demand for out-of-home and in-home activities, as well as revealing how the interrelationships between time use and trip generation are affected by household characteristics and residential accessibility. Further, by expanding the model to include the breakdown of travel by mode, and through adding exogenous network-based level of service variables, Golob noted that the forecasts from the model would be sensitive to link volumes and other measures of transportation system characteristics. With such extensions, the forecasts from this type of model system could be compared with those obtained from conventional travel-forecasting models.

In another recent application of the structural equation modeling methodology to time allocation and travel, Lu and Pas (1997) developed and estimated a model relating socio-demographics, activity participation (time allocated to subsistence, maintenance, leisure and other activities, separately for in-home and out-of-home activities) and travel behavior (measured by the number of trips, number of trip chains, daily travel time, and percent of trips by car) at the individual level. This research showed that a complex set of interrelationships among sociodemographics, activity participation and travel behavior exist, and can be captured by this modeling framework. Lu and Pas (1997) furthermore found that most of the estimated direct effects correspond with the findings previously reported in the literature with partial models. In addition, they found that travel behavior can be explained better by including activity participation endogenously in the model. Also, they found that relationships between in-home and out-of-home activity participation exist and can be estimated and interpreted. By examining the direct, indirect, and total effects in the model system, one can better capture and understand the relationships among sociodemographics, activity participation, and travel behavior, thereby demonstrating the usefulness of structural equations models in modeling the complicated relationships among sociodemographics, activity participation, and travel behavior. Lu and Pas (1998) also used the structural equation models framework to study the interrelationships between in-home and out-of-home activity participation in three different activity types (subsistence, maintenance and leisure) on two consecutive days of the week.

DISCUSSION AND CONCLUSIONS

This chapter has shown that while first-generation travel demand models treated time very narrowly and rather superficially, and in fact considered travel time as a measure of the spatial separation of traffic analysis zones, time has become an increasingly central concept in the field of travel demand analysis and modeling over the past 20 years. In particular, the activity-based approach to travel demand analysis and modeling has emphasized the importance of time to the understanding and modeling of travel behavior.

Time is currently recognized and incorporated in travel demand analysis and modeling in the following ways: (1) as an explanatory factor in models of destination, travel-mode, and route choice, with travel time being one of the attributes used to describe the indirect utility of the alternative choices in these models; (2) as a dimension along which we can describe when, and over what horizon, travel and related decisions are made, with recognition of within-day, day-to-day, and longer-term dynamics in travel and related behavior; (3) as a constraint on the allocation of other resources; (4) as an important dimension describing when activities and travel begin and end, that is. clock time or scheduling; and (5) as a resource to be allocated, in terms of choice of time devoted to participation in various activities and related travel.

During the period in which time was emerging from obscurity, time and activity participation were also assuming a central role in microeconomic theories of consumer choice, and hence in utility-maximizing models of consumer choice behavior. These latter developments can be traced to the seminal work by Becker (1965), which was extended by the work of a number of others including DeSerpa (1971) and Evans (1972). In Chapter 3 in this volume, Jara-Díaz describes his recent contribution to the this modeling direction, using Evans' (1972) as a starting point.

As Jara-Díaz notes, microeconomics is not concerned with the motivations that underlie activity participation. In that regard, a number of other disciplines, including sociology, psychology, and social psychology, can contribute to the further development of time allocation models (see Bhat and Koppelman, 1993; Chapin, 1974; Gärling 1992; Tonn, 1984a, 1984b for examples of other theories and conceptual frameworks concerning time allocation).

The chapter provides an overview of some of the recent and current research in modeling time use in the context of travel demand modeling, in order to illustrate the central role that time has assumed. Specifically, the chapter provides an overview of recent developments in daily and weekly time allocation modeling in the transportation field using both utility maximization and structural equation models. This overview reveals that a rich variety of such models has recently been developed, contributing to exciting new developments.

Kurani and Kitamura (1996) point to some of the multiple uses of time in travel demand models that are mentioned above, and they correctly identify that the variety of treatments of time is due, in part, to the lack of a unified conceptualization of time in activity analysis. However, rather than seeing the multiple uses of time in a negative light, the present author sees these multiple uses of time as exciting, and indicative of the important role that time could, and should, play in the development of a comprehensive theory of travel behavior. That is, it seems to this author that time could play a unifying role in the development of an integrated theory of travel behavior, a theory that would contribute greatly to the further development of travel demand models in support of policy analysis and planning studies.

While time and time use is likely to play an important role in the development of an integrated theory of travel behavior, it has also recently been suggested that time use data, analysis, and models, could play an even broader role in the next generation of transportation planning methodologies. In a recent paper, Kitamura, Pas and Fujii (1998) discuss the challenges set forth by contemporary transportation planning and policy analysis issues and they argue that time use data and models could be used to address many of the difficult current issues, ranging from induced travel to responses to travel demand management measures. They also discuss how time allocation can be employed as a measure of user benefit, thus broadening the scope of the evaluation of alternative transportation plans and policies, and they illustrate this notion with a simple example. The idea of time use being a measure of user benefit, or welfare, could no doubt enhance the central role of time in travel demand modeling, and in transportation planning and evaluation, as we enter the 21st century.

ACKNOWLEDGEMENTS

The author is grateful to the conference organizers, Duke University, and the National Institute of Statistical Sciences (NISS) for travel support to attend the very stimulating and enjoyable conference "Theoretical Foundations of Travel Choice Modeling", held in Stockholm in August 1996, at which these ideas were first presented. The content of this chapter reflect some of the discussion at the conference. Specifically, the author would like to acknowledge discussions with Chandra Bhat, Sergio Jara-Díaz, Ryuichi Kitamura, and John Polak. The conference organizers and book co-editors, Tommy Gärling, Thomas Laitila, and Kerstin Westin, are also sincerely thanked for their very helpful comments on an earlier version of this chapter. Of course, the author alone assumes all responsibility for the contents of this chapter. NISS is also gratefully acknowledged for partial support of the preparation of this chapter through an on-going National Science Foundation project entitled "Measurement, Modeling, and Prediction for Infrastructural Systems" (grant number DMS 9313013). Xuedong Lu's assistance in locating some of the references cited in this chapter is also very much appreciated.

REFERENCES

Allaman, P. A., Dunbar, F. C. and D. Steinberg (1981). *Impacts of demographic trends on time allocation for household activities*. Paper presented at the 60th annual meeting of the Transportation Research Board, Washington, DC.

Bain, J. H. (1976). *Activity choice analysis, time allocation and disaggregate travel demand modeling*. M.S. Thesis, Department of Civil Engineering, Massachusetts Institute of Technology MIT, Cambridge, MA.

Becker, G. S. (1965). A theory of the allocation of time. *The Economic Journal,* **75**, 493-517.

Ben-Akiva, M. E. (1972). *The structure of travel demand models.* Doctoral Dissertation, Massachusetts Institute of Technology MIT, Cambridge, MA.

Bhat, C. and F. S. Koppelman (1993). A conceptual framework of individual activity program generation. *Transportation Research A,* **27**, 433-446.

Bhat, C. R. and R. Misra (1997). *Discretionary activity time allocation of individuals between inhome and out-of-home and between weekdays and weekends*. Paper presented at the 8th international conference on travel behavior research, Austin, TX.

Burnett, P. and S. Hanson (1979). Rationale for an alternative mathematics approach to movement as complex human behavior. *Transportation Research Record, No.* **723**, 11-24.

Burns, L. D. (1979). *Transporation, Temporal, and Spatial Components of Accessibility.* Lexington Books, Lexington, MA.

Chapin, F. S. (1974). *Human Activity Patterns in the City.* Wiley, New York.

DeSerpa, A. (1971). A theory of the economics of time. *The Economics Journal,* **75**, 493-517.

Downes, J. D. and P. Emmerson (1985). Budget models of travel. In: *Proceedings of 13th PTRC Summer Annual Meeting - Seminar L*, pp. 63-80. London.

Ettema, D. F. (1996). *Activity-Based Travel Demand Modeling.* Doctoral Dissertation, Technical University of Eindhoven, Eindhoven, The Netherlands.

Ettema, D. F. and H. J. P. Timmermans (1997). Theories and models of activity patterns. In: *Activity-Based Approaches to Travel Analysis* (D. F. Ettema and H. J. P. Timmermans, eds.), pp. 1-36. Pergamon, Oxford.

Evans, A. (1972). On the theory of the valuation and allocation of time. *Scottish Journal of Political Economy*, **10**, 1-17.

Fujii, S. and R. Kitamura (1997). *A structural equation model system of commuters' time use and travel.* Paper presented at the 8th international conference on travel behavior research, Austin, TX.

Gärling, T. (1992). Determinants of everyday time allocation. *Scandinavian Journal of Psychology,* **33**, 160-169.

Golob, T. F. (1990a). The dynamics of household travel time expenditures and car ownership decisions. *Transportation Research A,* **24**, 443- 463.

Golob, T. F. (1990b). Structural equation modeling of travel choice dynamics. In: *Developments in Dynamic and Activity-based Approaches to Travel Analysis* (P. Jones, ed.), pp. 343-370. Avebury, Aldershot, UK.

Golob, T. F. (1997). *A simultaneous model of activity participation and trip chain generation by households*. Paper presented at the 8th international conference on travel behavior research, Austin, TX.

Golob, T. F., M. J. Beckmann and Y. Zahavi (1981). A utility-theory of travel demand incorporating travel budgets. *Transportation Research B,* **15**, 375-389.

Golob, T. F., M. A. Bradley, and J. W. Polak (1996). *Travel and Activity Participation as Influenced by Car Availability and Use*. Paper presented at the 75th annual meeting of the Transportation Research Board, Washington, DC.

Golob, T. F. and M. G. McNally (1997). A model of activity participation and travel interactions between household heads. *Transportation Research B,* **31**, 177-192.

Golob, T. F. and H. Meurs (1987). A structural model of temporal change in multi-modal travel demand. *Transportation Research A,* **21**, 391-400.

Goodwin, P. (1991). The right tools for the job: A research agenda. In: *Longer Term Issues in Transport* (J. H. Rickard and J. Larkinson, eds.), pp. 3-40. Avebury, Aldershot, UK.

Goodwin, P., R. Kitamura and H. Meurs (1990). Some principles of Dynamic analysis of travel behavior. In: *Developments in Dynamic and Activity-Based Approaches to Travel* (P. Jones, ed.), pp. 56-72. Avebury, Aldershot, UK.

Hägerstrand, T. (1970). What about people in regional science? *Papers in Regional Science,* **24**, 7-21.

Hanson, S. and J. Huff (1982). Assessing day-to-day variability in complex travel patterns. *Transportation Research Record, No.* **891**, 18-24.

Hanson, S. and P. Hanson (1980) Gender and urban activity patterns in Uppsala, Sweden. *Geographical* Review, **70**, 291-299.

Hanson, S. and J. Huff (1988). Systematic variability in repetitious travel. *Transportation* **15**, 111-135.

Hupkes, G. (1982). The law of constant travel time and trip-rates. *Futures,* **14**, 38-46.

Jones, P., M. Dix, M. Clarke and I. Heggie (1983). *Understanding Travel Behavior*. Gower, Aldershot, UK.

Jones, P., F. S. Koppelman and J.-P. Orfeuil (1990). Activity analysis: State-of-the-art and future directions. In: *Developments in Dynamic and Activity-Based Approaches to Travel* (P. Jones, ed.), pp. 34-35. Avebury, Aldershot, UK.

Juster, F. T. (1990). Rethinking utility theory. *Journal of Behavioral Economics,* **19**, 155-179.

Kitamura, R. (1984). A model of daily time allocation to discretionary out-of-home activities and trips. *Transportation Research B,* **18**, 255-266.

Kitamura, R., E. I. Pas and S. Fujii (1998). Time-use data, analysis and modeling: toward the next generation of transportation planning methodologies. *Transport Policy*. In press.

Kitamura, R. T. Yamamoto, S. Fujii and S. Sampath (1996). A discrete-continuous analysis of time allocation to two types of discretionary activities which accounts for unobserved heterogeneity. In: *Transportation and Traffic Theory* (J.-B. Lesort, ed.), pp. 431-453. Elsevier, Oxford.

Koppelman, F. S. and E. I. Pas (1984). Estimation of disaggregate regression models of person trip generation with multiday data. In: *Proceedings of the Ninth International Symposium on Transportation and Traffic Theory* (J. Volmuller and R. Hamerslag, eds.), pp. 513-529. VNU Science Press, Utrecht, The Netherlands.

Kraan, M. (1996). *Time to travel? A model for the allocation of time and money*. Doctoral Dissertation, Department of Traffic and Transport Management, University of Twente, Enschede, the Netherlands.

Kurani, K. S. and R. Kitamura (1996). *Recent developments and the prospects for modeling household activity schedules*. Technical Report, Institute of Transportation Studies, University of California, Davis, CA.

Lave, C. A. (1968). *Modal choice in urban transportation*. Doctoral Dissertation, Department of Economics, Stanford University, Palo Alto, CA.

Lisco, T. E. (1967). *The value of commuters' travel time: A study in urban transportation*. Doctoral Dissertation, Department of Economics, University of Chicago, IL.

Lu, X. and E. I. Pas (1997). *An examination of activity time allocation on two consecutive days.* Paper presented at the 8th International conference on travel behavior research, Austin, TX.

Lu, X. and E. I. Pas (1998). *A structural equation model of the relationships among socio-demographics, activity participation and travel behavior.* Unpublished manuscript, Department of Civil and Environmental Engineering, Duke University, Durham, NC.

Mahmassani, H. S. (1997). Dynamics of commuter behavior: recent research and continuing challenges. In: *Understanding Travel Behavior in an Era of Change.* (P. R. Stopher and M. Lee-Gosselin, eds.), pp. 279-314. Pergamon, Oxford.

Pas, E. I. (1980). *Toward the understanding of urban travel behavior through the classification of daily travel/activity patterns.* Doctoral Dissertation, Department of Civil Engineering, Northwestern University, Evanston, IL.

Pas, E. I. (1985). State-of-the-art and research opportunities in travel demand: Another perspective. *Transportation Research A,* **19**, 460-464.

Pas, E. I. (1986). Multiday samples, parameter estimation precision, and data collection costs for least squares regression trip-generation models. *Environment and Planning,* **18**, 73-87.

Pas, E. I. (1990). Is travel demand analysis and modeling in the doldrums? In: *New Developments in Dynamic and Activity-Based Approaches to Travel Analysis* (P. Jones, ed.), pp. 3-27. Avebury, Aldershot, UK.

Pas, E. I. (1997). *Time use and travel demand modeling: recent developments and current challenges.* Paper presented at the 8th International conference on travel behavior research, Austin, TX.

Pas, E. I. and F. S. Koppelman (1987). An analysis of the determinants of day-to-day variability in individual's travel behavior. *Transportation,* **14**, 3-20.

Quarmby, D. A. (1967). Choice of travel mode for the journey to work: Some findings. *Journal of Transport Economics and Policy,* **1**, 273-314.

Salomon, I. (1983). Lifestyles - A broader perspective on travel behavior. In: *Recent Advances in Travel Demand Analysis* (S. Carpenter and P. Jones, eds.), pp. 290-310. Gower, Aldershot, UK.

Stopher, P. R. (1967). *A behavioral approach to modal split forecasting.* Doctoral Dissertation, Department of Civil Engineering, University of London, London.

Stopher, P. R. and A. H. Meyburg (eds.) (1975). *Urban Transportation Modeling and Planning.* Lexington Books, Lexington, MA.

Tonn, B. E. (1984a). The cyclic process decision-heuristic: an application in time-allocation modeling. *Environment and Planning A,* **16**, 1197-1220.

Tonn, B. E. (1984b). A sociopsychological contribution to the theory of individual time-allocation. *Environment and Planning A,* **16**, 201-223.

Train, K. and D. McFadden (1978). The goods/leisure trade-off and disaggregate work trip mode choice models. *Transportation Research,* **12**, 349-353.

Worrall, R. D. (1967). The urban panel as a longitudinal data source. *Highway Research Record, No. 194,* 62-77.

Yamamoto, T. and R. Kitamura (1997). *An analysis of time allocation to in-home and out-of-home activities across working days and non-working days*. Unpublished manuscript, Department of Civil Engineering, Kyoto University, Japan.

Zahavi, Y. (1977). *The "UMOT" model*. Technical Report, Urban Projects Department, The World Bank, Washington, DC.

11

TWO COMPUTATIONAL PROCESS MODELS OF ACTIVITY-TRAVEL CHOICE

Ryuichi Kitamura, Kyoto University, Japan
Satoshi Fujii, Kyoto University, Japan

INTRODUCTION

The ultimate goal of travel choice analysis is to gain a full understanding of why people travel and to develop quantitative capabilities that facilitate the prediction of travel behavior. Toward this end, the individual's travel choice has been examined in connection with his or her activity engagement, with the recognition that travel demand is a derived demand. Studies from this viewpoint are referred to as activity-based analyses of travel choice (for reviews, see Axhausen and Gärling, 1992; Damm 1983; Ettema, 1996; Gärling et al., 1994; Jones et al., 1983; Jones et al., 1990; Kitamura, 1988; and Kurani and Kitamura, 1996). Along with the recent developments of micro-simulation model systems, the activity-based analysis is now entering the stage of large-scale application to travel demand forecasting and policy analysis. This is of particular significance because of shifts in the focus of urban transportation planning.

One shift is away from facility construction toward transportation systems management (TSM) and toward travel demand management (TDM) – attempts to resolve transportation problems by better operating existing facilities and managing travel demand, rather than building more facilities. Tools currently available for passenger travel demand forecasting and policy analysis are mostly derived from the trip-based, four-step procedure, developed in the 1950's and 1960's when urban population was growing rapidly, motorization was progressing, and suburban sprawling was starting. Not surprisingly, these tools, geared towards the planning needs of decades ago, are not well suited to address current planning issues. In particular, they are mostly not applicable to the evaluation of the effectiveness of

TDM measures that are extensive, sophisticated, and fine-tuned to target specific traveler segments.

Another shift is the recent emphasis on the environmental impact of vehicular transportation. Increasing skepticism surrounds the notion that pollutant emissions from road traffic can be reduced by expanding road capacity and thereby improving traffic flow. At the same time, there have been increasing concerns with new trips induced by capacity addition. If travel demand follows the economic demand-supply relationship and if the primary cost component of travel is the time cost, then reduced travel time due to capacity addition would increase the demand for travel; people would then travel more often and for longer distances.

If an increase of transportation capacity induces new trips, then they must be accurately reflected in the environmental impact assessment; otherwise environmental impacts will always be underestimated. No effective methods, however, have been established for the assessment of the magnitude and characteristics of induced trips, especially for induced trips in the narrower sense, that is, those trips that were not made before but are newly made due to improved travel condition. While one may be tempted to develop an aggregate demand function that expresses travel demand as a function of network service levels, it is doubtful if such a function can fully represent various constraints and thresholds that govern individuals' activity engagement and trip making.

Quite important is to recognize that capacity addition and TDM and other current policy measures both have far-reaching impacts on urban residents' daily life. This is more so with TDM measures than with traditional, supply-oriented measures. For example, flexible work hours, compressed work weeks, or telecommuting all affect how a worker allocates time to different activities within each day and across weekdays and weekend days. Trip-based models are inadequate for the assessment of the effectiveness of such measures. Activity-based models are called for.

One of the approaches taken to modeling activity engagement and scheduling behavior is to apply computational process models, of which production systems are one example, to the problem of how people formulate and execute schedules. The approach is defined in Kurani and Kitamura (1996) as:

> "Tasker and Axhausen (1994) describe the computational process approach as one in which the decision maker '... tries a sequence of possibilities and selects the first one that is suitable.' They differentiate between such models based on a search process from utility based models, stating:
> 'Firstly, the search process is a computational process, not a mathematical function. Thus, the order in which heuristic functions for various modes in the tree

> are evaluated affects the way the search is conducted, and its final result. Calculations involving utility, however, optimize in a sequence independent manner...
> Secondly, heuristics can be represented, not just as a scalar, but as a vector of quantities (time, money, safety, comfort, social obligation, etc., along with risk elements for each of these where appropriate), ..."

Production models were developed by Newell and Simon (1972) as models of human problem solving. The production system consists of, or condition-action pairs that specify an action to be executed when a condition is met. Computational process models are themselves an example of a physical symbol system:

> ... capable of inputting, outputting, storing, and modifying symbol structures, and of carrying out some of these actions in response to the symbols themselves. 'Symbols' are any kinds of patterns on which these operations can be performed, where some of the patterns denote actions (that is, serve as commands or instructions) (Simon, 1990).

Presented in this paper are two computational process models of activity-travel behavior: the Prism-Constrained Activity-Travel Simulator (PCATS) and Activity-Mobility Simulator (AMOS). PCATS (Kitamura et al., 1996a) is a model system of activity engagement and travel, which incorporates the concept of Hägerstrand's time-space prism (Hägerstrand, 1970) and generates activities and trips within prisms. Coupling constraints in an individual's daily itinerary are first identified in PCATS and prisms are established in the time-space dimension. The choices of activity type, duration, location, and travel mode are then simulated sequentially while considering the prism constraints. Model components to simulate these choices are estimated using the results of a time-use survey conducted in the Osaka-Kobe metropolitan area of Japan, where PCATS has been implemented using network and land use data from the area. The impacts on activity-travel behavior of (a) increased travel time due to worsening congestion, and (b) change in work schedules, are assessed using the model system.

AMOS (Kitamura et al., 1993, 1995, 1996b), on the other hand, focuses on the individual's adaptation behavior. It is a simulation model system which has been developed primarily to evaluate the effectiveness of TDM measures, by replicating how a traveler would modify her activity and travel when a change takes place in the travel environment, for instance, an increase in auto travel cost and reduction in travel time due to congestion pricing. AMOS takes an observed daily travel pattern of an individual; generates an adaptation option (e.g., change commute travel mode) that may be adopted by the individual when faced with the change in the travel environment; adjusts the baseline pattern (e.g., re-sequence activities, select new destinations) to produce a modified activity-travel pattern; evaluates the utility of the modified pattern; based on a satisficing rule, accepts one of the modified patterns so far generated and terminates the search, or continues to search for alternatives. An AMOS prototype has been implemented in the Washington, D.C., metropolitan area (Kitamura et al., 1995; RDC, 1995). Because all TDM measures considered in the implementation project are aimed at commuters, the current prototype is formulated for commuters only.[21]

In the next section, the outline of PCATS is presented. This is followed by the presentation of scenario analysis results. A brief summary of PCATS is given. The components of AMOS are described next. Discussed thereafter are AMOS model implementation efforts and results of scenario analyses to evaluate TDM effectiveness. A summary of AMOS is given next. A final section offers conclusions.

OVERVIEW OF PCATS[22]

The individual's activity engagement and travel are simulated by PCATS within constraints as embodied by Hägerstrand's prisms. In defining prisms for each individual, it is assumed that the simulation period, say a day, can be divided into periods of two types: open periods and blocked periods. Open periods are ones in which the individual has the option of traveling and engaging in activities. In blocked periods, on the other hand, the individual is committed to engage in certain activities at certain locations. Activities participated within a blocked period will be called fixed activities; those pursued in an open period flexible activities.[23] Given the speed of travel, the ending time and location of a blocked period and the beginning time and location of the subsequent blocked period together define a time-space prism in which the individual's activity and travel must be contained.

It is assumed that the individual's activity decision is dependent on the past, but not dependent on the future, except for the presence of prism constraints. It is further assumed that

[21] For details, see RDC (1995) and Pendyala et al. (1996). AMOS is currently extended to include both commuters and non-commuters.

[22] The description of PCATS draws on Kitamura et al. (1996a).

[23] The activity categories used in PCATS are: sleep, personal care (other than taking bath), taking bath, child care, meal, domestic chore, work and work-related, school and study, social, grocery shopping, comparison shopping, hobbies and entertainment, sports and exercises, TV viewing, reading, resting, medical and dental, and others. A set of assumptions are adopted to determine whether an activity is fixed or flexible. Sleep is always classified as a fixed activity. Personal care (other than taking bath), taking bath, TV viewing, reading, and resting, on the other hand, are always classified as flexible. Activities of the remaining types are classified as fixed if the respondent indicated in the survey that the activity was subject to both temporal and spatial constraints; otherwise they are regarded to be flexible.

activity engagement decision is made sequentially, conditioned upon past activity engagement. This depiction of decision structure in PCATS is based on the identity discussed below.

Overall Structure

The individual's decision within a open period is depicted in PCATS as shown in Figure 11.1. The characterization found is based on a decomposition of activity-travel decisions into a series of conditional choices. As aspects of activity-travel decisions, consider activity type, activity duration, location, and travel mode (if applicable). Let

$\boldsymbol{X}$	$= (X_0, X_1, \ldots, X_k)$	= type of activity
$\boldsymbol{D}$	$= (D_0, D_1, \ldots, D_k)$	= duration of activity
$\boldsymbol{L}$	$= (L_0, L_1, \ldots, L_k)$	= location of activity
$\boldsymbol{M}$	$= (M_0, M_1, \ldots, M_k)$	= mode of travel used to reach the location of activity
k	= total number of activities during the study period	

Then the vectors, $(\boldsymbol{X}, \boldsymbol{D}, \boldsymbol{L}, \boldsymbol{M})$, define the individual's activity-travel patterns, starting with the initial condition, (X_0, D_0, L_0, M_0), and ending with the final activity, (X_k, D_k, L_k, M_k). If one can define $\Pr[\boldsymbol{X}, \boldsymbol{D}, \boldsymbol{L}, \boldsymbol{M}]$ for any $(\boldsymbol{X}, \boldsymbol{D}, \boldsymbol{L}, \boldsymbol{M})$, then complete probabilistic characterization of activity-travel patterns is achieved. This probability, however, is overly complex. The approach taken in PCATS is to adopt the following decomposition:

$$\begin{aligned}\Pr[\boldsymbol{X}, \boldsymbol{D}, \boldsymbol{L}, \boldsymbol{M}] \equiv\ & \Pr[X_k, D_k, L_k, M_k | \tilde{\boldsymbol{X}}_{k-1}, \tilde{\boldsymbol{D}}_{k-1}, \tilde{\boldsymbol{L}}_{k-1}, \tilde{\boldsymbol{M}}_{n-1}] \\ & \times \Pr[X_{k-1}, D_{k-1}, L_{k-1}, M_{k-1} | \tilde{\boldsymbol{X}}_{k-2}, \tilde{\boldsymbol{D}}_{k-2}, \tilde{\boldsymbol{L}}_{k-2}, \tilde{\boldsymbol{M}}_{n-2}] \times \ldots \\ & \times \Pr[X_1, D_1, L_1, M_1 | X_0, D_0, L_0, M_0]\end{aligned}$$

where

$$\tilde{\boldsymbol{X}}_{n-1} = (X_0, X_1, X_2, \ldots, X_{n-1})$$
$$\tilde{\boldsymbol{D}}_{n-1} = (D_0, D_1, D_2, \ldots, D_{n-1})$$
$$\tilde{\boldsymbol{L}}_{n-1} = (L_0, L_1, L_2, \ldots, L_{n-1})$$
$$\tilde{\boldsymbol{M}}_{n-1} = (M_0, M_1, M_2, \ldots, M_{n-1})$$

Thus, the simultaneous probability associated with $(\boldsymbol{X}, \boldsymbol{D}, \boldsymbol{L}, \boldsymbol{M})$ is decomposed into a series of conditional probabilities associated with (X_n, D_n, L_n, M_n), given the past history, $(\tilde{\boldsymbol{X}}_{n-1}, \tilde{\boldsymbol{D}}_{n-1}, \tilde{\boldsymbol{L}}_{n-1}, \tilde{\boldsymbol{M}}_{n-1})$, $n = 1, 2, \ldots, k$.

To be exemplified with the time of day, consider the following expression:

$$\Pr[X_n, D_n, L_n, M_n | t_{n-1}, \widetilde{\boldsymbol{X}}_{n-1}, \widetilde{\boldsymbol{D}}_{n-1}, \widetilde{\boldsymbol{L}}_{n-1}, \widetilde{\boldsymbol{M}}_{n-1}, \widetilde{\boldsymbol{R}}_{n-1}], \; n = 1, 2, ..., k$$

where

t_{n-1} =the completion time of the $(n-1)$-th activity

$\widetilde{\boldsymbol{R}}_{n-1}$ =$(R_0, R_1, R_2, ..., R_{n-1})$

R_n =the travel time to reach the n-th activity location (0 if no travel is involved).

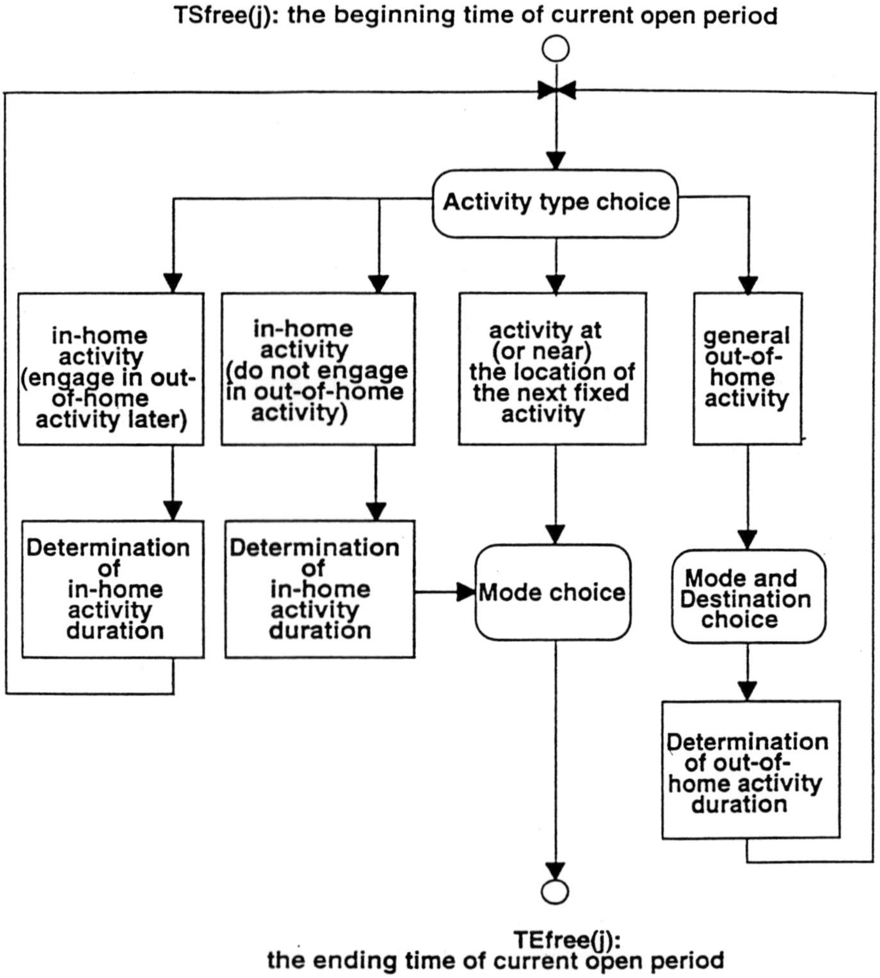

Figure 11.1. PCATS activity type choice model for simulation of activity and travel in the open period.

This probability is unity if the n-th activity is in a blocked period. No attempt is required to develop a model for the probability in this case. The initial state at time t_0 is defined by (X_0, D_0, L_0, M_0, R_0). The first activity may be the continuation of the 0-th activity (this is an exception and no subsequent activity for $n = 2, 3, ..., k$, will be a continuation of the preceding activity).

R_n is not treated as a random variable in PCATS; it is assumed that the travel time is deterministic given the travel mode used and the origin and destination of the trip (M_n, L_{n-1}, and L_n). The activity completion time, t_{n-1}, can be uniquely determined given $\tilde{\boldsymbol{D}}_{n-1}$ and $\tilde{\boldsymbol{R}}_{n-1}$.

The joint probability associated with the attributes of the n-th activity can be further decomposed into many (in this case 4! = 24) different sequences of conditional probabilities. For example,

$$
\begin{aligned}
&\Pr[X_n, D_n, L_n, M_n | t_{n-1}, \tilde{\boldsymbol{X}}_{n-1}, \tilde{\boldsymbol{D}}_{n-1}, \tilde{\boldsymbol{L}}_{n-1}, \tilde{\boldsymbol{M}}_{n-1}, \tilde{\boldsymbol{R}}_{n-1}] \\
&\quad = \Pr[X_n | t_{n-1}, \tilde{\boldsymbol{X}}_{n-1}, \tilde{\boldsymbol{D}}_{n-1}, \tilde{\boldsymbol{L}}_{n-1}, \tilde{\boldsymbol{M}}_{n-1}, \tilde{\boldsymbol{R}}_{n-1}] \\
&\quad \times \Pr[D_n | X_n; t_{n-1}, \tilde{\boldsymbol{X}}_{n-1}, \tilde{\boldsymbol{D}}_{n-1}, \tilde{\boldsymbol{L}}_{n-1}, \tilde{\boldsymbol{M}}_{n-1}, \tilde{\boldsymbol{R}}_{n-1}] \\
&\quad \times \Pr[L_n | X_n, D_n; t_{n-1}, \tilde{\boldsymbol{X}}_{n-1}, \tilde{\boldsymbol{D}}_{n-1}, \tilde{\boldsymbol{L}}_{n-1}, \tilde{\boldsymbol{M}}_{n-1}, \tilde{\boldsymbol{R}}_{n-1}] \\
&\quad \times \Pr[M_n | X_n, D_n, L_n; t_{n-1}, \tilde{\boldsymbol{X}}_{n-1}, \tilde{\boldsymbol{D}}_{n-1}, \tilde{\boldsymbol{L}}_{n-1}, \tilde{\boldsymbol{M}}_{n-1}, \tilde{\boldsymbol{R}}_{n-1}] \\
&\quad = \Pr[D_n | t_{n-1}, \tilde{\boldsymbol{X}}_{n-1}, \tilde{\boldsymbol{D}}_{n-1}, \tilde{\boldsymbol{L}}_{n-1}, \tilde{\boldsymbol{M}}_{n-1}, \tilde{\boldsymbol{R}}_{n-1}] \\
&\quad \times \Pr[X_n | D_n; t_{n-1}, \tilde{\boldsymbol{X}}_{n-1}, \tilde{\boldsymbol{D}}_{n-1}, \tilde{\boldsymbol{L}}_{n-1}, \tilde{\boldsymbol{M}}_{n-1}, \tilde{\boldsymbol{R}}_{n-1}] \\
&\quad \times \Pr[M_n | D_n, X_n; t_{n-1}, \tilde{\boldsymbol{X}}_{n-1}, \tilde{\boldsymbol{D}}_{n-1}, \tilde{\boldsymbol{L}}_{n-1}, \tilde{\boldsymbol{M}}_{n-1}, \tilde{\boldsymbol{R}}_{n-1}] \\
&\quad \times \Pr[L_n | D_n, X_n, M_n; t_{n-1}, \tilde{\boldsymbol{X}}_{n-1}, \tilde{\boldsymbol{D}}_{n-1}, \tilde{\boldsymbol{L}}_{n-1}, \tilde{\boldsymbol{M}}_{n-1}, \tilde{\boldsymbol{R}}_{n-1}], \text{ etc.}
\end{aligned}
$$

The identities imply that there will be many different sequential models for (X_n, D_n, L_n, M_n), and that these sequential models will all offer in principle the same statistical fit to the data. Then a particular sequential model may be preferred and selected considering theoretical support, policy sensitivity, and ease of modeling. The sequence adopted in the development of PCATS is $X_n \rightarrow L_n \rightarrow M_n \rightarrow D_n$.

Activity Type Choice Model

The activity type choice model is formulated as a nested-logit model with a two-tier structure. Exactly which alternatives can be included in the choice set is determined considering prism constraints. Namely, the formation of choice sets in PCATS is governed in part by prism constraints. In the first (upper) tier, one of the following three broad classes of activities is chosen: in-home activity, activity at (or near) the location of the next fixed activity, and general out-of-home activity. The second tier under "in-home activity" includes engage in out-of-home activity subsequently, and do not engage in out-of-home activity within the current open period. If the former is the case, then the duration of the in-home activity is determined, and the activity choice model is applied again to simulate the next, out-of-home activity, with the "in-home activity" alternative excluded from the choice set. If the latter is the case, then the travel to the location of the next fixed activity is simulated. Likewise, if the option of "activity at (or near) the location of the next fixed activity" is selected in the first tier, then the travel to the next fixed location will be simulated.

If "general out-of-home activity" is chosen, then the activity type is selected in the second tier. Activities classified into the following six activity types comprise the choice set in the second tier: meal, social, grocery shopping, comparison shopping, hobbies and entertainment, and sports and exercises.

The explanatory variables used to model the choice of out-of-home activity type include personal attributes: age, sex, home-maker or not, time of day, and probability that the activity duration fits within the open period.

The last variable is defined as $\Pr[Z \leq D_f | X_n = j] = F_j(D_f | \mathbf{S}_{ji})$

where

Z = activity duration
D_f = $T_s - t_{n-1} - R_s$
T_s = the beginning time of the next fixed activity
R_s = travel time to the location of the next fixed activity using the fastest travel mode that is available
F_j = the cumulative distribution function of activity durations for activity type j, and
$\mathbf{S}_{ji}$ = a vector of explanatory variables for activity type j and individual i.

Coefficient estimates of this variable indicate the anticipated tendency that activities which are less likely to fit within an open period tend not to be pursued (see Otsuka, 1996).

Destination and Mode Choice Model

The destination and mode choice model is also formulated as a nested-logit model. The first tier concerns the choice of destination, and the second tier the conditional choice of travel mode given the destination. In the current version of PCATS, one model is applied to all trips; this is restrictive and in future models will be differentiated by trip purpose. Municipalities are used as the unit of geographical aggregation in this study. Travel modes are classified into public transit, automobile, bicycle, and walk. The uncertainty in the choice set is represented during the model estimation using the method outlined in Ben-Akiva and Lerman (1985). The explanatory variables used in the model of destination choice are: zonal population, number of commercial establishments, intra-zone destination dummy, the possible minimum travel time to the destination zone, then to the location of the next fixed activity, and probability that the activity duration will fit within the open period given the activity is pursued at the destination zone, and in the conditional mode choice: person and household attributes (age, sex, employment status, driver's license holding, household income, number of vehicles available); time of day; travel time and cost by mode, and number of transfers; intra-zone trip dummy, and location type indicator (indicators of the combination of the current location type and the location of the next fixed activity).

This model is used in PCATS to generate a destination and mode for each trip. As is the case for activity choice, only those destination-mode pairs that are feasible under prism constraints and other coupling constraints (primarily for auto availability) are included in the choice set.

Activity Duration Models

The distribution of durations of flexible activities is determined by activity type, assuming
The distribution of durations of flexible activities is determined by activity type, assuming
that the parameters of the distribution (the mean and a shape parameter) are a function of personal attributes and other explanatory variables. Weibull distributions are exclusively used in the current version of PCATS. The explanatory variables used in the duration model are: person and household attributes (age, sex, employment, household income, driver's license holding, number of vehicles available) past activity engagement (the cumulative amount of time spent on the same type of activity since t_0) time of day (the mid-point of the time when the activity [including travel] starts [(t_{n-1})] and the beginning time of the next fixed activity) the time in-home activity can start if returned home; time availability (the amount of time between the completion time of the previous activity and the beginning time of the next fixed activity; and location type indicator (indicators of the combination of the current location type and the location type of the next fixed activity).

The activity duration models described above are used to determine the duration of the activity, given its type, location, the mode used to reach the activity location, and while considering prism constraints. The maximum possible activity duration is first determined based on the size of the prism, which is a function of the speed of travel, the location of the trip origin, the location of the activity, and the location of the next fixed activity. Then the distribution as given by the duration model for the activity type is truncated at the maximum, that is, a probability mass equaling to the probability that the activity duration will exceed that maximum is placed at the maximum. The resulting mix distribution is used to generate activity durations in the simulation.

Validation

A validation analysis was conducted with the intent of evaluating how well the simulation system replicates observed activity and travel patterns. In the analysis expected values obtained from the PCATS simulation are compared against observed values for several indicators of activity-travel patterns. Expected values are obtained by performing 100 simulation runs for each sample individual and taking the average of the results over a total of 374 sample individuals whose activity records are complete. Table 11.1 summarizes the results.

Table 11.1 indicates that total travel time, in-home flexible activity duration, and number of trips are relatively well represented by the simulation. The *t*-statistics indicate, however, that at $\alpha = 0.05$ predicted values and observed values are significantly different for all indicators. n particular, number of non-work destinations and number of non-work trip chains have very small correlation coefficients. These discrepancies between the observation and prediction point to possible deficiencies in the model components, especially the activity type choice models.

Table 11.1.
Results of the Validation Study

	Predicted		Observed			
	Mean	SD	Mean	SD	t	r^2
Total travel time	116.3	70.7	127.9	87.1	-2.00	0.622
In-home flexible activity duration	314.5	152.9	288.7	191.0	2.04	0.673
Out-of-home flexible activity duration	28.4	72.3	39.6	75.6	-2.07	0.329
Number of non-work destinations	0.071	0.61	0.31	0.58	-5.42	0.169
Number of non-work trip chains	0.059	0.28	0.013	0.11	2.86	-0.027
Number of trips	2.89	1.56	3.38	1.79	-4.00	0.576

SD: standard deviation across sample individuals
t: t-statistics associated with the difference between the predicted and observed values (not based on the standard deviations associated with "predicted" values)
r^2: Squared Pearson correlation coefficient between predicted and observed values

Scenario Analysis

PCATS is applied in this section to assess how changes in the travel environment may affect an individual's activity and travel. To this end, a sample individual's activity and travel after the completion of work activities are simulated using PCATS for the scenarios shown in Table 11.2. The sample individual has the following profiles: An employed male of 54 years old; a household income in the 1,500,000 to 2,000,000 yen range; has held a driver's license for 30 years; one vehicle available to the household; commutes to CBD Osaka; lives in Kaizuka City which is approximately 30 km to the south from Osaka along the Osaka Bay and has good freeway access to the CBD.

The individual is assumed to be at the work location when work ends (which is assumed to represent the ending point of a blocked period), and the next blocked period is assumed to begin at midnight. It is thus assumed in the simulation that the entire evening period, after work till midnight, is an uncommitted block of time. Simulation is repeated 100 times for each scenario.

The results of the simulation runs are summarized in Table 11.3. Since static travel time is used in the simulation, there is no random element in travel time (and therefore in in-home time) for the first travel pattern (W-H) where the individual returns home immediately after work and engages in no out-of-home activity. In the base case, out-of-home activities are engaged in 16 of the 100 runs (16%). The mean out-of-home activity duration is 109 minutes for the W-O-H pattern where out-of-home activities are engaged during the commute to home, while the mean duration is much shorter 52 minutes in the W-H-O-H pattern where activities are engaged in separate home-based trip chains.

The frequency of the simple W-H pattern increases to 91, 89 and 90, respectively, in the three scenarios. Quite notable in Scenario 1, where work ending time is moved to 6:00 PM, is the substantial reduction in out-of-home activity duration and the slight reduction in travel time

Table 11.2
Scenarios Used in the Simulation Analysis

Scenario	Description
Base case	Work ends at 5:00 PM. A car is used to commute.
Scenario 1	Work ends at 6:00 PM. A car is used to commute.
Scenario 2	Work ends at 5:00 PM. Public transit is used to commute.
Scenario 3	Work ends at 5:00 PM. Car commute takes extra 30 min.

Table 11.3
Results of Scenario Simulation with a Sample Individual

		After-work travel pattern[a]			
		W-H	W-O-H	W-H-O-H	Other
Base case	Frequency	84	8	7	1
	Travel time[b]	51	122	160	
	In-home time[b]	369	188	208	
	Out-of-home time[b]	0	109	52	
Scenario 1	Frequency	91	5	4	0
	Travel time	51	114	184	
	In-home time	309	177	115	
	Out-of-home time	0	69	62	
Scenario 2	Frequency	89	6	5	0
	Travel time	79	135	180	
	In-home time	341	190	155	
	Out-of-home time	0	94	85	
Scenario 3	Frequency	96	6	2	2
	Travel time	81	136	214	
	In-home time	339	195	178	
	Out-of-home time	0	89	29	

[a] W-H: work →home. W-O-H: work →other →home. W-H-O-H: work →home →other →home
[b] In minutes. Out-of-home time excludes travel time.

associated with the W-O-H pattern. In-home activity time does not show very much change. The shortening of the after-work open period caused by the change in work ending time has prompted the individual to engage in out-of-home activities less frequently. When the W-O-H pattern is engaged, the activity location is closer and the activity duration is much shorter, presumably to accommodate the tighter time constraints. These tendencies are not found for the W-H-O-H pattern, however. Yet, it is cautioned that the frequency of out-of-home activity engagement is small in the simulation results and the statistics presented under the W-O-H and W-H-O-H patterns contain large variations.

Similar reductions in out-of-home activity engagement can be found for Scenarios 2 and 3. The mean travel times associated with pattern W-O-H exhibit increases of less than 15 minutes from the base case, while the activity times decrease by 15 to 20 minutes. Much larger changes are associated with the W-H-O-H pattern. However, this is at least in part due to the small sample size.

Although the model system is still in its early stage of development and the analysis performed here is not extensive, this scenario analysis has demonstrated that PCATS facilitates the analysis of time-oriented policies such as changes in work schedules while explicitly considering time-space constraints in the analysis. PCATS also represents the repercussions of a change in the travel environment, including induced (or suppressed) travel and changes in activity location and duration.

Summary

PCATS is still in its early stage of development. There are many areas where development, extension, and refinement are needed. Nevertheless, it can be concluded that the study has demonstrated that activity-travel behavior in time-space prisms can be simulated reasonably well and that travelers' responses to changes in travel time or work schedules can be examined using the micro-simulation model system. The PCATS model system is readily applicable to other types of scenarios, such as changes in store hours or extended operating hours of public transit which are difficult to address with the conventional trip-based models which do not incorporate the time dimension and disregard time-space constraints.

A future task, is to incorporate into PCATS behavioral assumptions conceiving activity engagement. The "utility-maximizing," nested-logit model of activity type choice in PCATS captures the salient tendencies associated with activity type choice; however, it hardly captures the reason for activity engagement, or does it embody a behavioral mechanism of activity engagement. Effort is currently ongoing to introduce into PCATS the notion of time allocation to various activities, thus capturing the individual activity-travel scheduling effort as a resource allocation problem under prism constraints.

OVERVIEW OF AMOS

AMOS is a model which predicts changes in travel behaviour that will follow a change in the travel environment. Its development has been motivated by the recognition that the traditional, trip-based, four-step procedures are incapable of addressing the problems raised by TDM and other policy measures that are now the primary focus of urban transportation planning. An AMOS prototype has been developed and implemented in the Washington DC, metropolitan area with the intent of predicting traveler responses to selected TDM measures. The TDM measures considered for evaluation include: parking surcharge; bicycle/pedestrian facility improvements; parking pricing with employer-paid voucher; congestion pricing; the combination of the first and second measures; and the combination of the third and fourth measures. AMOS comprises five main components and a reporting routine (see Figure 11.2). More detailed discussions can be found in RDC (1995), Kitamura et al., (1995); and Pendyala et al., 1996).

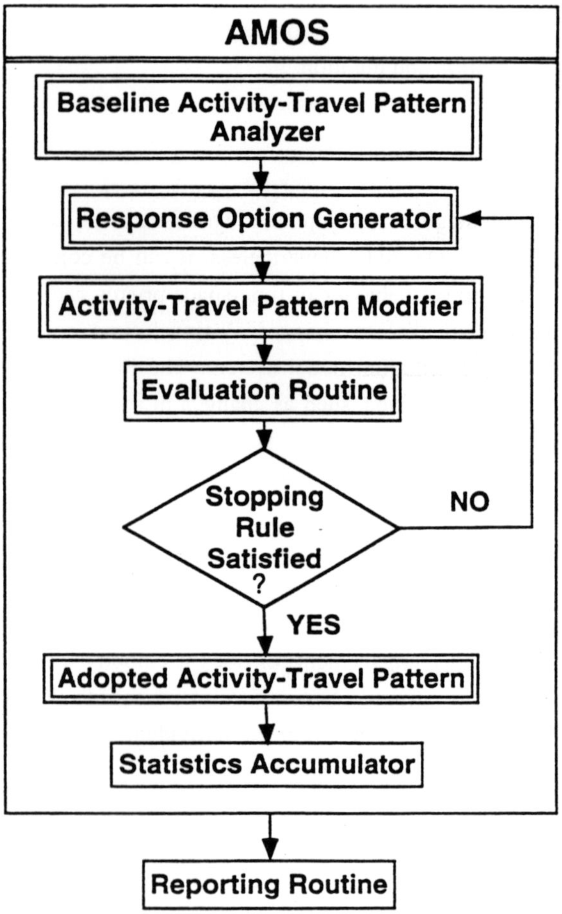

Figure 11.2. Outline of activity-mobility simulator (AMOS).

Baseline Activity-Travel Pattern Analyzer

The Baseline Activity-Travel Pattern Analyzer is summarized in Figure 11.3. It inspects daily travel diary data and determines whether they are complete with all trips and pertinent information intact. It also checks whether the sample trip maker and/or the travel pattern fall in the categories intended for analysis. Another major function the Analyzer performs is to develop indicators of travel pattern characteristics (e.g., whether or not there is a stop during the commute trip) that feed into the Response Option Generator. Trip diary data used are those typically available from metropolitan planning organizations (MPO's).

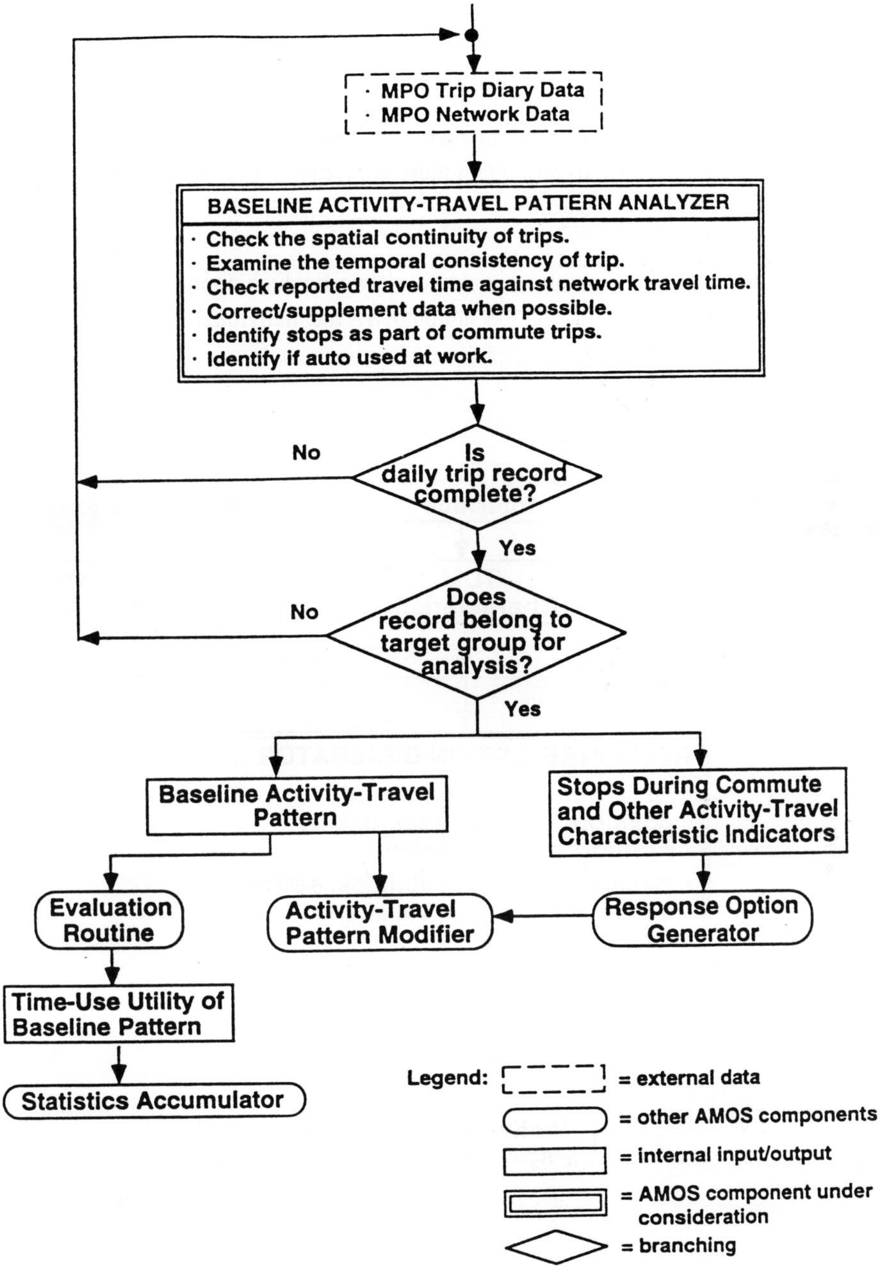

Figure 11.3. AMOS baseline activity-travel analyzer.

Response Option Generator

The Response Option Generator (Figure 11.4) is a key stochastic element of AMOS. The input to the Generator consists of: household and person attributes, network and land use characteristics, TDM attributes, and the indicators of the baseline activity-travel pattern characteristics prepared by the Analyzer. Given these the Generator simulates how the sample individual responds to the TDM measure.

The central component of the Generator is a neural network which computes the probability that each possible response option will be adopted by the individual based on the input

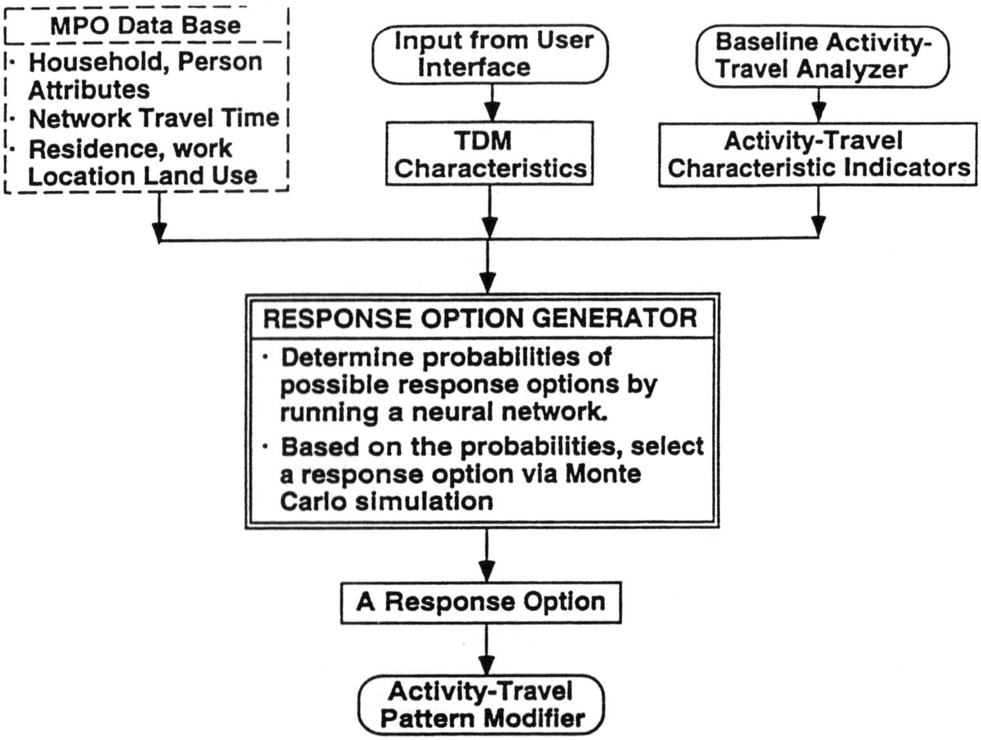

Figure 11.4. AMOS response option generator.

variables.[24] The use of a neural network draws from the connectionism, a branch of cognitive science. Connectionists postulate that humans process information by breaking it down into smaller elements that are inter-connected with different levels of intensity. In other words, human thinking is a process of connecting one informational element (e.g., a concept) to another. This idea can be depicted by a neural network, which can be "trained" to best replicate observed connection patterns between input (in this case TDM attributes and other variables) and output (response options).

Generating a response option alone does not automatically produce a complete and feasible new activity-travel pattern. Quite often the primary change implied by the response option triggers secondary and tertiary changes. For example, a solo-driving commuter who stops at a grocery store on the way home from work may choose to switch to public transit because of congestion pricing. To be able to do this, however, the commuter may also need to make a home-based trip chain to a nearby grocery store by auto after returning home by bus. The next module performs such adjustments.

Activity-Travel Pattern Modifier

The Activity-Travel Pattern Modifier (Figure 11.5) examines the baseline pattern and if the response option necessitates performs: (i) activity re-sequencing, (ii) activity re-linking, (iii) mode and destination assignment, and (iv) trip timing adjustment. Such adjustments are needed primarily when a travel mode change or a departure time change implied by the response option makes the baseline pattern infeasible or impractical. The modifier then examines the feasibility of the resulting modified activity-travel pattern using a rule base.

Activity re-sequencing refers to the re-arrangement of the order in which out-of-home stops are made. Re-linking, on the other hand, refers to the re-combining of out-of-home stops into trip chains. For example, consider a sequence of three out-of-home activities, A, B and W, where W is work. Suppose these three are pursued as A-W-B. This may be re-sequenced as W-A-B. Letting H be the home base, these activities may be linked as: H-W-A-B-H or H-W-H-A-B-H. In the former case, activities A and B are pursued on the way home from work; in the latter they are pursued in a separate home-based trip chain.

---

[24] The response options were originally obtained during a computer-aided telephone interview survey. They include: do nothing differently, change departure time to work, switch to transit, switch to car/vanpool, switch to bicycle, switch to walk, work at home, and others. Theoretically they may include the option, "switch to drive alone," which is not included in the prototype because no respondent volunteered this option.

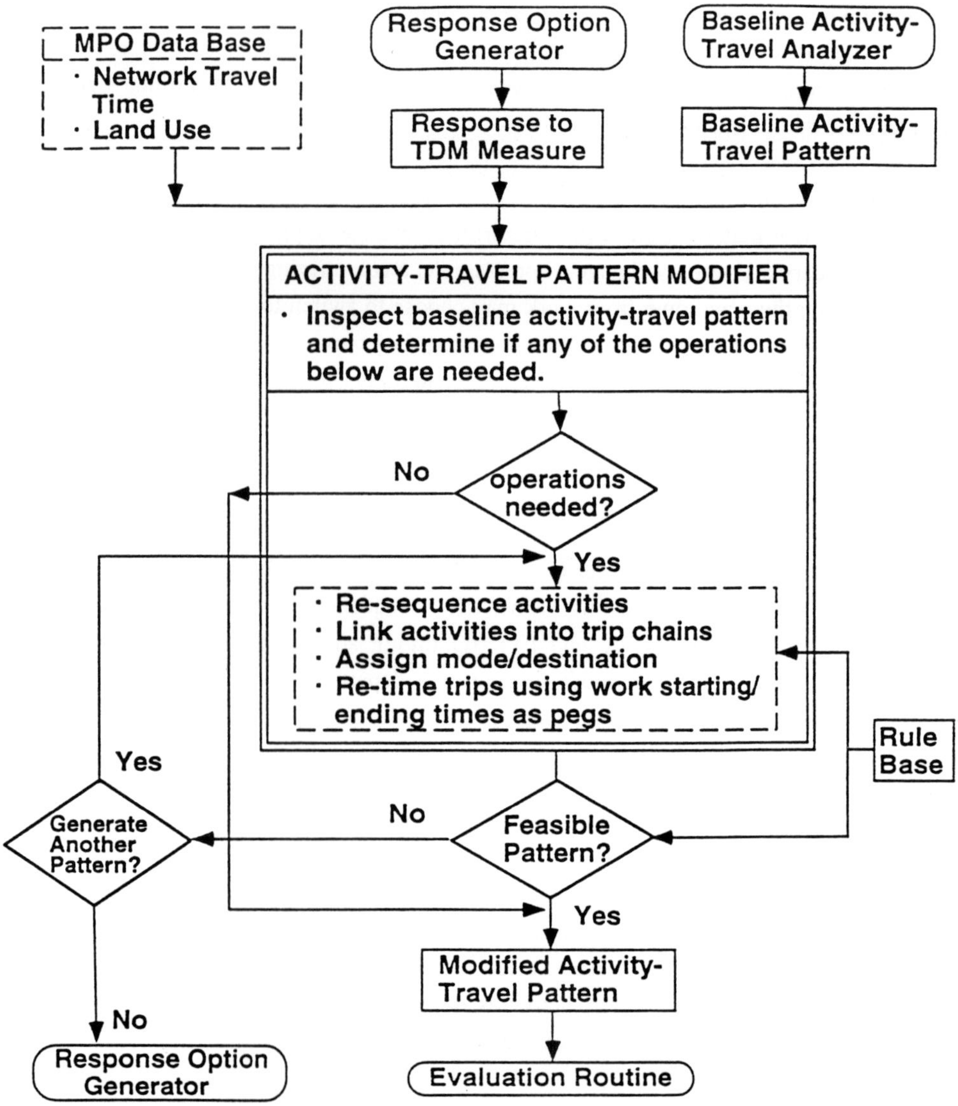

Figure 11.5. AMOS activity-travel pattern modifier.

While the response option from the generator may dictate which mode is to be taken for some trips, there may be some degrees of freedom associated with other trips. In the latter case the modifier simulates mode choice for the set of trips whose modes are not fixed. Likewise, new destinations may be chosen for certain types of activities when activities are re-sequenced or re-linked, and the destination locations in the baseline pattern are no longer suitable.

The timing of trips is determined while using the work starting and ending times as "pegs" (Cullen, 1972). For example, the starting time of a home-to-work commute trip by a new mode will be determined such that the commuter will arrive at the workplace at or earlier than the time observed in the baseline pattern. The duration of each out-of-home activity is fixed in this initial prototype.

After a modified activity-travel pattern is generated, the Modifier checks if it is feasible. Used in this check is a rule base, which contains several groups of rules that govern and constrain travel. For example shopping cannot be pursued outside the store hours. In fact this rule base is constantly referred to in the above process of modifying the baseline activity-travel pattern and producing an alternative.

Evaluation Routine

The Evaluation Routine (Figure 11.6) assigns a utility measure to the modified activity-travel pattern using time-use utility functions (see Kitamura et al., 1995). The attractiveness of the modified pattern is determined by the utility produced by allocating time to, and engaging in, the in-home and out-of-home activities contained in the pattern. The utility functions are being developed using the time-use data obtained from the survey conducted as part of the study. The effort includes the generalization of the utility functions to include non-time elements such as mode attributes and monetary expenses, and sequencing and timing of activities. Using the time utility concept, AMOS evaluates TDM measures while considering their impacts on the entire daily activity, not just on the commute trips which these measures often target.

Acceptance Routine

The Acceptance (Search Termination) Routine (also Figure 11.6) evaluates the set of time-utilities associated with the activity-travel patterns so far generated, and determines whether the search should continue or one of the patterns so far generated should be adopted. The routine is based on the assumption that the individual forms a subjective distribution of utilities associated with alternative patterns assesses the likelihood of obtaining a better activity-travel pattern; and terminates the search when the cost of search exceeds the expected gain of searching further. It is assumed that the individual can determine which of the alternatives so far evaluated has the largest utility and will choose that alternative. Experiments are being designed to validate this theoretical search termination model and to estimate the parameters.

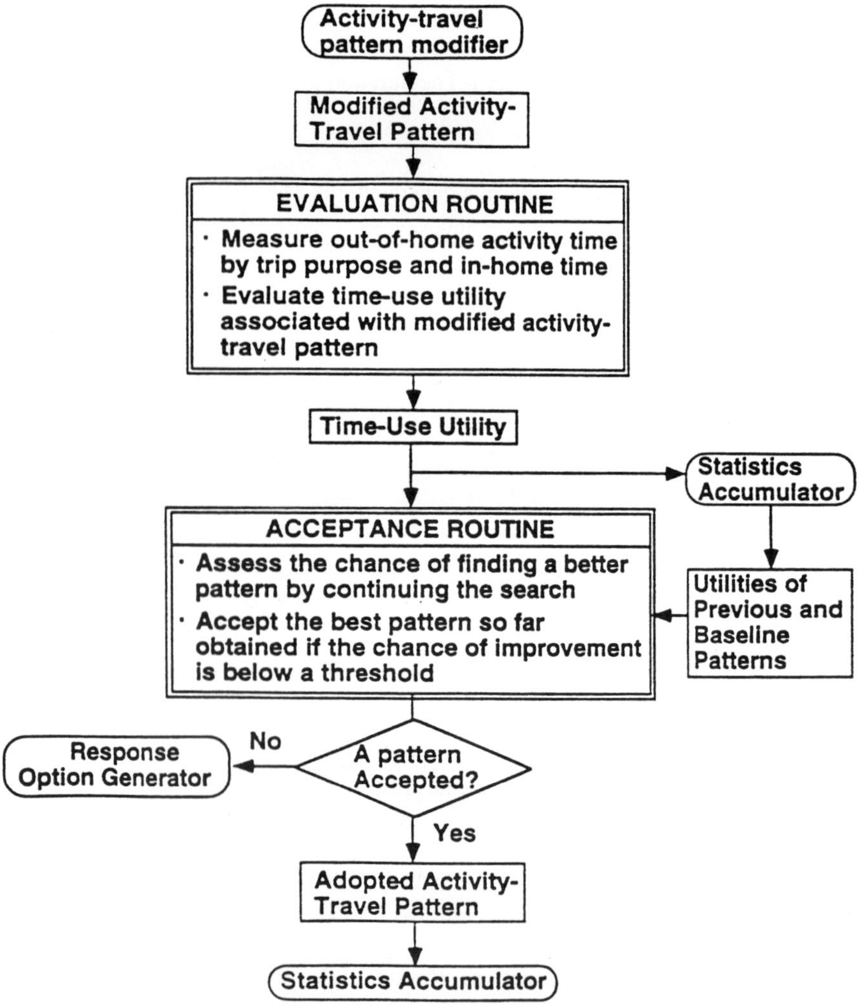

Figure 11.6. AMOS evaluation and acceptance routines.

Statistics Accumulator and Reporting Module

The Statistics Accumulator performs bookkeeping functions and produces two files. One is a temporary file and contains detailed information about the alternative activity-travel patterns generated for a sample individual in the simulation. This file supports the search process described above. The other file is a permanent file which contains the attributes of the pattern adopted by each sample individual. This file is later accessed by the Reporting Module to produce desired statistics and forecasts.

AMOS Prototype Results

Results of the AMOS prototype implementation in the Washington DC, metropolitan area are summarized in this section.[25] A small-scale telephone interview survey was conducted to obtain data to calibrate model components, especially to train the neural network in the Response Option Generator. The implementation effort adopts the traffic analysis zone (TAZ) system and zone-to-zone network travel time matrices by travel mode, maintained by the Metropolitan Washington Council of Governments (MWCOG), the MPO of the region. Network skim travel time data are available for: drive alone (SOV), ride-sharing (HOV), public transit with walk access, and public transit with auto access. Travel times by bicycle and walk are estimated by applying assumed speeds (6.5 mph and 2.5 mph, respectively) to the centroid-to-centroid distance. The implementation effort thus utilizes as much information as available from the MWCOG data base. Note that travel time data used are static, thus possible changes in network service levels due to TDM measures are not reflected in the current application.

AMOS Survey

A three-phase survey using computer-aided telephone interview techniques, was conducted to generate a data set to calibrate AMOS components. The survey which targeted commuters included a time-use section which collected data on both in-home and out-of-home activities as well as details of each trip made. Also in the survey was a set of customized stated-preference (or "stated adjustment") questions which asked respondents how they would respond to each of a set of TDM measures. The survey was conducted in November and December of 1994. Approximately 650 interviews were completed. Adults who commuted at

---

[25] An early predecessor where an activity-based model system can be found is van der Hoorn (1983) which was developed based on a weekly time-use data set.

least three days a week were the target of the survey. For further information, see RDC (1995) and Pendyala et al. (1996).

TDM Measures Considered

In the stated-adjustment section of the survey, respondents were given a description of a TDM measure (see Table 11.4 for the measures included), then asked in an open-ended format "What would you do?" if the measure had been in fact implemented. Follow-on questions were asked to probe into details of the stated behavioral adjustment (e.g., how to drop off a child at the day-care when public transit is used to commute). Commute travel time and other pertinent parameters were customized such that the hypothetical scenario would closely represent each respondent's commute situation.

Neural Network

Responses to the TDM section were used to train the neural network of the Response Option Generator. The trained network consisted of 45 input nodes and 8 output nodes with two hidden layers. The input nodes may be grouped to represent: personal and household attributes, work schedule characteristics, commute characteristics, trip chaining characteristics, mode characteristics, and TDM scenarios. The eight output nodes comprise

Table 11.4
TDM Measures Included in the AMOS Survey in the Washington, D.C. Metropolitan Area

TDM #1	Parking Tax. Incremental parking tax at work place at: $1 to $3 per day in suburbs[a]; $3 to $8 per day in D.C. and central areas
TDM #2	Improved Bicycle/Pedestrian Facilities. Well-marked and well-lighted bicycle paths and a secure place to park a bicycle wherever respondent went.
TDM #3	"Synergy" Combination of TDM 1 and TDM 2
TDM #4	Parking Charge Combined with Employer-Supplied Commuter Voucher. Employers provide employees with a commuter voucher while employees must pay for a parking surcharge: $40 to $80 per month for both voucher and surcharge
TDM #5	Congestion Pricing. Area-wide implementation of congestion pricing, effective from 6:00 AM to 9:00 AM and from 4:00 PM to 7:00 PM: $0.15 to $0.60 per mile; 10% to 30% travel time savings
TDM #6	"Synergy" Combination of TDM 4 and TDM 5

[a]Different parameter values are assigned to respondents randomly within the range shown.

change departure time, use transit to work, ride-share to work, ride bicycle to work, walk to work, work at home, do nothing different, and other (long-term responses treated as doing nothing in short-term policy analysis). A statistical procedure was developed to convert the output of the neural network (which lies between 0 and 1 but is not a probabilistic measure) into the probability that each response option will be adopted. The set of probabilities thus obtained is used in the simulation. Note that these probabilities vary from individual to individual depending on the person and household attributes and travel characteristics.

Simulation Results

The following scenarios are considered in the implementation study: (1) (TDM #1) parking pricing: parking surcharge of $8.00 per day; (2) (TDM #4) parking pricing with employer-paid voucher: parking charge of $80 per month and a commuter voucher of $60; (3) (TDM #5) congestion pricing: congestion charge of $0.50 per mile, travel time reduction by 30%; and (4) (TDM #6) a synergy combination of TDM #4 and TDM #5: parking charge of $80 per month, commuter voucher of $60, and congestion charge of $0.50 per mile.

The number of sample households from the MWCOG data base that were made available to the study was unfortunately very small and the results presented here are subject to sampling errors. It must be noted that this exercise has been made for illustrative purposes and the size of the sample used here, and some of the simplifying assumptions, warrant neither generalization of the results obtained nor general assessment of the relative effectiveness of the TDM scenarios.

A total of 20 simulation runs were performed for each TDM measure, and the average of the 20 runs are used in the analysis. In the rest of this section, the baseline case is first examined, then simulation results for TDM#1 and TDM#5 are discussed. For additional results, see RDC (1995).

Baseline. Table 11.5 presents the distribution of trip purposes (work vs. non-work), travel mode (auto-driver, auto-passenger, other), mean trip duration by mode, percent of hot starts, and average number of trips per person for AM peak, PM peak, and off-peak periods. Over 60% of the trips are work trips in the morning peak period (non-work trips include trips from work to home), and about three-quarters of the trips are made by auto. The large fraction of trips by other mode in the afternoon peak period represents walk trips made in this period by this sample of commuters. The average number of trips per person is 3.21.

Table 11.5

Baseline Travel Characteristics

	Total	AM Peak	PM Peak	Off-Peak
TRIP PURPOSE				
Work	42.2%	64.0%	31.1%	36.6%
Non-Work	57.8%	36.0%	68.9%	63.4%
TRAVEL MODE				
Auto – Driver	54.0%	65.1%	54.7%	45.5%
Auto – Passenger	18.4%	10.5%	18.9%	23.6%
Other	27.6%	24.4%	26.4%	30.9%
TRIP DURATION (minutes)				
Total	18.5	21.7	22.0	13.4
Auto-Driver	21.6	24.5	24.9	15.2
Auto-Passenger	17.0	16.4	21.4	14.2
Other	13.6	16.4	16.2	10.1
HOT STARTS (%)	37.7%	34.9%	35.9%	37.8%
PERCENT OF TRIPS	100%	27.3%	33.7%	39.0%
TRIPS PER PERSON	3.21			

Table 11.6

AMOS Simulation Results: Parking Pricing (TDM #1)

	Total	AM Peak	PM Peak	Off-Peak
TRIP PURPOSE				
Work	43.2%	63.2%	35.5%	35.4%
Non-Work	56.8%	36.8%	64.5%	64.6%
TRAVEL MODE				
Auto – Driver	47.5%	57.5%	48.6%	40.8%
Auto – Passenger	16.4%	10.3%	18.7%	23.9%
Other	36.1%	32.2%	32.7%	35.3%
TRIP DURATION (minutes)				
Total	19.4	21.6	22.8	13.4
Auto-Driver	21.2	23.6	24.8	15.3
Auto-Passenger	16.5	16.4	21.4	14.2
Other	18.4	19.7	20.7	10.5
HOT STARTS (%)	37.4%	34.5%	37.4%	39.2%
PERCENT OF TRIPS	100%	26.9%	33.0%	40.1%
TRIPS PER PERSON	3.31			

Parking Pricing (TDM#1). The most notable change with the parking surcharge of $8 a day is in modal split (Table 11.6). The fraction of auto driver trips decreased from 54.0% in the baseline case to 47.5%, and auto passenger trips from 18.4% to 16.4%. The fraction of other modes increased by 7.8% during AM peak, 6.3% in the PM peak, and 4.4% during off-peak periods, respectively.

The overall average trip duration (in minutes) shows only small changes between the two cases. Importantly, however, the mean trip duration for other modes increased from 13.6

minutes to 18.4 minutes suggesting that long-distance commuters tended to remain auto commuters while shorter distance travelers adopted other options. The distribution of trips across morning peak, afternoon peak, and off-peak shows only minor changes. The fraction of morning peak trips decreased slightly from 34.9% to 34.5%, while that of afternoon peak trips increased from 35.9% to 37.4%. The average number of trips per person increased slightly from 3.21 to 3.31. This reflects activity re-linking as a result of a commute mode change which resulted in slightly more trips per person.

Congestion Pricing (TDM#5). The results with congestion pricing at a level of $0.50 per mile with 30% reduction in travel time are summarized in Table 11.7. The fraction of auto trips 50.2%, is higher with this TDM than with parking pricing (47.5%), but is lower than the baseline result (54.0%). Other than mode shares, the results of this TDM are very similar to those of TDM#1.

Discussion. The exercise reported here has shown that AMOS, an activity-based micro-simulation model system, is capable of producing travel forecasts by simulating daily travel patterns. It has also been demonstrated that the TDM measures considered have certain impacts on travel demand. From a model development viewpoint, the results are very encouraging as they show that activity-based models can be implemented in a metropolitan region and can produce forecasts for policy analysis in a practical manner.

Table 11.7

AMOS Simulation Results: Congestion Pricing (TDM #5)

	Total	AM Peak	PM Peak	Off-Peak
TRIP PURPOSE				
Work	43.0%	64.4%	35.6%	36.5%
Non-Work	57.0%	35.6%	64.4%	63.5%
TRAVEL MODE				
Auto - Driver	50.2%	56.3%	51.9%	39.7%
Auto - Passenger	17.0%	10.3%	18.3%	22.2%
Other	32.8%	33.4%	29.8%	38.1%
TRIP DURATION (minutes)				
Total	19.0	23.0	22.6	13.5
Auto-Driver	21.4	23.5	24.5	16.1
Auto-Passenger	17.3	16.4	21.8	14.5
Other	16.2	24.2	19.9	10.2
HOT STARTS (%)	36.8%	34.5%	36.5%	34.9%
PERCENT OF TRIPS	100%	26.9%	32.2%	40.9%
TRIPS PER PERSON	3.30			

From transportation policy viewpoints, however, the results may seem less encouraging because the effects of the TDM scenarios examined are relatively small, and because there are only a few discernible differences among the impacts of the respective TDM scenarios. These results may be simply due to the small sample used in the exercise. It is conceivable that the commuters in the sample had very limited alternative commute options and were able to respond within very narrow ranges to whatever TDM scenarios being implemented. Whether this observation can be generalized or not needs to be determined in the future by analyzing a larger sample of individuals.

Another possibility is that the Response Option Generator has not been fine-tuned enough to be able to detect possibly minute differences in commuters' responses to different TDM measures. In particular, the results suggest that a neural network be developed for each TDM measure separately (in the current prototype, the neural network is designed to be able to handle all the examined TDM scenarios).

The invariance in simulation results across the TDM scenarios may also be due to the limitations of the prototype used for the analysis. Importantly, destination choice has not been implemented in the prototype. In addition, the simplistic evaluation and acceptance rules adopted in the prototype may have resulted in premature search termination for each commuter, possibly leading to the acceptance of the baseline patterns with a higher probability than it should receive.

Summary

This exercise has demonstrated that a micro-simulation model system of daily travel behaviour, which adheres to the principles of the activity-based approach, is not only feasible but also will be a practical tool for policy analysis. The implementation of the AMOS prototype in the Washington D.C., metropolitan area utilizes the data base maintained by the MPO of the area. The medium scale survey (about 650 respondents) used in this study can be modified to entertain a wide range of TDM measures, making AMOS a flexible and realistic tool for transportation policy analysis.

AMOS is a model of adaptation. It replicates the cognitive process of searching for the best activity-travel alternative given a change in the travel environment. The cognitive process is formulated as the individual selects a principal change in the activity-travel pattern in response to the change in the travel environment, then makes adjustments to the activity-travel pattern while considering various constraints and rules. The new pattern is then evaluated and its superiority relative to the other patterns examined is assessed. This process is repeated until a satisfactory pattern is obtained. This process may represent the individual's trial-and-error search effort which may take place over a period of several weeks, or it may be thought of as a replication of purely mental exercises of devising suitable ways of adapting to the change in the travel environment. In any event, underlying the development of AMOS is the intent of replicating how the individual "thinks" and decides what to do and how to travel.

Efforts are ongoing currently on several fronts to extend the scope of AMOS by incorporating vehicle transaction and utilization behaviour, vehicle allocation, and synthetic generation of households and their activity-travel patterns (see Kitamura et al., 1996b). Planned future research activities include the development and incorporation of models for search termination, activity engagement, time allocation, inter-person interaction, and multi-day behaviour. Incorporating of realistic search strategies is another area for future effort.

CONCLUSION

Two computational process model systems of activity-travel choice, PCATS and AMOS, are presented in this chapter. The two model systems are motivated by different objectives, are built with different emphases, and are suited for different analytical purposes. PCATS, which explicitly incorporates Hägerstrand's time-space prisms, is suited for the analysis of factors that affect time-space constraints such as store opening hours, work schedules, and public transit operating hours. PCATS also offers a coherent analytical framework for examining induced or suppressed demand and addressing the effects of added capacity or mounting traffic congestion on travel demand. AMOS, on the other hand, is a system of behavioral adjustment. It is a policy tool that predicts changes in demand by simulating how individuals respond to changes in their travel environments. As such AMOS is best suited for the evaluation of the effectiveness of policy options such as TDM measures. The scenario analyses presented using prototypes of these two model systems have demonstrated that activity-based model systems are practical tools for policy analysis and warranted the development of full-scale model systems.

ACKNOWLEDGEMENTS

Parts of this study were presented earlier at the U.S. Federal Highway Administration's Travel Model Improvement Program (TMIP) conference on Activity-based Travel Forecasting, New Orleans, June, 1996. Funding for the development of PCATS has been provided by the Ministry of Education of Japan. The development and implementation of AMOS has been funded by the U.S. Environmental Protection Agency (EPA), Federal Highway Administration (FHWA), and Metropolitan Washington Council of Governments (MWCOG). The authors are grateful to the AMOS development team which has included Cynthia Chen, Seyoung Kim, Keith Lawton, Clarisse Lula, Eric Pas, Ram Pendyala, Purasuna Reddy, and David Reinke.

REFERENCES

Axhausen, K. W. and T. Gärling (1992). Activity-based approaches to travel analysis: Conceptual frameworks, models, and research problems. *Transport Reviews*, **12**, 323-341.

Ben-Akiva, M. and S. R. Lerman (1985). *Discrete Choice Analysis: Theory and Application to Travel Demand*. MIT Press, Cambridge, MA.

Cullen, I. G. (1972). Space, time, and the disruption of behaviour in cities. *Environment and Planning*, **4**, 459-470.

Damm, D. (1983). Theory and empirical results: a comparison of recent activity-based research. In: *Recent Advances in Travel Demand* (S. Carpenter and P. M. Jones, eds.), pp. 3-33. Gower Publishing, Aldershot, UK.

Ettema, D. (1996). *Activity-based travel demand modelling*. Doctoral dissertation, Department of Architecture, Eindhoven University of Technolgy, The Netherlands.

Gärling, T., M.-P. Kwan and R. G. Golledge (1994). Computational-process modeling of household activity scheduling. *Transportation Research B*, **28**, 355-364.

Hägerstrand, T. (1970). What about people in Regional Science. *Papers of the Regional Science Association*, **24**, 7-21.

Jones, P. M., M. C. Dix, M. I. Clarke and I. G. Heggie (1983). *Understanding Travel Behaviour*. Gower, Aldershot, UK.

Jones, P., F. S. Koppelman and J. P. Orfueil (1990). Activity analysis: State-of-the-art and future directions. In: *Developments in Dynamic and Activity-Based Approaches to Travel Analysis* (P. Jones, ed.), pp. 34-55. Gower, Aldershot, UK.

Kitamura, R. (1988). An evaluation of activity-based travel analysis. *Transportation*, **15**, 9-34.

Kitamura, R., S. Fujii and Y. Otsuka (1996a). *An analysis of induced travel demand using a production model system of daily activity and travel which incorporates time-space constraints*. Paper presented at the 5th world congress of the Regional Science Association International. Tokyo.

Kitamura, R., C. V. Lula and E. I. Pas (1993). AMOS: An activity-based, flexible and truly behavioural tool for evaluation of TDM measures. *Proceedings of the 21st Summer Annual Meeting: Transportation Planning Methods*, pp. 283-294. PTRC Education and Research Services, London.

Kitamura, R., E. I. Pas, C. V. Lula, T. K. Lawton and P. E. Benson (1996b). The sequenced activity mobility simulator (SAMS): An integrated approach to modeling transportation, land use and air quality. *Transportation*, **23**, 267-291.

Kitamura, R., R. M. Pendyala, E. I. Pas and P. Reddy (1995). Application of AMOS, an activity-based TCM evaluation tool, to the Washington D.C. metropolitan area. In: *Proceedings of 23rd European Transport Forum: Transportation Planning Methods*, pp. 177-190. PTRC Education and Research Services, London.

Kurani, K. S. and R. Kitamura (1996). *Recent developments and the prospects for modelling household activity schedules.* Technical Report, Institute of Transportation Studies, University of California, Davis, CA.

Newell, A. and H. A. Simon (1972). *Human Problem Solving*. Prentice-Hall, Englewood Cliffs, NJ.

Otsuka, Y. (1996). *Development of a simulation model system of activity and travel which represents the individual's decision process*. M. S. thesis, Department of Applied Systems Engineering, Kyoto University, Japan.

Pendyala, R. M., R. Kitamura and D. V. G. Prasuna Reddy (1996). *A rule-based activity-travel scheduling algorithm integrating neural networks of behavioural adaptation.* Paper presented at conference on activity based approaches: Activity scheduling and the analysis of activity patterns, Eindhoven, the Netherlands.

RDC Inc. (1995). *Activity-based modelling system for travel demand forecasting. A travel model improvement program (TMIP).* Technical report, Metropolitan Washington Council of Governments, San Francisco, CA.

Simon, H. A. (1990). Invariants of human behaviour. *Annual Review of Psychology*, **41**, 1-19.

Tasker, M. P. and K. W. Axhausen (1994) *DynaMIT: A travel behaviour simulation using satisficing search methods.* Unpublished manuscript, University of London Centre of Transport Studies, Imperial College, London.

van der Hoorn, T. (1983). Experiments with an activity-based travel model. *Transportation*, **12**, 61-77.

[illegible]

[illegible] FTR [illegible]

[illegible]

[illegible] TRG Transport Research Services, London.

[illegible] development and the prospects for modelling [illegible]

[illegible]

12

Information Representation for Driver Decision Support Systems

Mei-Po Kwan, Ohio State University, Columbus, OH, USA,
Reginald G. Golledge, University of California, Santa Barbara, CA, USA and
Jon Speigle, University of California, Santa Barbara, CA, USA.

Introduction

Intelligent Transportation Systems (ITS) utilize advanced communication and transportation technologies to achieve traffic efficiency and safety. There are different components of ITS, including Advanced Traveler Information Systems (ATIS), Automated Highway Systems (AHS), Advanced Traffic Management Systems (ATMS), Advanced Vehicle Control Systems (AVCS), and Advanced Public Transportation Systems (APTS). Development of a system for ITS depends on our ability to deal with a vast amount of information about the locations of places as well as with the complex representation of the transportation network linking those places, and to incorporate these into a geographic database. The system therefore needs to be constructed based upon the foundation of an integrated and comprehensive Geographic Information System (GIS). As compared to the simplified node-link graph theory representations of transport networks used by current ITS, GIS are able to provide more realistic representations of elements of the complex environment.

Transportation science has an expressed goal of increasing accessibility for all groups of people with regard to the environments in which they live and interact. A significant component of these goals is to further develop Intelligent Transportation Systems (ITS) through multi-level and multi-modal research and testing. This includes contributing to research and transportation system architecture, technology development, policy formation, and operational tests of various systems including ATMS, ATIS, and APTS. In this paper we focus on ATIS.

ATIS CHARACTERISTICS

Traffic congestion is a problem that appears to be increasing world-wide. In recent years considerable effort has been paid to the investigation of methods to reduce such congestion and the hazards and accidents that are usually associated with it. Collectively these efforts come under the aegis of Intelligent Transportation Systems (ITS). A critical part of ITS is the Advanced Traveler Information Systems (ATIS). Essentially these consist of in-vehicle information and guidance systems which help the driver to select routes which will reduce congestion, to find parking in areas where it is sparse, and to facilitate rescheduling of activities when congestion makes this a feasible alternative. It is argued that such assistive information will benefit individual drivers in terms of helping to achieve their scheduled behaviours and activities as well as benefiting the system by improving traffic flow.

The objective of an in-vehicle guidance system is to assist the driver to select routes which will help reduce congestion. Before the information given by the in-route guidance system can be effective, however, it must be perceived as being valuable by the driver, the driver must recognize that information so obtained is valid and reliable, the driver must accept that the action if taken will not increase stress levels or route him through fearful or dangerous places, and the driver must be willing to execute an appropriate action promptly.

The continuing increase in vehicle miles of travel and the corresponding environmental degradation, personnel time loss, increased congestion, and decreases in safety, have lead to the suggestion that ATIS may represent a feasible solution to many of these problems. The ATIS is usually conceived of as an in-vehicle route guidance system with supplementary information that allows changes in activity scheduling. The objective of an ATIS is to reduce the impacts of congestion by offering information to drivers that will help them select alternate routes that should benefit them individually as well as benefiting the operation of the system as a whole. While there is no doubt that drivers may respond in markedly different ways to in-vehicle information and route-guidance data, it is still suggested that a versatile ATIS will be capable of being personalized by each driver by acting as a decision support system and informative supplement to the driver's knowledge base.

ATIS AS A DECISION SUPPORT SYSTEM FOR TRAVEL

Decision Support Systems (DSSs) are integrated sets of tangible and intangible information that are designed to supplement a decision maker's personal knowledge base during problem-solving activities. The principal objective of a DSS is to support decision making by humans, not to replace it, with computerized recommendations. Use of such a system is presumed to bring to bear on a problem the strengths of personalized expert knowledge and comprehensive exogenous knowledge that may not normally be available to the decision maker. The result should be an informed and intelligent decision.

An ATIS can be considered a DSS that is designed to provide a set of information to a driver while on-route to help solve problems (e.g., what to do when faced with congestion, hazards, or other barriers to movement such as construction). To achieve this goal, the ATIS should not provide just a single set of commands or directions but rather allow a driver to use whatever criteria are most acceptable in terms of deciding whether to wait out the delay effects, or to undertake travel changes such as rerouting, rescheduling activities, replacing activities, changing destinations, and so on. But ATIS should be considered as a supportive tool designed to help the user, without automating the total decision process according to previously established sets of objectives or by imposing solutions which may be unacceptable to the driver (e.g., recommending rerouting that takes the driver through what may be imaged as a dangerous neighborhood). When using an ATIS as a DSS, it must be designed to be easy to use, it must have a user friendly in-car interface, it must help drivers achieve travel objectives and not divert them from attaining those objectives, and they must be designed to enable a user to benefit fully from the types of information dispensed. The ATIS then becomes a decision aid which might allow a user to generate a series of alternatives prior to making a critical decision. It thus provides the traveler with an opportunity to find a good or satisfactory solution without imposing the need to seek an optimal solution.

In the domain of surface travel by road, implementing an ATIS as a DSS can be complicated. The first problem arises in determining whether the information dispensed is most relevant to solving a system problem (e.g., clearing a point of congestion, hazard, or traffic delay) in as timely a way as possible, or providing information to allow drivers to make their own decisions on rerouting, rescheduling, and so on. Up until this time, the best quality and most quantity of information that has been geocoded and stored has related to the environmental base (i.e., transport network). Given this emphasis, information has been available to handle a system problem in a timely way, by determining alternate routes around a barrier or obstacle.

In recent years, however, there has been an increasing effort to focus on the driver as a recipient of advanced traveler information by developing a series of simulators which allow manipulation of environments and observation of different driver behaviors (see Koutsopoulos et al., 1995). The use of such simulators is most helpful when attempting to decide the likelihood that drivers will accept information dispensed through their in-vehicle guidance system.

The development of driver simulators and the consequent modeling of driver behavior has been an attempt to enrich the potential of ATIS methodology. Simulators themselves range from PC based pseudo game systems in which a participant moves a cursor on the screen (representing a vehicle) through a two-dimensional network or maze in attempts to achieve some goal that is constrained by à priori determined barriers. Other simulators are more comprehensive and involve video or time-sliced slide displays of actual road conditions. Of significant importance, however, is the increased acknowledgment that these simulations work best if the environment is designed to have as many important real-world features as possible (e.g., traffic lanes instead of undifferentiated network arcs; signalized or otherwise controlled intersections instead of unconstrained nodes in a graph theoretic representation of a system; indicators of traffic speed and driving headway; etc.). Emphasis on driver behavior has also developed beyond simple observation of actions to the stage of providing supplementary information to the driver so that normally considered criteria can be used in decisions concerning rerouting, rescheduling, destination substitution, and so on. The most productive driver simulations are those in which the supplemental information allows decision makers to use their personal skills and expert knowledge in a flexible and adaptable way. But, while acknowledging these needs, it is also necessary to realize that an interactive system (e.g., in-vehicle guidance) can only be effective if it works in a realistic time frame and does not delay the decision-making process of a potential traveler. Thus, not only do drivers require timely decision support, but they must also be able to comprehend the information provided, integrate it with their goals and objectives as they exist at a given time and place, and make a decision that will benefit both the driver and the system.

Regarding an ATIS as a driver decision support system provides a framework for developing a potentially useful ATIS. The first filter provided by this framework is whether a driver is capable of synthesizing the information provided in such a way that personal and system-wide benefit will occur after a decision has been made. A second component of this framework concerns the system's construction, which involves selection of a type of database, a user interface, the type of display for distributed information, the amount and type of information, the range of alternatives that may be considered to replace current actions, the user friendliness of the display system, the complexity of the information provided, and the probability that a driver will use such information. With these guiding principles in mind, we now turn to suggest one way of developing such an ATIS using a Geographic Information System (GIS) as a host. We will also discuss the relative advantages and disadvantages of two different types of GIS - the longer existing relational based GIS which uses link-node network structures, and the more recently emphasized object-oriented database and data model.

HELPING THE TRAVELER RESOLVE CONFLICTS

We have suggested that one of the most common uses of an ATIS will be to provide information that will help a traveler overcome the problems associated with congestion. These problems can involve rerouting, rescheduling, destination substitution, activity compression, and activity deletion.

Rerouting decisions require information on the nature of the roads that make up the path selected as an alternative to that currently being traveled. The database must therefore have the capability of incorporating all levels of a road hierarchy, from highways to local streets. In addition, information must be available on the type of flow (e.g., one-way or two-way flows) and the directionality of the flow (i.e., towards or away from a potential destination). Data such as the number of lanes, whether on-street parking exists, the type of neighborhood the selected streets pass through, and the location of facilities (e.g., commercial, business, or other establishments) must also be accessible. In addition, access to information concerning the expected time of travel along the new route requires accumulation of expected travel times on each level and segment of the general road hierarchy making up the selected path. Other features are required that might be captured in a visual display (e.g., whether the ultimate route parallels the new one or proceeds orthogonally or diagonally away from it, as might be expected if the alternate route represents a shortcut).

The occurrence of congestion may cause gridlock. In that case rerouting may not be a feasible alternative. If the driver is locked in a tightly linked lane of traffic with no immediate chance of escape (e.g., on an interior lane of a multi-lane highway), it makes little sense to provide information about alternate routes that could be taken. Rescheduling of activities then becomes a possible alternative. Most drivers do this mentally. However, an ATIS option that assist in rescheduling. Such an option has been offered elsewhere by Golledge et al. (1994), and Kwan (1997) in the form of a Computational Process Model linked to a Geographic Information System (GIS). These devices, called SCHEDULER and GISICAS, respectively, allow conflict resolution in both temporal and spatial domains by rescheduling prioritized activities to give precedence to those with the greatest needs, providing temporal constraints are not violated.

Along with potential rescheduling of activities comes the possibility for substituting among destinations. Rescheduling, in fact, might encourage the traveler to exit the congested area at the earliest opportunity and to undertake travel to a new destination which is in a vastly different locale than the original one. To make these destination substitutions, the ATIS must contain sufficient information to allow choice of alternative places to go. Destination substitution may have only a limited possibility with respect to work trips, but for many other trips it can become a viable alternative (e.g., substituting a different restaurant for the one originally planned to patronize).

Congestion invariably involves time loss. Time loss may prevent extensive rescheduling, just as the nature of the surrounding road system may make rerouting a less desirable alternative. One option, therefore, is to undertake activity compression. This would simply maintain the original activity schedule but reduce the amount of time allocated to each remaining activity. Thus, a two hour recreational period might be reduced to forty-five minutes, a one hour shopping trip reduced to fifteen minutes, and so on. Often activity compression is also associated with destination substitution. For example, one may have planned to visit a specialty food store to pick up goods for an evening meal; after experiencing traffic congestion and the need for activity compression, a destination such as a fast-food restaurant may be chosen as an alternative to the original one.

If time loss because of congestion or barriers is significant, a final option is activity deletion. In this case the ATIS would need to provide sufficient information (or to have sufficient information contained within its programming) to allow the decision maker to delete low priority activities. For example, if at lunch time one was planning to get a haircut, but was caught up in congestion while traveling to the barber, that activity could simply be deleted from the day's schedule and reinserted at some later appropriate time. Again, however, information necessary to allow deletion to take place must be provided in a timely, clear, and usable manner.

DEVELOPMENT OF A GIS FOR ATIS

The first requirement of a GIS for ATIS is focused on the ability to represent the transportation network in detail in order to apply different routing algorithms for both modeling of movement and simulations of flows. In the real world, a transportation network has different types of roads (e.g., highway, arterials, local streets, etc.) and different types of

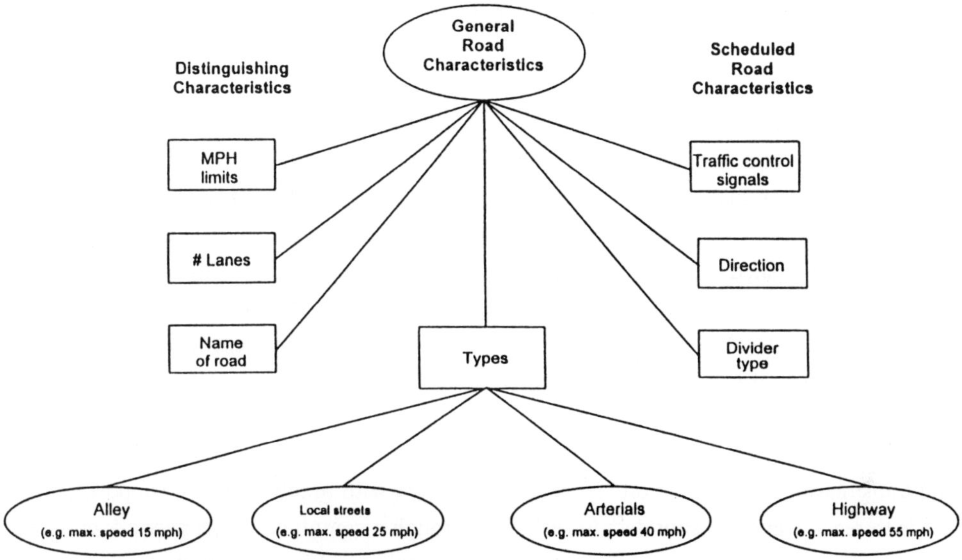

Figure 12.1. Sample of components of a road system

intersections (e.g., signalized and non-signalized) that are of interest to ITS builders (Figure 12.1). For some applications, information on intersections, lanes and lane changes such as turn lanes and merging lanes, highway exits, etc. may be important. We may also require geometric representation for road curvature and incline.

Current GIS data models usually represent the network as a collection of links and nodes. Whenever two links intersect, a node is created. Unless additional structure like a turn table is added, the link-node nature of topological data models is basically planar and cannot distinguish an intersection at grade from an intersection with an overpass or underpass which does not cross at grade. It is difficult to accurately represent overpasses/underpasses and this situation may lead to problems when running various routing algorithms.

There are other situations in which we may want to differentiate types of nodes. First, if there is a decision point that the traveler needs for movement, the node-link model usually snaps that point to the nearest node regardless of type. Thus, this approach ignores the different types of links and nodes making up the complex network. Moreover, when the traveler needs to make a decision to move in the network, there are many decision points that he or she needs to encounter in addition to intersections (e.g., merging lanes, divided lanes). Sometimes, different nodes may lead to different behaviors. For example, we usually make a U-turn at a dead-end node. This behavior is commonly not available in much of the link-node software.

Second, since ITS have to operate at various spatial scales, a multi-level representation of a transportation network is needed. At the local level (i.e., city or county), the transportation network needs to be represented in very great detail for navigational purposes, whereas only the major elements such as the interstate or state highways need to be represented in the case of regional, state, or interstate travel. Further, relationships between data at different levels have to be established (e.g., solving the aggregation problem). Only through the use of a more efficient geographic data model can we hope to overcome these and other related problems (Roberts et al., 1991). Several recent attempts to tackle this problem have focused on the object-oriented approach (e.g., Mainguenaud, 1995; van Oosterom and Schenkelaars, 1995).

A third problem concerns a system's ability to handle the real-time update of traffic patterns and transmission of information. To provide travelers with timely decision support, an ATIS needs to receive and transmit data in real-time through a communications network connecting a vast number of system elements, such as traffic sensors and location-tracking devices. Even for a small city, the real-time update and transmission of data is a very challenging problem. To keep information on the dynamic environment current and self-consistent even in the presence of concurrent updates and queries from around the network, we need to go beyond existing data handling technologies. The system has to be interoperated with other transportation software and requires certain spatial partitioning for effective access.

Geocoded data systems are complex. Incorporating such systems into a dynamic real-time multilevel ATIS system makes them even more so. Objects, when geocoded, have both geographically and temporally referenced geometry. That is, objects are located in time and in space. These spatio-temporal characteristics tie all the geocoded data together. But, spatial databases are usually extremely large. For example, in the Santa Barbara area, with a population of about 180,000, the street system consists of 14,000 arcs and 28,000 nodes. When we add descriptor and attribute information to every segment and node, the size of the database escalates by many orders of magnitude. In geocoded data global accuracy and relative accuracy are important. Relative accuracy is more important than global accuracy because the real-world activities that are modeled or simulated within the database are usually local in nature.

In databases of transportation networks that are tied to link-node representations, real-world accuracies (e.g., curved streets and multi-path intersections) are not always represented accurately but are stylized into graph theoretic renderings. Furthermore, with respect to the current state of knowledge, we have very little idea of the amount of inaccuracy and error that may be present in many of these large databases. The presence of such error will, of course, give unreliably bad performance when attempting to either model travel behavior or to send information that will cause revision of activity schedules and consequent changes in spatial behavior. However, it appears that, when considering these problems, the use of object-oriented database systems and object-oriented data modeling strategies now have the greatest potential for use when attempting accurate and usable information transmission and the consequent modification of travel behavior.

Perhaps the most fundamental requirement when considering building a GIS for use in an ATIS context is the degree to which it is a realistic representation of existing network elements. Such a representation would need to include a road hierarchy, directionality of traffic flow, number of lanes, turn restrictions at intersections, signaling and signage (e.g., traffic lights and stop signs), existence of pedestrian crossings (particularly school crossings), presence or absence of bicycle lanes and on-street parking, the presence or absence of dividing strips, the opportunity for making U-turns, the nature of the road's surface, allowable and safe speeds, street gradients, driveway access from streets, whether or not through traffic is possible, and so on. Many existing transportation networks used in ATIS simulations and model building have been developed simply as node-link graphs, and while some of the above attributes are included, many of them are not.

In addition to the realistic representation of the street network, information is needed concerning the multimodality of each system element. In other words, the potential for linking with other transportation modes (e.g., busses, trains) must be included. In addition, the ATIS and its GIS components must be amenable to spatial and temporal update as permanent changes occur in the environment or as significant temporary barriers or changing circumstances come into existence (e.g., closure of entrance or exit ramps on freeways for certain time periods to allow widening, change of grade, or other features).

Perhaps the main purpose, however, of having a good GIS in an ATIS is to provide alternative models for vehicular routing and navigation. Any existing ATIS, whether designed for use in practice or in simulated conditions, contains only a restricted number of routing algorithms. These are usually based on minimum distance considerations and are most easily implemented in a simple node-link network. However, unless a very comprehensive set of attributes are associated with each node or link, this simplified graphical solution to navigation and routing may prove not so simple after all. For example, if one is being allocated to a route segment which has a lane divider that prevents U-turning, and U-turning is impossible at the next intersection, how does one access a destination on the other side of the road? Similarly, in the simplified graph theoretic network representation, a route may require turning at a node which in reality is not an intersection but an overpass or underpass.

Obviously without knowing the constraints of traffic flow directionality, one may further come up with a routing solution that indicates travel should occur against the legal direction of flow. Or simply, the driver may wish to implement a routing algorithm that uses a path selection criteria other than shortest path. This alternative may depend on trip purpose (which may change if activity rescheduling occurs) (Golledge, 1995).

Obviously many of the existing GIS that rely on link-node structures have difficulty in meeting the various informational needs mentioned above. Their information base is incomplete, their potential for conflict resolution is limited, and their ability to use human perceptions is likewise limited.

In comparison to the link-node systems, object-oriented Geographic Information Systems (OOGIS) allow for class hierarchies among objects and provide for the possibility of the inheritance of attributes among different levels of a class. For example, a general level of the class "road" might be "highway." Many of the characteristics of a highway can be inherited by lower level members of the general class of roads - such as arterial streets or neighborhood streets.

An object-oriented database allows a map to be divided into object layers, each defined by a common set of attributes. Each object layer may consist of entities encoded as circles, lines, points, polylines, nodes, links and line attributes. Object-orientation also provides a means of partitioning a map and of displaying the zoomed-in partition. Partitioning allows the rendering of only those sections of the map which overlap with a currently defined viewing box. As the viewing box expands to incorporate more layers, it also incorporates adjacent partitions. Attribute lists relevant only to each partition (partitioned entity lists) can be displayed and consequently the time and effort of examining unnecessary data is conserved. Partitioning can take place in a number of different ways.

An object-oriented GIS can also form a class hierarchy. For example, if a road layer class is defined we may then be able to derive more specific types of roads from this class (e.g., highway, local street). This establishes a kind of "parent-child" relation between objects. The "child" class inherits the properties of the "parent" class including the parent class' attributes and its associated functions. For example, a relation might be established between a class labeled "car" and a class labeled "vehicle." It could then be said that "car" is derived from "vehicle" and that "car" is a type of "vehicle." As a type of vehicle it would inherit many of the attributes associated with the class vehicle.

An additional favorable characteristic of an object-oriented GIS is called "polymorphism." This represents the quality or state of being an object which is able to assume different forms.

In a programming context it means the same construct can be used to manipulate objects of different types. For example, one can conduct "overloading" in which the same operator ("+" or "print") can perform different functions depending on the recipient's object class.

In the link-node GIS structure there is an assumption that the transportation network is represented as links and nodes and there is no differentiation between these links and nodes. Consequently the delay function at intersections, for example, cannot be easily modeling differently for signalized intersections versus non-signalized intersections, or highways versus local streets. With polymorphism in an object-oriented context, these different elements in a network can be modeled using different functions but with the same operator ("cost") used to calculate the delay function in all circumstance (Khoshafian and Abnous, 1990). An object-oriented system also allows for parametric polymorphism. This provides for the construction of abstract classes with different type parameters. For example, a delay function can be formulated as COST(D) with D as a parameter of different type of distance measures. In transportation research, distance is often measured either as travel time or over-the-road distance. So the parameter D can be implemented as a type of integer (when used in terms of minutes of travel time) or as a type of real number (when used in terms of meters of physical distance). Both sets share the code as it is implemented in COST(D).

The relevance and reliability of the information contained in an ATIS is a key component of such travel aids. Thus, for both travel-behavior - simulations and for real-world applications, the need for development of a database and data model that reflects complexities of real-world conditions is essential. It is in this context that we argue for the development of Geographic Information Systems (GIS) as an appropriate host technology. In the remainder of this chapter we will examine some of the characteristics of existing and potential GIS databases for ATIS use. And, finally, we will discuss the theoretical implications and model building implications of having access to more comprehensive and more realistic databases that reflect transport-network conditions and driver-behavior conditions in ways that are closer to real-world situations.

TOWARDS AN OBJECT-ORIENTED GIS

Worboys (1992a, 1992b, 1994a, 1994b), and Worboys et al. (1990) surveyed the current state of the object-oriented paradigm as it applies to the handling of geo-referenced information. He outlined the major concepts behind the approach and its application in handling spatial information. These concepts have also been presented in much of the research, some of which will be discussed below.

In his conceptualization Worboys defined a geo-object which unifies spatial, temporal, graphical and textual/numerical objects. He also pointed out the little use of proprietary object-oriented systems, except in cases like Milne et al. (1993) and David et al. (1993). Research on the extended relational system has also faced difficulties. Projects in extended relation DBMS on applications of geo-referenced information include work by Lohman et al. (1991) and Rowe and Stonebraker (1987).

Some object-oriented systems have been implemented using object-oriented modeling and commercial OODBMS. Williamson and Stucky (1991) developed a generic GIS supporting earth resource imaging analysis. The system consists of (1) a graphic raster and text interface; (2) a database containing maps, images, and graphical and textual descriptors; and (3) a collection of processes which transform the representation and content of the data objects in the GIS. The system purportedly improved quality of reports, database updates, and analysis and more timely access to widely dispersed information. They argue that a more intuitive interface with the information in the database also reduces the necessary minimum training level for analysis.

COMPARISON OF OBJECT-ORIENTED AND RELATIONAL APPROACHES

Typical modeling approaches use relational databases for data modeling and as the base for programming languages they use to model different transportation processes. However, the object-oriented approach has been said to have superior modeling power because the DDL and DML are merged with the OO programming language. Because of this there is a tendency for the use of OODBMS to expand in comparison to relational models. While the latter models simply describe system states, object-oriented models can describe both system states for data and system processes or behaviors in an integrated context (Rumbaugh, 1991).

Some limitations of existing relational DBMS have been identified by various researchers such as Herring (1992). First, relational DBMS lack extensibility to provide for special application needs (e.g., provisions for the user to add new data types and methods, addition of user-defined code, design of new storage methods, and access to standard packages). A table is created for each entity type, a row corresponds to an entity and columns contain the attribute values. However, a relational schema ends up with many additional tables because an attribute is restricted to being a simple built-in type. An attribute may not be a set of values, and relationships are also modeled using tables. For GIS spatial and non-spatial data, they are complex enough that spreading entities and attributes into numerous tables is undesirable (Wiegand and Adams, 1994).

Some of the advantages of OO modeling are that related data can be kept together and relationships can be directly modeled (Medeiros and Pires, 1994). The new geographic data models based on a set of feature objects provide the needed framework for a scaleless and seamless database (Guptill, 1989; Mainguenaud, 1995). It also has the DBMS characteristic of extensibility needed by GIS (Haas and Cody, 1991). For example, the ability to add new data types (e.g., points) and operations on them (e.g., distance functions) and the ability to have a new set of operations as part of the query language (e.g., overlay) are added benefits (Gunther and Lamberts, 1994).

The use of relational databases has been popular in the past and because of this, a set of reliable and working software tools have been developed. This software includes the core database engine, modeling tools, and application development utilities. It usually incorporates a powerful query language (SQL) which has a sound mathematical basis in relational calculus (Date, 1985) and usually operates on collections of fixed-format tables. As opposed to this, the tools in an object-oriented approach draw on a semantically richer background. Unfortunately, this has allowed greater personalization of the approach and a standard query language has not yet emerged. As a result, two different types of object-databases have developed, one of them following many of the ideas of the relational model and the other designed to be integrated with an OO programming language (such as VERSANT). The goal of VERSANT's type of OODBMS is to transparently provide persistence to the classes set in the OO programming language. Ties to the relational model provide enhanced relational databases with an object interface. Those tied to VERSANT or other similar databases emphasize persistent objects. Other tools include translators between objects on the program side and relational sets on the storage side. Regardless of the tools selected, there is a need for more application oriented technology. The development of this software, however, must focus on questions of its modularity, its performance, its scaleability, its openness, its robustness, and its ease of use (Göllü, 1995).

The advantages and capabilities of object-oriented systems include: (a) The representation of elements in the object-based environment can be categorized into a hierarchy of object classes. The ability to perform various functions on different classes through polymorphism can further differentiate network elements and modal choices in ITS applications. This ability is not readily available without adding additional structure in the existing relational GIS. (b) An object-oriented representation will be more congenial to the perceived environment of an individual. It will therefore provide a more intuitive and user-friendly environment for the implementation of an ITS. (c) It overcomes the difficulties of the planar data structure and will allow for a finer differentiation of various geographic objects in the system. Running routing and spatial search algorithms will become less problematic. (d) It will facilitate the representation and processing of a multi-level transportation network through introducing new classes across different levels and new functions for these classes; spatial search and queries will be handled more efficiently. (e) It will facilitate the integration of a wide variety of geographic objects and relations within a comprehensive geographic database.

There are also important advantages when implementing such object-oriented systems in the real world: (a) Less costly data integration. Because of the costs of acquiring and maintaining geographical data, the cost of developing an ITS can be greatly reduced if part of the data can be shared and integrated across many other applications. An object-oriented approach can greatly facilitate this through a unified scheme of object abstraction and classification. (b) Less costly maintenance and expansion. The modularity of object-oriented systems renders them highly extensible, reusable and maintainable. (c) Higher data access efficiency and reliability of the system.

The major objectives when building an object-oriented system include the establishment of an object-oriented data model through constructing the key abstractions, class structure, and functions that are required. Thus, we suggest that it is possible to build up an object-oriented GIS for ITS applications by generating a set of high-level abstractions of the ITS environment through domain analysis. These abstractions may include the transportation network, the mobile and non-mobile users of the systems, and the activity schedules and the adjustment strategies of travelers. Key mechanisms involved in the system would include processes like data acquisition, message parsing, activity scheduling, routing, spatial search, and information display. The class structure and the module architecture of the system will then be constructed based upon the above analysis. Specifically, elements of the complex transportation network (including routes that utilize various modes such as transit routes) at various spatial scales could be represented in terms of abstract data types supporting inheritance, polymorphism, and dynamic binding.

EXPERIMENTING WITH AN OBJECT-ORIENTED ATIS

The first step in an experiment we conducted was to conceptualize the class hierarchy that could be applied to an ATIS, and develop an object-oriented GIS that could handle inheritance, polymorphism, and dynamic binding. We then developed a script language associated with this object system using C++ and C language; it can handle various mathematical and conditional statements, as well as defining classes and functions. The system is designed to be fully functional within the IBM PC environment.

Based on this preliminary work, we tried to utilize a real-world transportation network and operate it in our object-oriented system. We used an ETAK database of the Santa Barbara area as the primary host for our efforts. This is a comprehensive database available commercially.

Our first step was to read the ETAK mapbase file structure for the Santa Barbara network into our object class hierarchy. Next, we undertook the following steps: (1) completing modification of the ETAK database to include attributes such as travel times and splits, weights, or penalties on turns at signed or lighted intersections, one-way streeting, lane changes, and temporal summaries of traffic volumes; and (2) expanding the class hierarchy in order to accommodate these realistic traffic characteristics.

Methods

Object Layering. The first task was to divide the base map of the Santa Barbara area into layers. Most of the layers can be defined according to the attribute types defined within the ETAK MapBase format. Each record attaches information to a point, line, or region and has a code which determines the type of attribute. An object layer is created for each type of attribute; and each layer contains lists of the following types of entities: circles, lines, points, polylines, nodes, links, and line attributes. The first four are the graphical entities common to most CAD and GIS systems. The nodes and links are the entities used to define the street network and will be described below. The line attributes are records used to describe the line and link entities. They contain a subset of the information attached to each line element within the MapBase format: they include the left and right street addresses, census block numbers, topological information, etc.

Because the MapBase format defines many more attribute types than actually occur within the Santa Barbara county map, most of the layers ended up empty. An important work that was to add in some of the detailed attributes in the ETAK database and build the layers. Transit routes were stored in another layer with pointers referring to the street network in order to perform routing. Space will be conserved by allocating only the entity lists which actually contain items in order to minimize storage.

Partitioning. The major purpose of partitioning is to enhance the efficiency of spatial search. Some means of partitioning the map is required if displaying the zoomed-in map takes nearly as long as displaying the entire map. The reason is that storing the map as a "flat" file requires comparing each entity to the viewing box. The extent of the viewing box is determined by a translatable origin and axes scale factors. This problem can be alleviated by partitioning the contents of each layer. Partitioning allows the rendering of only those sections of the map which overlap with the viewing box without "touching" all of the entities. The increase in performance resulting from accessing only the necessary entities can be observed whenever a spatially localized subset of the map is to be referenced. This will occur not only during the tasks of displaying the map but during many other analyses, including route computation.

The description given above of a "layer" stated that each layer contained a set of entity lists, one for each type of entity. The partitioned entity lists replace that description. Each layer consists of either the lists of entities or a set of partitioned entity lists. A hierarchical spatial tree can be used to implement the partitioning. The tree consists of a set of hierarchically arranged bounding boxes. The bounding box for the entire map is used as the boundary of the "root." In a sense, the unpartitioned entities are stored at the root of the tree because all of the entities are contained within its associated bounding box (by definition). The root bounding box is divided into *N*-by-*N* quadrants to define the 1st level. The original box is the "parent" and the smaller *N*-by-*N* boxes are the "children." Each child is itself subdivided into *N*-by-*N* boxes. This recursive subdivision is repeated until a specified "depth" is achieved. The bounding boxes at the first level define the tree's "leaves." The contents of each layer is partitioned by deciding which leaf's bounding box contains the entity. This decision is made using the entity centroid. The entity is copied to the appropriate entity list in the leaf and is subsequently removed from the unpartitioned list. Although it is possible to use a different spatial partitioning for each layer, the same spatial quadtree is used for all layers.

In the first phase of building the system, only the leaves of the tree will be used to determine whether to draw each partition's entities. The more sophisticated and efficient approach would be to compare the bounding boxes at each level of the tree to the viewing box. A saving in processing time would arise whenever the remaining depth of the tree could be ignored because the bounding box at the given level did not overlap the viewing box.

Object Hierarchy. We require the capability to hierarchically relate different road types for use in route selection. In a multi-modal situation, we also require the different object layers of routes to be related. Code can be written to create these relations at the "layer" level. We build on the recent work of Göllü (1995) for this purpose. The first step is to define a "road-layer" class and then to derive more specific types of roads from this class. "Derive" is the object-oriented term meaning to establish a "kind-of" or "parent-child" relation between objects (Egenhofer & Frank, 1989). The derived or "child" class is said to "inherit" the properties of its "parent" class, including the parent class's attributes and its associated functions. A relation might be established between a class "car" and a class "vehicle." It could then be said that "car is derived from vehicle" and that "a car is a type of vehicle."

The class definitions are stored in text files which, when processed, define the object hierarchy. The class definitions describe both the hierarchical relation as well as the attributes attached to "instances" of each class. In our case, the "instances" correspond to items placed into the given layer. Defining the class hierarchy is taken care of by the first line, "class road: layer." It is useful at the same time to define the potential attributes of entities contained by the hierarchy. The attributes are only "potential" because the entities are not required to have defined attributes.

Associations Between Entities and Attributes. There seem to be several approaches to associating entities and attributes, each having different drawbacks. One is for each entity to contain a pointer to its attribute. When the entity possesses an attribute, the pointer would be set to the attribute's memory address; when the entity does not, the pointer would be set to NULL. A second method, similar to the first, would be to place a pointer to an entity within each attribute. A third method would be to assign a unique identification code (i.e., "handle" or "object identifier") to each entity. Attributes would then refer to their entity using this code. A single-bit flag might be set within an entity to indicate that an entity possesses an attribute.

The first two of these approaches are pointer-based. They have the drawback that they allow only a single attribute to be associated with each entity (unless a more complicated scheme is implemented to allow lists of attribute pointers for each entity). One or the other approach would be preferred depending on from which side most of the computation would be done. If most computation involved only the attributes with infrequent reference to the entities, then the pointers might be better placed within the attributes. An additional consideration is whether the entity-attribute relations will be sparse or dense. The space consumed by NULL pointers under sparse conditions might be unacceptable. For the third method, actually associating an entity with its attribute requires searching the attribute list for the attribute having the entity's identification code. This required computation is what is made implicit within the pointer-based approaches.

Generic Attribute Hierarchies. The capability to define class hierarchies is not specific to layers, but may also be used to define hierarchies of any information (e.g., types of activities, types of destinations). The use of the word "layer" in the road-class definition causes some special handling that creates internal data structures related to the map. Another way to say this is that the keyword "layer" acts as a built-in "base" class. One goal of our continuing research will be to decide what we want hierarchies of, what the attributes should be, and how we will fill in the attributes.

THEORETICAL AND APPLIED IMPLICATIONS OF USING OOGIS IN ATIS

Transportation science encompasses a variety of theories concerning (among others) network structure, routing algorithms, traveler activity patterns, mode choice, demand forecasting, vehicle or traffic assignment, trip allocation, and traveler behavior. This chapter has direct relevance for trip allocation and prediction, activity scheduling, and traveler behavior, and indirect relevance for other theories.

Most transportation models derive their basic assumptions from econometric models. These include assumptions of economic rationality and utility maximization of travel-related choices. Simplified spatial assumptions include those of shortest-path route selection and selection of closest alternatives for interaction purposes. In addition, travelers are assumed to be low risk takers, creatures of habit (i.e., invariably repeat behaviors), and to have perfect information. Such assumptions are often necessary to produce normative or optimal solutions for transportation problems.

In practice, information is usually partial or incomplete, is not available instantaneously, and may be difficult to interpret or integrate into an existing knowledge base. Knowledge bases are imperfect, as are our mental models (i.e., cognitive maps) of environments through which travel takes place. Environments are partially represented. Drivers often have well formed habits that are difficult to change. But much driver behavior is not habitual, but is better described in terms of problem solving. Driver decisions may be rational, but often we do not know the appropriate criteria by which to judge them. Consequently, we limit those criteria to a few that are mathematically tractable. Sometimes we develop theories or models and assumptions that are sensible for dealing with freight traffic, or for dealing with fleet vehicular movements that have little or no driver independence involved, then assume that individualized vehicular traffic can be likewise constrained.

When discussing ATIS, we must remember that we are dealing with individual drivers pursuing their own activity schedules, rather than fleet operators seeking cost minimization solutions to driving problems. There is, therefore, a need to work in a more realistic rather than a more abstract and simplified environment; there is a need to incorporate more realistic assumptions of driver behavior and even network structure; there is a need to consider individual differences in risk taking propensity and preferences for travel. Allowance must be made for both active movers (those seeking solutions to traffic problems) and passive stayers (those waiting *in situ* for someone else to solve the problem).

A viable ATIS must be more realistic than normative. As such it is difficult to imagine a good usable and practical ATIS being driven by existing theoretical principles or concerns. The potential for developing a real-world real-time ATIS may be significantly increased by using a GIS as a host.

A host GIS can allow different levels or classes of objects. Starting with the most basic feature - the transport network - a GIS can share an accurate geocoded database complete with curves, gradients, non-planar intersections, and movement controls; or it could develop cognitively warped and viewpoint specific systems. This introduces versatility into the network representation.

A GIS should provide alternative path selection algorithms with a variety of route selection criteria (e.g., shortest path, fastest path, longest leg first, most aesthetic route, etc.). Such criteria may be most important in selecting alternate routes and alternate destinations after experiencing congestion, hazards, construction, or other travel delay scenarios.

By more faithfully and realistically encoding routes and linking sets of attributes to them, rerouting and changing destinations should become more acceptable procedures. There is little sense in recommending an alternative route that is restricted to local streets with numerous stop signs, if a traveler's criteria for path selection is "fastest route." It makes little sense to recommend an alternative route that crosses many arterials if a traveler wishes to avoid traffic signals or wants to minimize left turns. Most ATIS to date provide only one alternative routing model (often a modified Dijkstra algorithm); however, some do offer a set of alternative route selection models which allow individuals to express their preferences for different criteria.

Object-oriented GIS in particular allows for detailed attributes to be attached to routes. Routes can be conceptualized as continuous lanes over the entire length of a street rather than as multiple arcs with beginning and ending intersection nodes. Furthermore, as shown previously, features such as hierarchical layering, class structures, inheritance of attributes, polymorphisms, and encapsulation processes give much versatility and considerable replication of how we perceive the reality of transportation systems.

Thus, as technology (such as GIS) allows us to model and describe transportation systems more completely, there emerges a need to re-think existing models and theories which were developed in times of sparser knowledge and understanding of both system complexity and human behavior. While not denying the continued usefulness of (normative) theories and models as ways to formalize the state of our knowledge, there is also little doubt that as we enter this new age of information distribution, new sets of constraining assumptions are needed. As we develop more comprehensive tools for obtaining environmental information (e.g., remote sensing, satellite imagery) our need to simplify for the sake of mathematical elegance may be reduced, and as we develop technologies for detailed examination of driver activities and behavior (e.g., traffic and driver simulators) our knowledge of the range of feasible and likely behaviors should escalate - again requiring revisions of limiting behavioral assumptions.

Thus, as we explore the use of current and future technology, there should be direct and immediate feedback to the domains of theory and model building.

CONCLUSION

We have conceptualized and developed an object-oriented geographic information system. Our continuing effort will focus on the data modeling issues of a multi-modal network, and the implementation of a multi-strategy travel decision support system built on this object-oriented system. We have discussed the advantages and disadvantages of an object-oriented approach to transportation modeling. Apart from its improved capability of handling the transport network as more of a perceptually accurate system, the object-oriented data model allows us to incorporate hierarchical layering within the basic network that ties to the normal engineering way of interpreting road systems. Polymorphism allows us to perform ITS functions on various classes of objects in the transportation network, an ability that is not easily obtained in existing systems. The approach also appears likely to substantially decrease the time involved in interacting with the database particularly by using partitioning and inheritance characteristics. Cost of operation should consequently be reduced. Both these factors are important when considering the primary aim of the ATIS component of an ITS is to get useful information to travelers in as timely a manner as possible so that on-route decision making can be undertaken.

There are many basic research problems relating to the development of workable object-oriented data models for use in transportation planning and our research has examined some of these. We also expect that the data model as developed will greatly facilitate the implementation of ITS by more quickly resolving conflicts with respect to ultimate selection of routes, substituting destinations, changing activity patterns, and rescheduling activities.

Our project was designed to contribute to the next generation of traffic management technology, particularly in terms of dispensing information to travelers in a pre-planning or on-route phase via an ATIS. ITS generally appears to be moving more towards Object-

Oriented data structures and models and we believe our work is in line with these nationwide trends.

Some of the critical features involved in evaluating the worth of any object-oriented data model or object-oriented data structure include its ease of use, the relevance of the attributes defined in the system, whether or not the model deals with real or artificial concepts, the degree to which there is a clear translation between model entities and the actual objects, and whether or not there are acceptable matches between those activities undertaken in the real world and those activities incorporated into the model. Other criteria relate to the number of modules embedded in the system that have to be changed in order to work in a real environment, the number of steps that operation of the system requires, the degree to which one must know and accept a process model of the system, and the time that is required to integrate changes in the system to ensure it is a dynamic real time one.

ACKNOWLEDGEMENT

This research was funded by grant DRTS92-G-0009 from the University of California Transportation Center.

REFERENCES

Date, C. J. (1985). *An Introduction to Database Systems.* Edison Wesley, Reading, MA.

David, R., I. Raynal, G. Schorter and V. Mansart (1993). GeO2: Why objects in a geographical DBMS. *The Third Symposium Databases*, 264-276.

Egenhofer, M. J. and A. U. Frank (eds.) (1989). *Object-oriented modeling in GIS: Inheritance and propagation.* Paper presented at the American congress on surveying and mapping, American Society of Photogrammetry and Remote Sensing. Falls Church.

Golledge, R. G. (1995). Path selection and route preference in human navigation: A progress report. In: *Spatial Information Theory: A Theoretical Basis for GIS* (A. U. Frank and W. Kuhn, eds.), pp. 207-222. Springer, New York.

Golledge, R. G., M.-P. Kwan and T. Gärling (1994). Computational-process modeling of household travel decisions using a geographical system. *Papers in Regional Science*, **73**, 99-117.

Göllü, A. O. (1995). *Object management systems.* California PATH Research Report UCB-ITS-PRR-95-19.

Gunther, O. and J. Lamberts (1994). Object-oriented techniques for the management of geographic and environmental data. *Computer Journal,* **37,** 16-25.

Guptill, S. C. (1989). *Speculations on seamless, scaleless cartographic data bases.* Paper presented at Auto-Carto 9, 9th international symposium on computer-assisted cartography, Baltimore, MD.

Haas, L. M. and W. Cody (1991). Exploiting extensible DBMS in integrated Geographic Information Systems. In: *Advances in Spatial Databases* (O. Gunther and H. Schek, eds.), pp. 423-450. Springer, New York.

Herring, J. R. (1992). Tigris - a data model for an object-oriented geographic information system. *Computers & Geosciences,* **18**, 443-452.

Khoshafian, S. and R. Abnous (1990). *Object-Orientation: Concepts, Languages, Databases, User Interfaces.* Wiley, New York.

Koutsopoulos, H. N., A. Polydoroupolou and M. Ben-Akiva (1995). Travel simulators for data collection on driver behavior in the presence of information. *Transportation Research C,* **3**, 143-159.

Kwan, M.-P. (1997). GISICAS: An activity-based travel decision support system using a GIS-interfaced computational-process model. In: *Activity-Based Approaches to Travel Analysis* (D. F. Ettema and H. J. P Timmermans, eds.), pp. 117-134. Pergamon, Oxford.

Lohman, G. B. Lindsay, H. Pirahesh and K. B. Schiefer (1991). Extensions to Starburst: Objects, types, functions and rules. *Communications of the Association for Computing Machinery,* **34**, 94-109.

Mainguenaud, M. (1995). Modelling the network component of geographical information systems. *Journal of Geographical Information Systems*, **9**, 575-593.

Medeiros, C. and F. Pires (1994). Databases for GIS. *SIGMOD Record,* **23**, 107-115.

Milne, P., S. Milton and J. L. Smith (1993). Geographical object-oriented databases - a case study. *International Journal of Geographical Information Systems, 7,* 39-55.

Özveren, C. M. (1989). *Analysis and control of discrete event dynamic systems: A state space approach.* Doctoral dissertation, Massachusetts Institute of Technology, Cambridge, MA.

Ramadge, P. and W. Wonham (1987). Supervisor control of a class of discrete event processes. *SIAM Journal of Control Optimization,* **25**, 206-230.

Roberts, S. A., M. N. Gahegan, J. Hogg and B. Hoyle (1991). Application of object-oriented databases to geographic information systems. *Information and Software Technology,* **33,** 38-46.

Rowe, L. A. and M. R. Stonebraker (1987). *The Postgres data model.* Paper presented at the 13th VLDB conference, San Mateo, CA.

Rumbaugh, J. (1991). *Object-Oriented Modeling and Design.* Prentice-Hall, Englewood Cliffs, NJ.

Van Oosteron, P. and V. Schenkelaars (1995). The development of an interactive multi-scale GIS. *Journal of Geographical Information Systems*, **9**, 489-507.

Wiegand, N. and T. M. Adams (1994). Using Object-oriented database management for feature-based Geographic Information Systems. *Journal of the Urban and Regional Information Systems Association,* **6**, 21-36.

Williamson, R. and J. Stucky (1991). An Object-Oriented Geographical Information System. In: *Object-Oriented Databases with Applications to CASE, Networks and VLSI CAD,* (R. Gupta and E. Horowitz, eds.), pp. 297-311. Prentice-Hall, Englewood Cliffs, NJ.

Worboys, M. F. (1992a). A generic model for planar geographical objects. *International Journal of Geographical Information Systems,* **6**, 353-372.

Worboys, M. F. (1992b). Object-oriented models of spatiotemporal information. *GIS/LIS Proceedings,* **2**, 825-834.

Worboys, M. F. (1994a). A unified model for spatial and temporal information. *Computer Journal,* **37**, 26-34.

Worboys, M. F. (1994b). Object-oriented approaches to geo-referenced information. *International Journal of Geographical Information Systems,* **8**, 385-399.

Worboys, M. F., H. M. Hearnshaw and D. J. Maguire (1990). Object-oriented data modeling for spatial databases. *International Journal of Geographical Information Systems,* **4**, 369-383.

13
CAN WE EVER OBTAIN THE DATA WE WOULD LIKE TO HAVE?

Kay W. Axhausen, Leopold-Franzens University, Innsbruck, Austria

INTRODUCTION

Past practise in transport planning never had any doubts that it had the data it needed for its tasks: OD data, travel diary data, screen line and other counts, and supply data was assumed to be correct, complete, and sufficient for the models used to give the answers which the planners had been asked to give. While one is justified to doubt the models and the data used, they probably sufficiently matched the level of sophistication of the questions asked which mostly concerned rather crude changes to the network and other supply parameters, such as costs or timetables. It is only recently, when the frequent blockage of the application of these past solutions has led to a search for new, less crude solutions, that transport planning has started to doubt its modelling and its ability to obtain all the data required. The new solutions are well known, if seldom implemented:

1. (market) pricing of:
   - road space by link, time-of-day, day-of-week, and season
   - parking space by location, time-of-day, day-of-week, and season
   - public transport by line/location, time-of-day, day-of-week, and season
   - residential and office space by location and quality
2. innovation in the pricing regimes:
   - discount instruments for public transport, such as the German *Bahncard*, which gives a 50% discount on all tickets for a year for a one-time payment of DM 220 (second class) and DM 440 (first class).
   - loyalty schemes, such as frequent flyer cards

   - season tickets for whole groups, such as the *Semesterticket* at German universities providing all students with a season ticket for the term for a single heavily discounted price
3. regulation or deregulation of:
   - time-space regimes, for instance, shopping hours, as in Germany or Austria,service times, as public houses in the UK, etc.
   - working hours, including options such as telework, telecommuning, etc.
   - service provision, for instance regulation of public transport, car sharing, or telecommunications services
   - land use patterns
4. information provision, such as
   - current measurements/estimates for traveller and network operators
   - short-term/long-term forecasts for travellers and network operators
   - traces for the network regulator
   - access to shopping and services, for instance travel agencies, etc.

Some of the older ones remain popular but are often impossible to fund or implement:

5. supply change of:
   - network capacities, in particular also telecommunication capacities
   - network services, that is lanes open, busses running, green times, etc.
   - residential or office space

What is not well known, is how these new measures will interact with the activity systems of the travellers, that is what primary long-term and secondary short-term effects these changes will have. Even the primary short-term effects are difficult to assess, as most models used to evaluate them do not offer the full range of behavioural flexibility actually required to make that full assessment. Good examples are the effects of urban congestion pricing, pricing of parking or of radio-based information systems on the overall activity patterns. The studies have generally focused on the primary effects, but have not attempted obtain a complete assessment of all effects (see, e.g., Polak et al., 1993). Even for the traditional solutions some basic questions are still without a satisfactory answer: the amount of induced demand or the interaction between telecommunication and the demand for physical travel come to mind (Claisse and Rowe, 1993; SACTRA, 1995).

The main problem for most existing model types is that they are trying to capture policies aiming to modulate demand in time and space in a static framework. The frameworks are static as they assume that the whole period they treat, be it an hour or a whole day, is planned and optimized in one step and that all activities are optimized with reference to that period only. If the reverse is true, that is, that persons do not construct their behaviour in one step and that all of their activities have meaning at longer time frames, then new types of models are necessary to describe the effects of policies with confidence. Recent developments are trying to address these challenges from a variety of directions (see Axhausen and Gärling, 1992):

- expansion of the coverage of *joint models of the choice of activity pattern*, timing, destinations and mode employing utility maximising principles (Ben-Akiva and Bowman, 1995, or similar work undertaken by the Hague Consulting Group and its collaborators).
- *explicit scheduling* models which try to model the process of the sequencing of given activities employing a mixture of heuristic and utility maximising principles (Ettema et al.; 1995Gärling et al., 1989).
- *activity pattern execution* models which supplement given activity pattern with information about destination, mode and timing, again employing a mixture of heuristic and utility maximising principles (Axhausen and Herz, 1989, McNally, 1996) or much more sophisticated models such as AMOS (RDC, 1995) and TRANSIMS (Smith et al., 1995).

With the exception of Axhausen (1990), all of these have separate models for the generation of the travel demands and the behaviour of the network. These approaches cover to a smaller or larger extent the behavioural dimensions of interest (Table 13.1).

The purpose of this chapter is twofold: the first is to step back to scrutinize the assumptions about the activity stream which are built into the modelling approaches mentioned above; the second is to speculate about the data needs and modelling frameworks resulting from such a query and the resulting conceptualization of the activity stream. At this stage, I can only speculate given the absence of suitable empirical work - assuming it is possible - and the absence of implemented modelling frameworks. It is hoped that these speculations will foster an interest in the empirical and modelling work implied by these discussions.

The next section will discuss the activity stream and the different possibilities to conceptualize it, while the following section will discuss the data items needed to capture it and the possibility to actually survey them in the field. The form of suitable modelling frameworks will be sketched in final section before some concluding remarks.

Table 13.1.
Behavioral Dimensions Generated by Different Models.

Long-term	**Medium-term**	**Short-term**
Participation in employment	Participation in "projects"	Participation in out-of-home activities
Choice of employment location	Allocation of tasks to household members	Formulation of activities
Participation in education and other long-term commitments		Activity selection
Choice of residential location		Activity sequencing
Vehicle ownership		Choice of information systems
Season ticket ownership		Amount of information acquisition
Ownership of discount instruments		Choice of activity duration
Licence ownership		Choice of persons involved
Ability of cycle		Choice of location
		Choice of mode
		Route choice

THE ACTIVITY STREAM

The traditional view in transport planning is that human behaviour consists of clearly defined elements, called activities, which have a start time, duration, location, and purpose. The purpose does not change for the duration which is fixed à priori. The number of purposes distinguished is normally limited to six to ten. The categories range currently from the general catch-all "At home" to the specific "Picking somebody up/dropping somebody off", although earlier schemes were content with an even less specific set: "home-based work", "home-based other", and "non-home-based".

This paucity of description reflects the past focus of transport planning on the morning commute, that is, a single type of trip. The activity following the trip was actually rather irrelevant to the modelling, as none of its further characteristics, in particular exact timing and duration, were used in the modelling. The method used matched the question asked (Weiner, 1988).

Time budget research has a richer description of human behaviour, as it accommodates the idea of secondary activities, that is, things done or persons attended to in parallel with a main activity, among a small number of others. Still, time budget research has in the past not added any further dimension to its description of the activity stream beyond those used by the transport planners. It has only been recently, that further dimensions, such as person(s) with whom the activity was undertaken or for whom the activity was undertaken have been added (Ås, 1978; Grønmo and Harvey, 1982; Szalai, 1972).

While time budget research encourages respondents to be as accurate as possible in their description of the activity undertaken, it also encourages them to think of activities as something happening in 15, 30 minutes, or even longer blocks depending of the survey format used.[26] The accuracy desired is specified for the respondent either by examples or by coding lists. The current EUROSTAT coordinated time budget survey undertaken by the EU member states employs for example a coding list of about 120 activities which is an order of magnitude higher than usual in transport planning. Nevertheless, the codes include rather undifferentiated categories, such as "work" or "education", which summarize a multitude of different activities. Introspection and observation show that the reality of our behaviour is more complex. It is rather less a series of neat packages than a stream of interwoven acts and purposes. The speed of change between acts/purposes varies from uninterrupted sleep to the frantic business of a secretary in a busy office or of a mother in a household with many children. This stream is aggregated naturally into higher-order categories or block/activities by the respondents, as the multitude of different acts cannot be remembered, but the researcher should be aware that the importance of the acts varies and is not constant over the duration of the constructed activity (Gärling and Garvill, 1993).

It should be stressed then that the activities we obtain from the respondent are a construct which the respondent builds in response to the instructions of the researcher. The instructions state or imply the level of abstraction and the expected duration. It would be interesting to see if the level of abstraction with which daily life is reported in normal story telling is the same in different cultures or if some cultures produce markedly more or less detail.

Transport planning and time budget research focus on "purpose" to describe the intent of the activity, but purpose is only loosely tied to the actual acts performed. One only has to think of the type of acts covered by the purpose work. Still, from a modelling perspective these different types of acts will have a major impact on the choices of the worker. Equally, there is

---

[26] There are alternative survey designs but the roostered time interval design is the most popular.

no clear link between an act and a purpose: preparing a meal can be paid work, pure leisure, or part of homework. The description by one is not complete without the other.

There are two more aspects which should be considered in the description of an act. One is the meaning this act or activity has for the current longer-term aims of the person. This longer-term aims can be rather short term in absolute terms but it is longer then the time frame of the act/activity. It should be clear that multiple meanings of equal rank are possible, as well as multiple higher order meanings/aims. At a relatively long-term perspective, meanings can be as mundane as earning a living or as lofty as to educate oneself. At a shorter term perspective, the equivalents could be to keep a tidy house or to keep informed about the politics of the day.

The other aspect is the project to which the act/activity belongs. A project is a set of acts/activities which belong together to produce a specific object or to stage a specific event: writing a book, furnishing a house, financing a holiday, having a dinner party. They have a starting point and an end point, which is either the successful achievement of the aim or the final abandonment of the project. They also have a history: they are formulated, interrupted, delayed, transformed, rested, slowed down or accelerated, abandoned and picked up again. Projects are the concrete realisations of the aims of the person, but they can in some cases become an aim in their own right. This has happened when the person has to ask himself "Why have I done this", after a project has been concluded. It should be stressed that against a specific time frame not all acts belong to a specific project. There are then many activities which form the background for those specific projects. They are the "overhead" for the various projects. In contrast, all acts have meaning, even if the aims they fulfil are very short term, such as watching TV to kill some left-over time.

A fourth major element, next to type, purpose, and meaning, is the value or utility derived from it an activity. The pleasure of doing something or achieving something.

To summarize the discussion so far, the individual acts of the interwoven stream should be described by: type (what is being done), purpose (for what is it done), meaning (why is it done), utility (how enjoyable is it/having done it), project (to which, if any, does it belong), participants (who else is involved), time (when did it start), duration (how long did it last), location (where did it happen), and tools (what was needed to make it happen, especially money, vehicles, and telecommunication devices). Consider the example of a telephone call which could be for work (purpose) done for a living (meaning) to set up a meeting (project) with a client (participant) using a mobile phone (tool) in the traffic jam on the motorway (location) starting at 8:45 (time) and lasting four minutes (duration). Equally, it could be done for sport (purpose), to keep fit (meaning), to set up a game (project) with a squash partner (participant) using the office phone (tool, location) from 10:00 to 10:15 (time, duration).

The discussion of projects and aims highlights the fact that the acts/activities we observe during a day have many different time horizons forming a complex system of temporal constraints. In addition, the existence of these projects and aims implies that the person has always a backlog of activities which could be undertaken to further the current projects or advance new ones. The long-term calendar of activities which have been already been identified but not scheduled, should hardly ever be empty. It also implies that the activities observed on any one day are as a rule not random for a particular person but reflect the needs and schedules of his or her projects. Clearly, the schedules can be interrupted by unforeseen events, such as surprise visits or sudden opportunities. These random elements have to be identified in a survey and not assumed, just as one cannot assume that the behaviour reflect an optimal solution constructed by the person at the start of the day.

Furthermore, if we allow for the fact that some activities do not achieve their purpose, for instance, we could not obtain the item we wanted to buy, we did not meet the person we had arranged to meet because we were delayed by traffic; the movie became unbearingly silly instead of funny, then it is possible that the schedule outlined at the beginning of the day will be changed dynamically: activity durations are shorter than originally planned; activities are brought forward from future days; activities are rescheduled or rearranged. The same applies for activities for which we achieve the purpose faster than expected: the meeting comes to a resolution early; the item is found immediately; we travelled faster than we expected.

The projects, their associated activities, sequences, and time horizons, are worked out at any one time in different levels of specificity by the travellers. The detail known will vary with the importance of the project, the time until it has to be finished, and the completion of the other projects. Also, the detail will vary over the time horizon of the project, but not necessarily will the detail be highest for those activities planned for the near future. For examples the dates for a holiday including the details of the flights and the trips to the airports will be known exactly, but none of the preparatory work needs to be scheduled until about a week before departure. In contrast, a lecturer might know exactly when he has to prepare the next lecture, but has no idea about the timing of the preparations for the following ones.

The calendar of the institutions and of the year impose rhythms on projects. The seasons encourage and discourage certain activities. The school year and the legal holidays define opportunities for long-distance travel or local festivals. The terms of a university defines the cultural programming of a town. The time table of a sporting league tie its fans to the radio or television. The unique combinations of these various rhythms make the idea of a "normal" day or "week" moot, even an "average" day might be average for the wrong reasons.
The idea of projects and aims also highlights the importance of the commitments persons have to particular activities and by implication other persons, locations and times, as belonging to

that larger frame. What we observe can only be understood in relation to these longer time frames.

These self-imposed constraints are in addition to those socially imposed. Hägerstrand's (1970) classification of authority, coupling, and capability constraints is still useful here but should be given a dynamic dimension. In particular, the coupling and capability constraints can be modified by the individual over time. The coupling constraints involving other persons can be renegotiated or some times, at a cost, be ignored. We phone up to announce a delay or to rearrange a meeting. We stop visiting our parents each weekend, etc. The capability constraints are changeable over a longer time frame by training (to walk faster, to cycle longer distances, but also by getting to know the environment better) or by acquiring new tools (a vehicle, a mobile phone, a new home at a different location, etc.). It should be pointed out that these resources and capabilities have their own time horizons and rhythms. They take time to acquire, they take time to maintain, and because of the transaction costs they are not immediately exchanged. They also form consistent sets which gives change a discontinuous profile (Jones and Clark, 1988). A good example is a house in the exurbs which requires a car for every driver, in-home entertainment facilities, etc., or, equally, the urban flat with no car parking space and no space for even a TV set. To the extent that travellers can select their ownership/membership of a set, our observations will be of self-selected samples.

The time horizons for the renewal or disposal of the various tools a household/person owns generate additional rhythms which have to be considered in the analysis of activities. The ownership of a season ticket binds most travellers to public transport for the whole season. All discussions and preparations for possible change are likely to be concentrated in the time just before the end of the season.

In the discussion of tools and resources, one should not forget that travellers have not only means, time, and money (Heidemann, 1990), but also social means, that is, the ability to draw on the means, time, and money of other persons. This allows the traveller to relax constraints by getting other people to achieve certain aims, but it in return means that at other times the traveller might be constraint to do favours for others. The degree of reciprocity involved might range from the straight tit-for-tat to the completely altruistic. It might involve straight "loaning" of time to any mixtures of means, money, and time (in the form of gifts, services, or loans). The length of the settling periods will vary from the relatively short-term, lets say with the neighbours, to the long-term for members of the family. These exchanges will therefore not necessarily be describable by an instrumental utility maximizing framework (Hargreaves-Heap, 1989), which in turn means, that different logics of behaviour might structure a day.

The discussion has sketched a view of the activity stream as dominated by the constraints of the season and the available tools, money, and time, divided between specific projects and a background stream of activities. The classification of a specific activity to either class will depend on the time frame of analysis. Daily work could be considered part of the background stream for a short-term analysis, while it could be part of a project to acquire a house or other item when considered over the longer term.

In addition, while the activities reflecting basic needs, sleeping, eating and hygiene, will in general be part of the background stream, under certain circumstances they might be part of a project, for example, if the dinner involves non-household members as guests. We know about the rhythms of these basic physiological needs, even if they can be manipulated and suppressed to a certain extent, but we have no empirical data about the choice and rhythm of the various project people engage in.

It is useful to distinguish activity needs and activities/projects when talking about the modelling of activity patterns. Activity needs are the normally unconscious product of the physiological and emotional needs of the person, although in planning the person can anticipate them (Dennett, 1991; Edelman, 1994). These needs become aware when certain thresholds are reached. A person does not have to think about hunger. The person is hungry. In contrast, activity scheduling is a conscious process by which the activity needs are translated into distinct activities. It is at this stage that projects of different time horizons get formulated. The choice of the particular time horizon is crucial, as the scheduling of the next week or month requires different levels of detail from the scheduling of the next hour or half-day. The person has always to maintain the consistency between these different horizons, although we know from observation and introspection that sometimes certain time horizons are given precedence over other: going swimming on a fine summer day instead of writing a paper is a good example.

Activity scheduling will be triggered, among other situations, when the performance of the activity does not meet the expectations. While in many cases it is impossible to change the schedule immediately, there are enough times when it can be changed. Consider for example an outing with the wrong kind of weather, a boring book or TV show. In these cases, one can return early, change the book or TV show. In contrast, it is rarely possible to leave a committee meeting early or send the in-laws away.

These discussions add further items to the activity description, as well as adding further types of data. The data items concern the meaning of the activity as performed and their planning history: When was it first conceived? What was the original aim/context/meaning? When was it scheduled for the first time? When was this successful execution scheduled? When did the original aim/meaning/context change? When could it have been performed last and still achieve its current aim/meaning? Who/what was the beneficiary of the activity? What could have replaced it and still achieve the same aim? How easy/at what cost could it have been replaced? Could it have been dropped and how easily/costly? How important was it in comparison to the other activities performed and still planned for the day? To what extent were the aims of the activity achieved? When was this type (type could be defined at various levels of complexity) of activity performed last? What was the earliest/latest time for the performance of the activity today? or, What were the alternative venues for it?

Next to the activities performed, the further data needs are: the known activity needs; the activities formulated; and the activities planned for a particular time period, described with the same items as the activities performed, but with the questions asked in the subjunctive.

Aim achievement is crucial since it indicates the success and satisfaction with the activity. Transport planners, as a rule, have not concerned themselves with this issue. However in the context of activity scheduling it becomes essential to know whether the activity performance was as expected or whether the traveller had reason to extend the activity or cut it short.

To complicate matters, the achievement can be usefully divided into two components: goal achievement utility and process utility (utility of having done something versus doing it) (see Gärling et al., 1996; Winston, 1982). How these two components accrue will be dependent on the type of activity, its purpose, and its meaning. The goal achievement utility can accrue at the beginning of the activity, if the participation alone has value to the traveller. Consider, for example, the presence at a celebration, where the purpose is achieved when the person has congratulated the guest of honour. Equally, for the chair of a meeting the goal achievement utility accrues at the very end of the meeting, when he or she has been able to achieve all of his or her agenda. The process utility is normally assumed to be increasing, but there is no reason why that should be the case in the presence of a goal achievement utility: for example, being at a dentist is a thoroughly unpleasant experience only endured for the removal of the pain or the improved dental health.

DATA COLLECTION

The conceptualisation of human behaviour sketched above as activity needs of time and situation varying in importance, which are then first translated into lists, then schedules of planned activities, and finally into actions, implies a model of continuous dynamic change.

The matching ideal data collection method is the multiday observation/selfreport of the plans and the executed activities.

This would require the respondent to report 25 items instead of the current 6-7 (location, time, duration, purpose, group size, vehicles used/money spent, type of location), not only for one type but for up to four types of abstractions (needs, activities formulated, schedules, and executed activities), of which the first three could be reported at multiple times during a day or other period. In addition, not only one day with the average three out-of-home activities would be covered but multiple days and more activities, including in-home and at-work activities, as well as telecommunication usage.

It is unlikely that a sample of persons willing to accept such a task is still representative of the population. The level of literacy, introspection, conscientiousness, commitment, and free time seem unlikely to be available in many cases. In addition, these respondents would have to be willing to change reported plans, that is, not to feel bound by what is known to a stranger. Also, the respondents would have to be willing not to plan in too much detail or to omit things, that is, not to respond with changed behaviour to the survey situation. Given the intrusiveness of the requirements, this seems unlikely in most cases.

In this sense, transport planning will never obtain all the data we need consistently from one person. It will be necessary to divide the items across different instruments. The following come to mind: multiday activity diary expanded with some additional items; multiday activity schedule survey, survey of weekly/monthly activity plans/schedules, diary of activity schedule and activities, stated preference survey of scheduling; and gaming simulation of rescheduling.

Other combinations are also possible. This is a first suggestion and needs to be refined and revised.

Multiday Activity Diary Expanded with Some Additional Items

The current one-day diaries of travel behaviour give us no idea of how the activities reported fit into the on-going projects of the respondents. Most of the diaries do not give an idea of the type of in-home, at-work, or telecommunication activities which might be converted into out-of-home activities, although some recent American surveys have made progress with respect to in-home activities. Equally, some recent German work has shown that it is possible to obtain information about telephone calls (incoming and outgoing) from a diary instrument.

The task of this survey would be to span a substantial number of consecutive days, ideally weeks such as the 1973 Uppsala survey (Hanson and Huff, 1982), and to cover the in-home, out-of-home, and at-work activities (private calls, lunch or other breaks, banking or shopping, if such facilities are available at the work place). Ideally, the use of telecommunication tools should also be covered. Each activity should be described with respect to: location (start location, if a movement), start time, duration, tools used (vehicles, telecommunication), party (size and composition), money spent, and purpose, achievement of aims, when first planned, and associated project, if any. Given the workload implied, the coding of the items should be relatively crude. Also, one could consider the use of a computer-based instrument which the respondent could carry with him or her.

Such a survey would give transport planning for the first time a partial understanding of the longer-term structures of everyday life. The impoverished understanding based on single days could be overcome.

The questions of achievement requires special care, as achievement has purpose and meaning specific dimensions, while also being divided in goal achievement and process elements. The possibilities of computer-assisted telephone/personal interviewer (CATI/CAPI) technologies allow appropriate context sensitive questioning but experimentation is required.

Matching the first survey, this survey would focus on the plans for the day, recorded either the evening before or in the morning. The day as a unit for the plans and as recording interval is arbitrary but natural given the many rhythms which work on a daily basis. The description of the activities would be identical to the one above but the respondents would be encouraged to use the "Don't know" category in preference to omitting planned activities for embarrassment. Clearly, it would be best to have the same person perform both surveys, but given the extraordinary work load this would imply and the various possible biases, a separation seems required for long-duration surveys.

The comparison of planned patterns and realized patterns and the degree to which plans are detailed would give us insights into the degree to which the transport/activity system fulfills the desires of the populations and to what extent the travellers can change or adapt their behaviour.

Survey of Weekly/Monthly Activity Plans/Schedules

Extending the scope of the last survey, this survey should obtain an idea of the overall structure of the various projects, in which a person is engaged at any one time. This survey should also probe for the various dimensions of meaning (temporal, personal, household-oriented, etc.). The description of the activities should be as in the fist two but solicit ranges of values and make it clear to the respondent that empty space in the calendar is normal and does not need to be filled at any price.

The results would give us an idea about the amount of advance planning and timing of the detailing of the projects/activities. It would also reveal the number of projects and their relative sizes and importance in terms of time, money, and number of persons involved.

Diary of Activity Schedules and Activities

For a one-day diary it is conceivable that respondents would be able to deliver both planned and executed activities. Obviously, the main objective is to see how the plans are transformed and changed by failures, delays, or unexpected opportunities.

It is very hard to say á priori what the best format for such a survey would be. While mail out/mail-back surveys might be possible, a mixture of a telephone interview and a written instrument might be best with the interview establishing the plans the evening before and the written instrument recording the actual performance. The description of the activities would be the same as in surveys discussed above.

SP Survey of Scheduling

The success of Ettema's MAGIC (Ettema et al., 1993) survey has suggested that it is possible to obtain information about activity scheduling and schedules from stated-preference type surveys. Ettema used a list of activity types to prompt the respondents to formulate activities and to schedule them subsequently. His focus was both on scheduling as a process, as well as the quality of the resulting schedule.

Two types should be performed. The first type should focus as Ettema's has on schedule construction from scratch, that is, from a list of activities drawn from the long-term calendar including background activities, while the second type should focus on rescheduling existing schedules under the impact of unexpected changes (delays, faster achievement of goals, etc.).

The context of the stated-preference interview should also be used to record a completed day with all the items discussed above. Both as a database in its own right, as well as an exercise to alert the respondent to all the various elements to be considered.

The observations from the stated-preferences would allow the modelling of the individual scheduling process and of the schedule quality aimed for while the traveller is constraint for the day/period.

Gaming Simulation of Rescheduling

The stated-preference method described above is restrictive in dealing with only one person at a time and in focusing the attention of the respondent very strongly on "plan achievement". In the tradition of HATS (Jones et al., 1983) or CUPIC (Lee-Gosselin, 1996), these gaming simulation exercise should stress the collective element of the decision making process and the trading of responsibilities, resources, and tasks in the household group/social network. In this setting, it would be useful to make a multiple day period the subject of the experiment, as exchanges between household members are less likely during the daily time frame due to the difficulty of communication and coordination. The first phase of the experiment should construct a week, while the second phase should introduce an element of stress (a member of the family visiting, a sick person, a broken car, a public transport strike, extra work at the place of employment, etc.). The participants should be encouraged to think about as wide ranging changes as appropriate, in particular about longer-term resource commitments and changes.

The analysis of the gaming simulation would give insights into the trading of tasks and responsibilities in addition to the opportunity to model the scheduling process and its outcome.

MODELLING DIRECTIONS

The description of the data collection methods implies a disaggregate modelling approach. Before considering two conceivable disaggregate modelling frameworks, it is worthwhile to mention an alternative. Given the data problems, in particular that the whole data set required for disaggregate models will be assembled from various sources, it might be preferable to abandon such disaggregate approaches and to explore the possibilities of aggregate and rather more restrictive approaches instead which address only particular and limited questions.

The two frameworks to be discussed are the extremes along a continuum of possibilities. These extremes could be labelled an evolutionary approach and an equilibrium approach. The evolutionary approach would describe the persons individually as the agents of a simulation and how they change under the impact of the various policies implemented at different times during the simulation run. Each simulated day would imply a specific date in the development of the system with all the specific characteristics of a specific day (season, weather, events, social time-space regime for the days, random incidents). The simulated persons would learn, that is, change their memories of the system performance, and they would age and change with their household. The simulation of the behaviour in the network is part of the demand simulation. The agents respond directly to the network performance by rescheduling the day, week, etc.

The scheduling model would, most likely, be a satisficing heuristic and would explicitly address the issue of project generation and abandonment. The background activities would be generated through a system of competing hazard-type functions indicating the increase in urgency over time. Past solutions would be reused, if possible. The agents make their choices on the basis of expected performance or information provided and adapt their behaviour on the next day or at the next opportunity. The results of the simulation would be less the mean performance of the system, but the variability of the system and the paths the system takes during its evolution. No "specific" state, such as an equilibrium or only steady state, is achieved or aimed for. Multiple runs of the system would be required. It is obvious that such an evolutionary system could be made to produce a steady state or even equilibrium solution by either removing elements of variability or by respecifying the objectives of the scheduling model. A steady state solution will be required as the initial solution and used to validate the model against observed conditions in the urban area modelled. It is likely that different initial solutions reproduce the aggregate criteria used for validation.

The equilibrium solution at the other extreme will generate a consistent solution with prespecified characteristics, that is, those of a dynamic Wardrop equilibrium suitably extended to address the choice of activity sequence, timing, and duration. Assignment or simulation of the network will be separated from the adjustment/calculations of the demand structures, like in the current TRANSIMS model (Smith et al., 1995). The rules for the calculation/adjustment of the demand structures can be without relation to human decision processes, as long as they are able to produce a valid solution. Memory processes will also be ignored. The "days" of the iterations have no meaning beyond tracing the process of convergence.

Both approaches face the challenge of showing that their status-quo solutions are consistent with the behaviour data and aggregate network behaviour. Given their claim to model scheduling, it will not be enough to reproduce aggregate statistics about link flow or speeds, or even activity durations and the distributions of the activity chain. For both purposes the developers will have to develop methods to compare the joint sequence, timing, and duration distributions. To the best of my knowledge, there has been no substantial work in this area since the work of Recker and McNally (1985), Pas (1984), and Schmiedel (1984) in the early 1980's.

Obviously, both approaches will be extremely computationally intensive and it is not clear when the computing times can be brought down to a practicable level. The consequence of these demands is that the models will have to be used in a focused manner and in combination with less comprehensive models. In analogy to the approach of SATURN (Van Vuren and Vliet, 1992), for example, which models a core of the network in great detail while abstracting the surrounding network, one could consider running the model for a particular area or social group with all detail while modelling the remaining area/social groups more crudely.

The inclusion of "memory" raises enormous questions, as our knowledge of the dynamics of the mental mapping of the performance characteristics of networks and of activity opportunities is virtually non existing. Initial experiments (Axhausen et al., 1994) show complex learning mechanisms already for small networks. Work for real-life situations is unknown to the author. This is an area in which collaboration with cognitive psychologists is urgently needed.

CONCLUSIONS AND FUTURE DIRECTIONS

The challenges of the current policy discussion require that transport planning models accommodate the generation, definition, and scheduling of activities in a household context. Otherwise, the models are wide open to accusations of incompleteness.

This chapter has sketched a conceptualisation of the human activity stream as a series of projects which at different time frames either retain their distinct project character or become "overhead", everyday life. Modelling can progress by either by trying to trace the dynamics of the projects, spanning in many case multiple days or periods, or by developing tools which generate schedules consistent with real behaviour for any one day essentially ignoring the dynamics of the projects (see also Axhausen, 1991).

The data needs for the first type of model are prodigious and it will be nearly impossible to establish a population sample which returns all the required information from the same persons. The information needs will have to be split among different instruments. The majority of these instruments are new and will require testing and piloting before they can be used on a large scale.

Three topics require special attention: methods to establish the consistency of generated activity schedules with observed activity schedules models of network learning and attributation and finally algorithms to extract decision heuristics from SP-experiments or gaming simulations.

ACKNOWLEDGEMENTS

The author is grateful for the comments received during the conference and in particular those of Tommy Gärling. This chapter develops and expands ideas set out in Axhausen (1996).

REFERENCES

Ås, D. (1978). Studies of time-use: problems and prospects. *Acta Sociologica*, **21**, 125-141.

Axhausen, K.W. (1990). A simultaneous simulation of activity chains and traffic flow. In: *Developments in Dynamic and Activity-based Approaches to Travel Analysis*, (P. Jones, ed.), pp. 206-228. Avebury, Aldershot, UK.

Axhausen, K.W. (1991). *The role of computer-generated role-playing in travel behaviour analysis*. Paper presented at the 69th annual meeting of Transportation Research Board, Washington, DC.

Axhausen, K.W. (1996). The data needs of activity scheduling models. In: *Activity-Based Approaches to Travel Analysis* (D. F. Ettema and H. J. P. Timmermans, eds.), pp. 229-241. Pergamon, Oxford.

Axhausen, K. W., E. Dimitrakopoulou and I. Dimitropoulos (1994). *The simple exponential smoothing learning model: some empirical evidence about the smoothing parameter.* Working paper, Centre for Transport Studies, Imperial College, London.

Axhausen, K. W. and R. Herz (1989). Simulating activity chains: a German approach. *Journal of Transportation Engineering*, **115**, 316-325.

Axhausen, K. W. and T. Gärling (1992). Activity based approaches to travel analysis: Conceptual frameworks, models and research problems. *Transport Reviews*, **12**, 323-341.

Ben-Akiva, M. E. and J. L. Bowman (1995). *Activity-based disaggregate travel demand modelling system with daily activity schedules*. Paper presented at the international conference "Activity-based approaches: Activity scheduling and the analysis of activity patterns." University of Technology, Eindhoven, The Netherlands.

Claisse, G. and F. Rowe (1993). Domestic telephone habits and daily mobility. *Transportation Research A*, **27**, 277-290.

Dennett, D. C. (1991). *Consciousness Explained.* Penguin, London.

Edelman, G. S. (1994). *Bright Light, Brilliant Fire: On the Matter of the Mind.* Basic Books, New York.

Ettema, D., A. Borgers and H. J. P. Timmermans (1993). Using interactive computer experiments for investigating activity scheduling behaviour. In: *Proceedings 21st PTRC Summer Annual Meeting*, pp. 267-282. PTRC Education and Research Services, London.

Ettema, D. A. Borgers and H. J. P. Timmermans (1995). *SMASH (Simulation model of activity scheduling heuristics): empirical test and simulation issues.* Paper presented at the international conference "Activity-based approaches: Activity scheduling and the analysis of activity patterns". University of Technology, Eindhoven, The Netherlands.

Gärling, T., K. W. Axhausen and M. Brydsten (1996). Travel choice and the goal/process utility distinction. *Applied Cognitive Psychology*, **10**, 65-74.

Gärling, T., K. Brännäs, J. Garvill, R. G. Golledge, S. Gopal, E. Holm and E. Lindberg (1989). Household activity scheduling. In: *Transport toward 2001,* Vol. IV, pp. 235-248. Western Periodicals, Ventura, CA.

Gärling, T. and J. Garvill (1993). Psychological explanations of participation in everyday activities. In: *Behaviour and Environment: Psychological and Geographical Approaches* (T. Gärling and R.G. Golledge, eds.), pp. 270-297. North-Holland, Amsterdam.

Grønmo, S. and A. S. Harvey (1982). *Time-diaries and trip-diaries: a comparison.* Paper presented at the 10th world congress of sociology, Mexico City.

Hägerstrand, T. (1970). What about people in Regional Science? *Papers of the Regional Science Association*, **24**, 7-21.

Hanson, S. and J. O. Huff (1982). Assessing day-to-day variability in complex travel patterns. *Transportation Research Record, No.* **891**, 18-24.

Hargreaves-Heap, S. (1989). *Rationality in Economics.* Blackwell, Oxford.

Heidemann, C. (1990). *Regional planning methodology.* Technical Paper #16. Institute for Regional Science, University of Karlsruhe, Germany.

Jones, P. M. and M. I. Clarke (1988). The significance and measurement of variability in travel behaviour: a discussion paper. *Transportation*, **15**, 65-87.

Jones, P. M., M. C. Dix, M. I. Clarke and I. G. Heggie (1983). *Understanding Travel Behaviour.* Gower, Aldershot, UK.

Lee-Gosselin, M. E. H. (1996). Scope and potential of interactive stated response data collection methods. In: *Household Travel Surveys: New Concepts and Research Needs, Conference Proceedings 10*, pp. 115-133. Transportation Research Board, Washington, DC.

McNally, M. (1996). An activity-based microsimulation model for travel demand forecasting. In: *Activity-Based Approach to Transportation Modelling* (H. J. P. Timermans, ed.), pp. 37-54. Pergamon, Oxford.

Pas, E. (1984). The effect of selected socio-demographic characteristics on daily travel activity behaviour. *Environment and Planning A*, **16**, 571-581.

Polak, J. W., P. Vythoulkas, P. M. Jones, R. Sheldon and D. Wofinden (1993). *Travellers' choice of time of travel under road pricing.* Paper presented at the 21st PTRC summer annual meeting, Manchester, UK.

RDC (1995). *Activity-based modelling system for travel demand forecasting.* Report to the Federal Highway Adminstration, DOT-T-96-2, San Francisco, CA.

Recker, W. W. and M. G. McNally (1985). Travel/activity analysis: pattern recognition, classification and interpretation, *Transportation Research A*, **19**, 279-296.

Schmiedel, R. (1984). *Bestimmung verhaltensähnlicher Personenkreise für die Verkehrsplanung.* Doctoral dissertation, Institute for urban and regional planning, University of Karlsruhe, Germany.

Smith, L., R. Beckman, D. Anson, K. Nagel and M.E. Williams (1995). TRANSIMS: Transportation analysis and simulation system. In: *Proceedings of the 5th National Conference on Transportation Planning Methods Application.* Transportation Research Board, Washington, D.C.

Standing Advisory Committee on Trunk Roads (1995). *Trunk roads and the generation of traffic.* HMSO, London.

Szalai. A. (ed.) (1972). *The Use of Time.* Mouton, The Hague, The Netherlands.

Van Vuren, T. and D. van Vliet (1992). *Route Choice and Signal Control.* Ashgate, Brookfield, UK.

Weiner, E. (1988). *Urban transportation planning in the United States: A historical overview.* Report DOT-T-88-26, US Department of Transportation, Washington, D.C.

Winston, G. C. (1982). *The Timing of Economic Activities.* Cambridge University Press, Cambridge.

14

BEHAVIOURAL DECISION THEORY: PROBLEMS AND PROSPECTS

John Polak, Imperial College, University of London, UK

INTRODUCTION

The desire to improve the behavioural credibility of travel choice models has been a recurrent theme in transport modelling research over the past 30 years and it remains one of the central motivations underlying a broad range of contemporary modelling work. The preceding seven chapters illustrate a number of the ways in which the research community is currently tackling this challenge. The chapters can usefully be grouped into four broad categories as follows. The contributions by Stern and Svenson discuss fundamental conceptual and behavioural issues concerned with the modelling of travellers' decision making process. The chapters by Kitamua and Fujii and by Mahmassani and Jou present dynamic and activity-based approaches to modelling complex, multidimensional travel choices, and the chapter by Pas reviews the contribution of time-use studies to the modelling of travel demand. The chapter by Kwan, Golledge and Speigel focuses on developments in modelling technology, in particular the role of Geographical Information Systems and object-oriented programming in the representation of spatial processes and spatial decision making. Finally, the chapter by Axhausen addresses the vital question of what types of data we should and can collect and how data collection and modelling in transport and related fields might in the future be more closely integrated.

In the remainder of this overview I discuss in more detail the contribution of the chapters to these topic areas and seek to identify common themes, unresolved questions, and fruitful directions for future research.

CONCEPTUAL AND BEHAVIOURAL ISSUES IN DECISION MAKING

Svenson's chapter reviews a number of approaches to understanding human decision making and traces the development of behavioural theories of decision making, drawing mainly on research in psychology and the other behavioural sciences. He proposes a four-level typology of decision making processes ranging from, at one extreme, those which are largely habitual in nature and may take little account of the actual characteristics of a particular set of the choice alternatives to, at the other extreme, those that involve careful deliberation in the construction of relevant choice sets and the evaluation of choice alternatives. It is interesting to note that the idea that the same individual may use different decision-making procedures in different contexts (which is both intuitively plausible and finds empirical support in the behavioural science research) has received relatively little attention in research on travel behaviour. There may therefore be important work to be done by transport researchers in identifying the incidence of different decision making regimes and in particular, the conditions under which these regimes change (e.g., under what circumstances does a hitherto habitual decision become the object of active reconsideration).

Svenson draws an important distinction between approaches to modelling decision making that are primarily structural and those that are primarily process-oriented. According to this classification, almost all existing approaches to travel choice modelling are broadly structural, in that they concentrate on modelling the final decision but pay little attention to the cognitive and psychological processes leading to that decision. By contrast, process-oriented approaches are explicitly concerned with how the decisions are made, as well as which decisions are made. The debate between advocates of structural and process explanations has a long and bloody history in the social and behavioural sciences, with little sign of victory on either side (Polak, 1997). Whilst on the one hand, structural approaches tend to be relatively parsimonious and tractable, they shed little light on the mechanisms of decision making and are therefore vulnerable to the charge of superficiality. Moreover, if viewed from a psychological perspective they can often be construed as making wholly unreasonable assumptions regarding, for example, decision makers' perceptual or cognitive abilities. Process explanations, on the other hand, typically offer a more compelling narrative explanation of decision making and may be associated with more fundamental regularities in decision making but they often rest upon assumptions regarding perceptual and cognitive processes which are inherently very difficult to test or calibrate. Moreover, experience has shown that they are not necessarily more accurate than structural approaches in predicting final choices.

The final section in Svenson's chapter presents a brief overview of Differentiation and Consolidation (Diff Con) Theory (Svenson, 1990), which is a relatively new approach that seeks to unify a number of existing process-oriented theories of decision making. An important feature of Diff Con theory is that it acknowledges that decision makers frequently have it within their power to modify key parameters of the decision problem, including their goals, the decision rules and regimes, and the (perceived) attributes of the choice alternatives. This may provide important insights into how travellers are often able to simply "absorb" changes in service levels and other attributes without there being any manifest behavioural response (even when viable choice alternatives exist). Svenson outlines how Diff Con Theory might potentially be applied to modelling mode and other travel choices as a multi-level and multi-regime decision processes.

Stern's chapter touches on many of the same issues as Svenson's but does so from the starting point of a detailed consideration of drivers' decision making in highly congested travel conditions. Stern notes that in such conditions, decision making is inherently dynamic and explores the implications of this for behavioural modelling. His discussion highlights a number of key issues. The first of these is the concept of a decision maker's knowledge. Knowledge underpins all aspects of decision making from the setting of goals to the evaluation of choice alternatives. However, there are very few models indeed that provide anything approaching a satisfactory representation of the accumulation and use of knowledge. This is of course not surprising, since representing travellers' knowledge clearly poses a number of substantial challenges including dealing with spatial, topological, and temporal complexity, integrating experiential and other sources of information, and representing the operation and use of memory. Notwithstanding these obvious difficulties, it remains difficult to see how truly credible models of dynamic decision making can be developed without broaching this topic.

Part of the difficulty in achieving an adequate representation of travellers' knowledge relates to the second of Stern's dynamic decision making concepts, which is the issue of learning. Without a concept of learning, any representation of knowledge is rendered somewhat sterile; it essentially becomes just another source of inter-personal heterogeneity. And likewise the treatment of learning is meaningless without a concept of what is being learnt. Thus the treatment of knowledge implies the treatment of learning and vice versa and in my view both issues are central to the treatment of dynamic decision making. Despite the complexities involved, this must surely be one of the most important topics for future research efforts.

Stern also emphasises the fact that highly congested contexts give rise to choices that are made under uncertainty. As congestion has increased so has the unpredictability of many aspects of transport system performance, conferring upon many travel decisions a significant element of uncertainty. For example, in congested conditions if a driver is faced with the decision of whether or not to divert, this choice will typically be between two uncertain prospects; the perceived distribution of travel time on the current route compared to the perceived distribution of travel time on the alternative route. It is a curious fact that although behavioural scientists (including economists) have over the past decades devoted an enormous effort to studying decision making under uncertainty and have developed a wide range of different approaches, the efforts of transport researchers in this area remain sparse. It would therefore appear that there is both the need and the opportunity for further useful work on this topic.

Stern also points out that many travel choices are made under circumstances in which the time available to deliberate upon the decision is limited. This phenomenon is perhaps most obvious in the case of en-route decisions, but may also apply to other aspects of travel choice, such as departure time decisions. Stern argues that the consequence of this type of constraint acting on the decision making process itself is potentially profound, influencing the nature and amount of information that can be processed in arriving at a decision as well as how it can be processed. Stern develops further the idea that the time taken to reach a decision is an important aspect of the decision making process in his exposition of Decision Field Theory (DFT) (Busemeyer and Townsend, 1989). The essential idea in DFT is that during the decision making process, attention fluctuates between choice alternatives as relevant information is retrieved, processed, and integrated into the decision making process. This process continues until one alternative emerges as sufficiently superior to the others to justify being chosen. The output of the DFT process is a decision or response time and a set of choice probabilities. Stern discusses the potential effect of a wide variety of personal, situational, and transportation-system related factors on the development of the DFT process.

Both Diff Con Theory and DFT offer illuminating insights into the decision-making process and alert us to the fact that utility maximisation need not be the final word in such matters. Confronted with the generally lamentable explanatory performance of most existing choice modelling approaches, I suspect that many travel behaviour researchers may in principle be sympathetic to the arguments of behavioural scientists who favour more process-oriented approaches, such as Diff Con and DFT. However, I doubt that theoretical arguments, no matter how eloquently delivered, will be sufficient to ultimately persuade even the sympathisers, let alone the skeptics. Rather, it will be necessary to demonstrate empirically that the inclusion of process elements in transport models can indeed result in tangible benefits in model performance (preferably but not necessarily in terms of improved predictive ability). This type of work calls for much closer cooperation between the behavioural scientists who advocate process oriented approaches and the travel behaviour community. The papers included in this volume illustrate the potential that exists for fruitful and creative dialogue between these research communities and I should like to see such cooperation become much more commonplace in the future.

MODELS OF COMPLEX TRAVEL BEHAVIOUR

The chapter by Mahmassani and Jou presents the latest in a series of studies undertaken by Mahmassani and colleagues over the past decade or more which have investigated the day-to-day dynamics of commuters' travel behaviour (e.g., Mahmassani, 1997). Like several previous studies in the series, this chapter investigates route and time of day switching behaviour but it extends past work both by explicitly taking into account the influence of trip chaining on the propensity of travellers to switch and by carrying out the simultaneous estimation of all model parameters within a consistent multinomial probit framework. The effect of trip chaining is represented by classifying the commute movement according to the nature of the activities undertaken at the stops (if any) between home and the work destination. The data used in model estimation come from a sample of almost 200 travellers who participated in a 10 day commuting diary survey. The chapter is therefore also of interest because it is one of the few examples in the field of travel behaviour of an approach that has been largely pioneered in a laboratory context being applied using data from real transport systems.

A particular feature of this chapter (and the wider corpus of work it represents) is that it eschews a conventional (random) utility maximising theory of travel choice and instead works with the assumption of a boundedly rational decision maker. The notion is operationalised by means of the concept of indifference bands expressed in terms of schedule delay relative to the preferred arrival time at work. Travellers are assumed to continue with their existing choice of route or departure time on day $t+1$ unless they experience a schedule delay on day t that falls outside their current indifference band, in which case they switch. Separate indifference bands are estimated for early and late schedule delays and for departure time and route switching behaviour. The specification of the model permits the use of an error structure that includes both contemporaneous and serial correlation between route and travel time indifference bands and also for the value of the indifference bands themselves to vary over time. In addition, the network travel times in the model are not constant but vary from day-to-day, according to prevailing traffic conditions. Together these features mean that this study provides a valuable empirical perspective on the important issues relating to goal adaptation and choice under uncertainty discussed by Svenson and Stern.

The estimation results indicate that travellers tend to relax their indifference bands (thereby reducing the likelihood of switching) if they experience high variability in travel time or if they have previously experienced conditions which led to switching on earlier days. This suggests that travellers are prepared to dynamically modify their goal (in this case, to arrive at work within the time window implied by the indifference band), if the attainment of that goal is found to be problematic. This type of behaviour is consistent with the notion of variable aspiration levels, originally proposed by Simon (1955) and also discussed by Svenson in the context of Diff Con Theory.

The framework presented by Mahmassani and Jou is simple, elegant, and has proven itself to be robust across a number of different laboratory and real-world applications. However, there are several aspects that might usefully be developed further in the future. At present the sole criterion upon which the boundedly rational choice mechanism operates is schedule delay. It might be desirable to extend the model specification to include a more general definition of the switching criterion. After all, it is certainly likely that schedule delay is not the only factor that affects network choice behaviour. Moreover, such a development would also make it easier to extend the concept of bounded rationality to other dimensions of choice, such as the choice of mode. A second potential area of development concerns the modelling of learning. The current models describe the day-to-day evolution of the indifference bands, but do not include an explicit model of travellers' knowledge or the learning mechanism. Although a learning model would be highly desirable, developing such a model is clearly very challenging, not least because of the need to collect a long time series of data on anticipated and experienced travel characteristics, at the individual level. However, recent developments in passive tracking and monitoring using Global Positioning System (GPS) and Geographical Information System (GIS) technologies may be able to make an important contribution in the collection of such data.

The model presented by Mahmassani and Jou is essentially focused on day-to-day network-level adjustments in behaviour and does not deal with the wider behavioural issues associated with the generation and adaptation of travel demand (e.g., the generation of trip chains is treated as exogenous). By contrast, the chapter by Kitamura and Fujii focuses explicitly on the behavioural processes leading to the generation of travel demand. They present two computational process models (PCATS and AMOS) that approach the modelling of travel demand from an activity-based perspective.

The essential characteristic of the activity-based approach is that it assumes that individuals' demand for travel is derived from a more fundamental demand to participate in spatially and temporally distinct activities. Advocates of this approach argue that by directly modelling these more fundamental activity participation and scheduling decisions models are obtained that are better specified, make better use of the information available in typical travel behaviour data and are better suited to the task of predicting travellers' response to contemporary transport policy measures. In comparing the motivations and concerns of behavioural scientists and activity-based researchers it is interesting to note that whereas the former group tend to focus their attention on the mechanisms of the decision-making process, the work of the latter group is to a much greater extent concerned with achieving an improved conceptualisation of the behavioural process being modelled.

PCATS is a good example of a model that concentrates on providing a plausible structural representation of the behavioural process, rather than the details of decision making. Drawing on Hägerstrand's notion of space-time prisms (Hägerstrand, 1970), the model generates activities and associated travel for each individual taking into account the individual's feasible space-time prisms. The problem of simultaneously modelling the spatial and temporal characteristics of activity participation and scheduling is decomposed into a sequence of decisions concerning the choice of activity type, activity location, destination, mode of travel, and activity duration with each decision dimension being conditional on the predecessor dimensions. Thus chains of activities and associated travel are built up sequentially activity by activity. Nested logit and duration models are used to derive choice probabilities and activity duration at each stage. This sequential decomposition simplifies considerably the computation and estimation of the model, but also limits the extent to which the model is able to represent the effect of factors that are global to the activity-travel chain or overall activity programme of the individual, such as the overall balance of time use. The authors report that work is currently underway to address this and other limitations of the current model.

The second model presented by Kitamura and Fujii is AMOS. Unlike PCATS which synthesises complete activity-travel chains, AMOS is concerned with modelling the modification of existing activity-travel chains in response to policies for managing travel demand. For each individual in the application sample, AMOS takes the existing activity-travel chain and in the light of a particular set of demand management policies, generates a set of possible response options, which are expressed in terms of various modifications to the existing activity-travel chain. It then evaluates the alternative activity-travel chains using a utility function based on time use in in-home and out-of-home activities and on the basis of this criterion either accepts one of the modified chains or triggers a further round of response option generation and activity-travel chain modification. The crucial elements in AMOS seem to me to be the mechanisms used for response option generation and the activity-travel chain modification. Researching these mechanisms of search and adaptation clearly poses considerable difficulties, the principal one probably being that of how to capture information on the operation of processes that for the individual may be substantially a cognitive rather than a behavioural experience. This is yet another area in which a greater interaction between travel-behaviour researchers and behavioural scientists may prove valuable.

Both PCATS and AMOS are examples of a growing trend towards the use of microsimulation modelling in the estimation of travel demand (Miller, 1997). In principle, microsimulation lends itself naturally to the development of dynamic and evolutionary models, yet at present virtually all existing microsimulation models (including PCATS and AMOS) are essentially "single-day" models. Given the established importance of multiday behaviour in the context of activity scheduling, the extension of current models beyond the "single-day" mode would seem to be an important area for future research.

The chapter by Pas reviews the development of time-use studies and their contribution to recent work in the field of activity-based modelling. The chapter covers the theoretical background to time-use studies, the application of time-use concepts to activity-based modelling, and several issues related to the collection of time-use data. Pas points out that transport researchers have traditionally had a very narrow conception of time, treating it principally as an attribute of alternative destination or modes. However, time in fact plays a fundamental role in the organisation of human activities and hence in travel. In particular, the decisions we make concerning activity participation are decisions about how to allocate our time, considered as a scarce resource, and the decisions we make regarding activity scheduling are essentially exercises in temporal planning, undertaken over a variety of planning horizons. Activity based researchers are now beginning to appreciate this wider role and Pas builds on this recognition by proposing that time use be seen as the unifying theme for a theory of travel behaviour.

Pas also highlights the fact that there exists a substantial body of existing theoretical and empirical research undertaken that has been carried out by time use research community (e.g., Harvey, 1997) and that this work contains important lessons for the transport research community. For example, time-use research has emphasised the importance of time use within the home and the increasing scope for substitution between in-home and out-of-home activities. Time use research has also alerted us to the fact that many activities may be conducted simultaneously (e.g., childcare and housework or working during the course of travel). And at a methodological level, time-use diaries, which typically involve a sampling in time of activities, provide an interesting alternative to conventional travel-diary instruments. All in all, it seems to me to be highly desirable that the tentative level of interaction we are already seeing take place between time-use researchers and transport researchers should continue in the future and that this interaction would greatly benefit both research communities.

DEVELOPMENTS IN MODELLING TECHNOLOGY

The chapter by Kwan, Golledge, and Speigle discusses alternative models for the representations of spatial (network) information in the context of Advanced Traveller

Information Systems (ATIS). The chapter compares the conventional link-node representation used in the vast majority of existing transport analysis and modelling projects with an object oriented GIS (OOGIS) representation. The chapter also raises a number of more general questions regarding the comparative merits of relational and object-oriented databases in the context of model development. Kwan et al. put forward a compelling case for the advantages of OOGIS representations. These include greater flexibility in the representation of spatial detail, improved economy and transparency of representation (principally by means of the use of inheritance mechanisms), and greater flexibility in the viewing, editing, and manipulation of spatial entities. The importance of these factors is likely to increase further in the future, especially when the need arises to link detailed activity-based models of travel demand to network-based routing models. In these circumstances, it will be necessary to considerably expand the range of information encoded at the network level, for example to include information about activity opportunities and their attributes. OOGIS may significantly reduce the problems associated with this.

Kwan et al. also make the important observation that there is enormous diversity across the population in the perception and processing of spatial information. Taking account of this diversity is clearly crucial from the point of view of developing successful ATIS but it is no less relevant to the development of successful transport models. Indeed, it would be fair to say that to date transport researchers have largely ignored the implications of differences in spatial perception. Yet with the increasing prominence of ATIS and related systems, the need for a better treatment of this aspect of spatial behaviour has become more pressing.

One further question that is raised by the discussion of modelling technologies concerns standards. At present there are no commonly agreed standards within the transport research community for the representation of network level information (or, for that matter, for the representation of other transport data). The lack of such standards inhibits the exchange of information, models, and methods within the research community. It may be the case that OOGIS technology can provide a suitable foundation upon which to develop widely agreed standards in the future.

DATA SOURCES AND DATA COLLECTION

The chapter by Axhausen raises a number of important questions both regarding our current conceptions of activity participation and scheduling behaviour and the related question of how we go about collecting useful data on these processes. Axhausen argues, somewhat controversially, that our current concepts of activity participation and scheduling are more an artefact of the current methods we use to collect activity-travel data than necessarily a true reflection of the underlying behavioural and cognitive processes involved. Rather than being neat, crisply defined entities that we juggle and shuffle in a manner akin to coloured blocks on a board, Axhausen argues that activities should be understood as much more complex, dynamic, and mutable entities. He argues that individual activities need to be understood not as isolated events but as part of a stream of interwoven and mutually interdependent activities that are continuously being updated and modified over a variety of different timescales. At any point in time, some characteristics of the future activity stream may be clearly defined and well-planned whereas others may be only loosely defined, even right up to the point at which the activity is executed. Likewise, at any point in time, we may be engaged in a number of different activities that serve a variety of different purposes. Furthermore, the objective or motivation of one activity may be dependent upon other activities, leading to the concept of projects understood to be groups of related activities that contribute to the achievement of a particular objective. A central element in Axhausen's conception of activity participation and scheduling is the idea of planning, which he sees as operating continuously on different timescales and different strategic and tactical levels.

If this conception is correct then, as Axhausen points out, there are important implications for our current and future data collection efforts. The first is that the task of collecting relevant and adequate data on the activity scheduling process may be considerably more demanding (both for the analyst and the respondent) then we presently think, requiring, for example, more data items per activity, substantially improved coverage of in-home activities, and, as a prerequisite, multiday data. However, not only would the scale of the data required substantially increase, but the nature of the data would change. In particular, it would be necessary to collect data on cognitive processes such as the planning and modification of activity schedules and on discrepancies between planned and realised schedules. Axhausen discusses these requirements in relation to current data collection methodologies and makes a number of concrete suggestions for potentially fruitful research directions. One point that emerges clearly from the discussion is that it is almost inconceivable that researchers will be able to obtain all the data they require from a single survey instrument or a single individual. This means that methods of combining data from a variety of different sources to produce a consistent and plausible synthetic "base" population are likely to emerge as an important topic of methodological research in the future.

Summary and Conclusions

The seven chapters included in this section embrace a wide variety of disciplinary and professional interests and approaches. Yet despite this diversity there are a number of common themes that emerge. Perhaps the most striking commonality is the increasing emphasis that is now being placed on the development of process oriented theories and models, both by behavioural scientists and travel behaviour researchers. Linked to this, it is also apparent that there exists a growing interest in the relationship between cognitive and behavioural processes, whether this be in the context of unpacking the "black box" of structural decision making theory or exploring the role of the planning-execution-update cycle in the activity scheduling process. A further element present in several of the contributions is the notion of using simulation modelling techniques (rather than wholly analytic procedures) as a framework within which to animate process oriented models.

All these developments bear witness to the increasing influence of the behavioural and cognitive sciences on travel behaviour research. I have no doubt that the infusion of new thinking from these disciplines opens up exciting possibilities for the field, just as in previous decades the influx of ideas and methods from the fields such as physics, economics and geography provided the spur for a whole range of significant advances (Pas, 1990). Yet whilst welcoming these new approaches and the enormous potential they hold, I feel it is important that we should to retain the capacity to appreciate what is good in our existing approaches. Moreover, it is vital that our critical faculties should apply with no less rigour to new theories and approaches than they do to the old. It therefore seems to me that a key issue for the future centres on how best to empirically evaluate and test alternative theoretical and empirical approaches, since it is only through meaningful empirical appraisals the true merits of competing approaches can be established.

References

Busemeyer, J. R. and J. T. Townsend (1989). Decision field theory: A dynamic-cognitive approach to decision making in an uncertain environment. *Psychological Review*, **100**, 432-459.

Hägerstrand, T. (1970). What about people in regional science? *Papers of the Regional Science Association*, **23**, 7-21.

Harvey, A. S. (1997). From activities to activity settings: Behaviour in context. In: *Activity-Based Approaches to Travel Analysis* (D. F. Ettema and H. J. P. Timmermans, eds.), pp. 209-228. Pergamon, Oxford.

Mahmassini, H. S. (1997). Dynamics of commuter behaviour: recent research and continuing challenges. In: *Understanding Travel Behaviour in an Era of Change* (M. Lee-Gosselin and P. R. Stopher, eds.), pp. 279-314. Pergamon, Oxford.

Miller, E. J. (1997). *Activity-based travel behaviour modelling in a microsimulation framework.* Resource paper presented at the 8th international conference on travel behaviour research, Austin, TX.

Pas, E. I. (1990). Is Travel Demand Analysis and Modeling in the Doldrums? In: *Developments in Dynamic and Activity-Based Approaches to Travel Analysis* (P. Jones ed.) pp 3-27. Gower, Aldershot, UK.

Polak, J. W. (1997). A new direction for stated preference research: The case for an experimental approach. In: *Urban Travel Survey Methods: Measuring the Present, Simulating the Future* (P. Bonnel, R. Chapleau, M. Lee-Gosselin and C. Raux, eds.), pp 319-335. Cartier, Lyons.

Simon, H. A. (1955). A behavioural model of rational choice. *Quarterly Journal of Economics,* **69**, 99-118.

Svenson, O. (1990). Some propositions for the classification of decision situations. In: *Contemporary Issues in Decision Making* (K. Borcherding, O. I. Larichev and D. M. Messick, eds.), pp. 17-31. Elsevier, Amsterdam.

PART 3. STATISTICAL THEORY

15

MEASURING WILLINGNESS-TO-PAY FOR TRANSPORTATION IMPROVEMENTS

Daniel McFadden, University of California, Berkeley, CA, USA

INTRODUCTION

A fundamental premise of economic science is that consumers have well-formed, stable preferences, and that their choices reflect these preferences. If this is true, then it is possible to deduce from choice behavior, or from carefully phrased direct questions about preferences, whether transportation improvements or other public policy initiatives are socially desirable. In practice, consumers make errors in perception and judgment that violate classical economic rationality. It is often still possible to use observed behavior to judge the social desirability of improvements, but additional problems arise for measurement and analysis. This paper discusses these problems for the estimation of willingness-to-pay (WTP) from choice observations in real or hypothetical markets, or from direct preference elicitation. Following common use, I will refer to revealed preference (RP) analysis of real market choice behavior and stated preference (SP) analysis of hypothetical market choice behavior. I will use the term preference scaling methods (PSM) from applied psychology to refer to various methods for measuring preferences, including SP analysis, scaling of perceptions, attitudes, and risk judgments, and direct elicitation of WTP. Included in PSM are overlapping techniques known in marketing as conjoint analysis (CA) and in resource economics as contingent valuation (CV).

The second section of this chapter reviews the random utility model (RUM) that forms the conceptual linkage from preferences to choice, and discusses some issues that arise in applying this model to transportation behavior. The third section reviews the problem of computing WTP from RUM-consistent choice probabilities. The fourth section considers the conditions under which choice behavior in real or hypothetical markets will be consistent with a RUM that accurately represents underlying preferences, and examines the major issues in designing reliable PSM experiments.

RANDOM UTILITY MODELS OF TRAVEL BEHAVIOR

Travel Demand Models

Travel demand models are used to forecast the impacts on the transportation system of changes in attributes of travel modes or destinations, and can also be used to estimate demands for visits to alternative sites. Travel demand models are particularly useful for analyzing the demand for quasi-public goods such as recreation that are not priced through user fees, but are indirectly priced through the time and dollar cost of travel; this is called the travel cost method (TCM) for measuring WTP. I will use the example of anglers choice of fishing sites, and the value of restoring sites that are injured by pollution, to motivate the discussion in this paper. However, the issues that arise here are the same in more general applications such as determining WTP for new or improved shopping destinations, travel modes, or vehicles.

The problem of travel demand analysis and valuation of attribute changes is simple in concept and complex in practice. The demand for trips to a destination is influenced by the availability of travel and non-travel substitutes. When consumers are utility maximizers and improvements to a site can be expressed as an equivalent reduction in travel cost, a consumer's WTP is measured by the Hicksian consumers' surplus attached to the equivalent price change (see Diamond and McFadden, 1974; Hicks, 1939; McFadden, 1997; Varian, 1984). Mean WTP in the population of consumers, weighted if necessary to adjust for differences in private and social marginal utility of income, is a measure of social value.

The first complication is that different individuals facing the same menus of alternatives make different choices. Then, the theory of utility maximization and of equivalent and compensating variations that provide the foundation for consumer surplus measures must accommodate unobserved heterogeneity in tastes. It is now conventional to tackle this using the RUM. The basic ideas of applying the RUM to transportation decisions and using it in cost-benefit analysis were introduced by McFadden (1973, 1974, 1981), Diamond and McFadden (1974), and Domencich and McFadden (1975).

A conventional RUM for recreational travel looks something like this:

$$\begin{bmatrix}\text{Utility}\\ \text{of a site}\end{bmatrix} = \left\{\begin{bmatrix}\text{Disposable}\\ \text{Income}\end{bmatrix} - \begin{bmatrix}\text{Travel}\\ \text{Cost}\end{bmatrix} - \begin{bmatrix}\text{Value}\\ \text{of Time}\end{bmatrix}\right\} \cdot \alpha + \begin{bmatrix}\text{Hedonic}\\ \text{Attributes}\end{bmatrix} \cdot \gamma + \varepsilon \quad (1)$$

where α and γ are parameters; ε is a disturbance that captures all unobserved heterogeneity; costs and income are deflated by a price index so that they are in real terms; and "sites" include both the no-travel option and alternative recreation destinations. The hedonic attributes of a site include components whose improvements we wish to value, such as fish abundance in the example of anglers' choice of site.

An important assumption is that the disturbance ε for an individual consumer is fixed, and is not altered by changes in attributes. Thus, randomness in individual tastes is absent, or suspended for the purpose of determining choice under counterfactual scenarios where the quality of some sites is improved.[27] This is an assumption of consumer sovereignty, requiring that tastes are defined prior to market circumstances and invariant with respect to these circumstances. This assumption is a fundamental prerequisite for economic welfare calculations, and is also essential for identification of preferences from observations on choice behavior.

A consumer's WTP for a quality improvement in a site attribute, such as an increase in fish abundance when injury due to pollution is removed, is obtained as an equivalent variation - the adjustment in income required to make the consumer exactly as well off after removal of the injury as he or she was when the injury was present. If a consumer visits an injured site despite the injury, he or she incurs its full impact, and his or her WTP for a change in the degree of injury is the ratio γ/α times the change. If a consumer visits an injured site only after the injury is eliminated, his or her WTP is less, since he or she was sheltered from the full impact of the injury by use of a substitute. The exact WTP calculation depends on the behavior of the disturbances ε; in multinomial logit and nested multinomial logit models, these calculations are enshrined in what are called "log sum" formulas. Economically, this is a consumer surplus calculation, estimating the area behind the demand curve for the injured sites between their equivalent prices for the injured and non-injured states.

There are subtle modeling issues in the handling of hedonic site attributes. If attributes that are correlated with the degree of injury are omitted, then the coefficient γ spuriously attributes

---

[27] The WTP calculation is not harmed by intra-consumer randomness in tastes, provided the distribution of tastes among consumers facing the same market circumstances is *invariant* with respect to changes in measured attributes. If, alternately, tastes shift with market circumstances, there is a fundamental failure of the economic welfare calculus, as well as a fundamental failure of identification that leads to confounding of taste changes and changes in attributes of alternatives.

their impact on behavior to the injury. On the other hand, included attributes whose levels are influenced by the injury will capture part of the overall effect of the injury, and may not be handled appropriately when the WTP calculation is done.

Travel Times and Costs

Consider the term in the indirect utility function involving income, travel cost, and travel time. The level of the estimated coefficient α, which is inversely proportional to WTP, depends on how travel costs are measured and how travel time is handled. Economic theory is clear: Travel costs should include all components that vary at the margin with recreational choices, and time spent traveling to and engaging in recreation should be valued in terms that reflect the individual's opportunity cost of time. Implementation of this principle introduces complications.

Individuals value time spent in different activities differently. In-vehicle travel time is not as onerous as working for most people, and they value it at a substantial discount relative to their wage. Another way of saying this is that travel time to a recreation site conveys some benefits that offset the opportunity cost of wages foregone. For time spent actually engaged in recreation, this offset may be complete. The structure of indirect utility is most easily seen by starting from a model of the direct utility associated with a fixed recreational choice,

$$\begin{bmatrix}\text{Direct Utility}\\ \text{of a Site}\end{bmatrix} = \alpha \cdot c + G(h + (1-\psi)\cdot TT + (1-\lambda)\cdot RT) + \begin{bmatrix}\text{Hedonic}\\ \text{Attributes}\end{bmatrix}\cdot \gamma + \varepsilon \quad (2)$$

where c is consumption of goods other than recreation trips, h is hours of non-recreational leisure, TT is travel time, RT is on-site recreation time, and G is an increasing concave function. The coefficient ψ is less than one, reflecting the fact that recreational travel is less onerous than working, and the coefficient λ is near zero, reflecting the fact that recreational activities convey approximately the same benefits per hour as other forms of leisure. (If RT is the result of consumer optimization of the allocation of time between recreation and other leisure, then λ will be exactly zero.) The consumer will choose c and h to maximize this expression subject to the budget constraint $N + w \cdot l = OPC + IVC + c$ and the time allocation constraint $H = l + h + TT + RT$, where N is non-wage income net of fixed costs, w is the wage rate, l is hours worked, OPC is out-of-pocket travel cost, IVC is the variable part of indirect vehicle cost (e.g., depreciation and maintenance due to "wear and tear"), and H is total time available. Let $z = h + (1-\psi)\cdot TT + (1-\lambda)\cdot RT$. Using the budget and time constraints to solve for l and c in terms of z, the kernel of the consumer's problem is to choose z to maximize $-\alpha wz + G(z)$. This is accomplished by a function $z = g(\alpha w)$ that satisfies $\alpha w \equiv G'(g(\alpha w))$, and correspondingly $l = H - \psi{\cdot}TT - \lambda{\cdot}RT - g(\alpha w)$ and $c = N + wH - w{\cdot}g(\alpha w) - OPC - IVC - \psi w{\cdot}TT - \lambda w{\cdot}RT$. Substituting these in the direct utility function gives

$$\begin{bmatrix}\text{Utility}\\ \text{of a Site}\end{bmatrix} = \alpha\cdot\{Y - OPC - IVC - \psi\cdot w\cdot TT - \lambda\cdot w\cdot RT\} + \begin{bmatrix}\text{Hedonic}\\ \text{Attributes}\end{bmatrix}\cdot\gamma + \varepsilon \quad (3)$$

where $Y \equiv G(g(\alpha w))/\alpha + N + wH - w{\cdot}g(\alpha w)$ is "disposable income". This implies that one needs to distinguish travel time components. In particular, different travel time components such as in-vehicle time and walking time may have different coefficients, and on-site recreation time (*RT*) will have a much smaller coefficient than travel time.

Travel cost also causes difficulties. Economic theory says a marginal dollar is a marginal dollar, independent of the category in which it is spent. However, in reality consumers pay more attention to visible travel cost dollars, such as out-of-pocket gas and lodging expenses, than they do to less visible costs such as the variable components of maintenance and depreciation (see Hensher, 1985; Manering, 1983; Manering and Winston, 1985; Keeler and Small, 1977; Train, 1986). This behavior appears to arise because people keep different categories of expenditures in different mental accounts (see Thaler, 1990). In the utility formula above, replace *IVC* by $\kappa{\cdot}\mu{\cdot}IC$, where *IC* is average indirect cost (including both fixed and variable components), μ is the fraction of average indirect costs that are variable, and κ is a behavioral parameter. For the conceptual experiment of calculating the equivalent variation in non-wage income *N* that keeps the consumer at the utility level with injury when the injury is removed, WTP should be calculated in terms of high-powered out-of-pocket dollars (e.g., multiplier α/β) rather than low-powered indirect cost dollars (e.g., multiplier $\alpha/\beta{\cdot}\kappa{\cdot}\mu$).

One would like to estimate the travel demand model as a function of *OPC* and *IC* to obtain consistent estimates of both β and $\beta{\cdot}\kappa{\cdot}\mu$. In practical travel demand analysis, *OPC* and *IC* are nearly collinear, as the detailed vehicle information necessary to measure relative variation in these variables is not collected. Then, it is necessary to use an external estimate of $\kappa{\cdot}\mu$; popular choices are zero and one, while consensus estimates from the travel demand literature are roughly $\mu = 0.3$ and $\kappa = 0.4$. At current price levels, *OPC* and *IC* in the U.S. are both about 14 cents per mile. If the estimates above are correct, then omitting *IC* from the analysis ($\kappa{\cdot}\mu = 0$) biases the WTP estimate downward by 12 percent, while combining *OPC* and *IC* into an overall average cost ($\kappa{\cdot}\mu = 1$) biases the WTP estimate upward by 100 percent.

Model Structure and Income Effects

The key restrictive features in the RUM for travel demand in equation (1) are that it is (i) additively separable into terms involving income, hedonic attributes, and the disturbance; (ii) linear in income; and (iii) linear in the hedonic attributes. An additional restrictive feature is introduced when the disturbances are specified to have distributions that do not depend on income or hedonic attributes. These restrictions may be relatively harmless for describing the empirical features of travel demand, as even poor specifications will often provide good within-sample fits. However, these restrictions can have a substantial impact on the correct formulas for WTP. To investigate these issues, consider a more general nonlinear specification of the direct utility function given a trip to a specific site, $U(c,h+\psi\cdot TT,RT,\mathbf{x},\ \varepsilon)$, where c is consumption other than the recreation trip, h is leisure time, TT is recreation travel time, RT is time spent in the recreational activity, $\mathbf{x}$ is a vector of site attributes including those that are impacted by injury and those that are not, and ε is a disturbance that is heterogeneous across individuals. In this specification, TT has a coefficient ψ reflecting the benefit of recreational travel relative to leisure. The consumer faces an income budget constraint $N + w\cdot l = OPC + IVC + c$ and a time budget constraint $H = l + h + TT + RT$, as before, where N is non-wage income net of fixed costs, w is the wage rate, l is hours worked, OPC is out-of-pocket travel cost, IVC is the variable part of indirect vehicle cost, and H is total time available. The first-order conditions for maximization of direct utility subject to these constraints are

$$\nabla_c U(c,h+\psi\cdot TT,RT,\mathbf{x},\ \varepsilon) = \alpha$$

$$\nabla_h U(c,h+\psi\cdot TT,RT,\mathbf{x},\ \varepsilon) = w\cdot\alpha \tag{4}$$

and assuming RT can be optimized,

$$\nabla_{RT} U(c,h+\psi\cdot TT,RT,\mathbf{x},\ \varepsilon) = w\cdot\alpha \tag{5}$$

The solution of these equations yields c, h, and RT as functions of TT, real non-wage income $y \equiv N/w$, real marginal travel cost $t = (OPC + IVC)/w$, $\mathbf{x}$, and ε. Substituting these functions into the direct utility function yields an indirect utility function $U^{\#}(y - t,TT,\mathbf{x},\ \varepsilon)$. Note that this function does not depend on RT, which is optimized out. This is not to say that duration of recreational activity does not influence utility level, but rather that this is already captured in the function $U^{\#}$. In practice, other constraints may prevent full optimization in RT. In econometric analysis of choice behavior, consistency will require modeling these constraints explicitly, rather than simply introducing the endogenous observed RT as an explanatory variable in the indirect utility function.[28]

Assuming that utility is scaled so there are no problems in forming moments, one can always write

$$U^{\#}(y - t, TT, \mathbf{x}, \varepsilon) = V(y - t, TT, \mathbf{x}) + \eta \qquad (6)$$

where $V(y - t, TT, \mathbf{x}) = E_{\varepsilon/y,t,TT,x}\, U^{\#}(y - t, TT, \mathbf{x}, \varepsilon)$ and η is the residual. Then, the disturbance is additively separable, but with a distribution that will in general depend on $y, t, TT, \mathbf{x}$. In the analysis below, I will often concentrate on the case of an additively separable disturbance whose distribution does *not* depend on y or $\mathbf{x}$. This is quite restrictive, but essential if one is to isolate attribute changes from tastes, identify individual WTP, and give WTP an economic welfare content.

Consider the dependence of the indirect utility function on income, and the impact of this on demand behaviour and WTP calculations. First, consider an indirect utility function that is additively separable and linear in income with a common coefficient across alternatives, and with a disturbance whose distribution does not depend on income,

$$U^{\#}(y - t, TT, \mathbf{x}, \varepsilon) = \alpha(y-t) + S(TT, \mathbf{x}, \varepsilon) \qquad (7)$$

This is a case of the Deaton and Muelbauer (1980) "almost ideal demand system", since y is income divided by a price index. An important, and restrictive, behavioral implication of this system is that income drops out of the comparison of the utilities of different alternatives. Then, income changes cannot change the probability of the event that a specific alternative has maximum utility. (Wage rate changes which alter the opportunity cost of time can change probabilities, and wage rates are empirically correlated with income, but transfers and non-wage income weaken the link.) In consumer surplus terms, this implies that there are no income effects, so that Hicksian and Marshallian surplus measures coincide. A related consequence of this linearity in income is that it is possible to aggregate preferences into a social or representative preference that generates the market demand functions using Roy's identity (see Chipman and Moore, 1980,1990).

Alternatives to the almost ideal demand system in which demand for trips or for particular sites *are* income-sensitive necessarily involve non-additive interaction between income and

---

[28] One method for handling RT consistently when time allocation between recreation and other leisure activities is incompletely optimized is to make trip duration a specific feature of choice alternatives, along with site, and model joint choice of site and duration.

trip attributes in the indirect utility function. McFadden and Train (1976) discuss the direct utility structures required to obtain various patterns of income effects.

There is no difficulty in principle in developing empirical RUM models containing non-linear and/or interactive income effects. For example, using the decomposition $U^{\#}(y - t, TT, \boldsymbol{x}, \varepsilon) = V(y - t, TT, \boldsymbol{x}) + \eta$ and assuming that the disturbances have standard distributions that do not depend on variables of the problem, such as extreme value or generalized extreme value, leads to tractable choice probability functions. However, these income effects complicate considerably the calculation of WTP. Even additive non-linear income effects, or additive linear income effects with coefficients that differ by alternative, are sufficient to upset "log sum" formulas for WTP.

WTP FROM RUM-CONSISTENT CHOICE PROBABILITIES

Mean WTP

In this section, I discuss exact formulas and bounds for mean WTP when choices are RUM-consistent. Suppose consumers face $i = 1,...,J$ discrete transportation alternatives, interpreted to include the no-travel option as well as travel to alternative sites. The consumer has income y. If alternative i is chosen, there is a "generalized transportation cost" (t_i) and a vector of attributes ($\boldsymbol{x}_i$). Included in $\boldsymbol{x}_i$ are any user fees that are collected. There may be other commodity prices in the background. If restricted to alternative i, the consumer will maximize preferences subject to this restriction and subject to the constraints imposed by the availability of time and income, as in the second section. The result will be a level of satisfaction that can be characterized by an indirect utility function $U(y-t_i, \boldsymbol{x}_i, \varepsilon_i)$ that is homogeneous of degree zero and quasi-convex in income and prices (including those in the background), increasing in income, and non-increasing in prices. The equation (1) commonly used for transportation behavior is a special case of this general RUM (see, for example, McFadden, 1981). Consumers are heterogeneous in unobserved factors that influence their behaviour (e.g., tastes). The effects of these unobserved factors are summarized by ε_i. Let $\boldsymbol{t} = (t_1,...,t_J)$ and $\boldsymbol{x} = (\boldsymbol{x}_1,...,\boldsymbol{x}_J)$. Consumers choose i to maximize $U(y-t_i, \boldsymbol{x}_i, \varepsilon_i)$. The proportion $P_i(y, \boldsymbol{t}, \boldsymbol{x})$ of consumers choosing i is given by the probability of the event

$$U(y-t_i, \boldsymbol{x}_i, \varepsilon_i) > U(y-t_j, \boldsymbol{x}_j, \varepsilon_j) \text{ for } j \neq i \qquad (8)$$

Choice of recreational fishing trips is a typical application of this setup. The travel cost t_i will be transportation and living expenses for the trip, including opportunity cost of time; and $\boldsymbol{x}_i$

will include the observed site attributes that matter to anglers, such as site size measured by the number of accessible fishing spots, ease of access, fish abundance, facilities at the site, aesthetics, user fees, and marginal equipment costs.

Suppose a policy is adopted that raises the quality of some alternatives from injured to restored conditions. In the recreational fishing example, this might be a damage remediation program that removes pollutants and increases fish abundance. Use the notation $\boldsymbol{x}' = (\boldsymbol{x}_1',\ldots,\boldsymbol{x}_J')$ for the attributes of sites in the *as is* case with injury, and the notation $\boldsymbol{x}'' = (\boldsymbol{x}_1'',\ldots,\boldsymbol{x}_J'')$ for the attributes of sites in the *but for* case with restoration. The *principle of equivalent variation* is that the *as is* income of each consumer is *lowered* in the *but for* case with restoration so that maximum utility *after* the quality improvement equals maximum utility *before* the quality improvement. Then, mean WTP is the average of these equivalent variation income reductions in the target population.[29]

For *any* pair of alternatives j, m, define C_{jm} to be the *net* reduction in income combined with the quality improvement that equates the utility from alternative j with injury and the utility from alternative m without injury; that is,

$$U(y-t_j, \boldsymbol{x}_j', \varepsilon_j) = U(y-t_m-C_{jm}, \boldsymbol{x}_m'', \varepsilon_m) \tag{9}$$

Define $C = C(y,\boldsymbol{t}, \boldsymbol{x}', \boldsymbol{x}'',\varepsilon)$ to be the income adjustment C that equates *maximum* utility at injured conditions to *maximum* utility at restored conditions:

$$\text{Max}_j\ U(y-t_j, \boldsymbol{x}_j', \varepsilon_j) = \text{Max}_m\ U(y-t_m-C, \boldsymbol{x}_m'', \varepsilon_m) \tag{10}$$

Then, C is the WTP of the individual for the quality improvement. If i is chosen before the improvement and k is chosen after the improvement, then $C = C_{ik}$ and these definitions imply $C_{ii} \le C(y,\boldsymbol{t}, \boldsymbol{x}', \boldsymbol{x}'',\varepsilon) \le C_{kk}$. Thus, the income compensation actually required to keep a consumer's maximum utility constant is bracketed by the income compensation required to keep the utility of the original alternative constant and the income compensation required to keep the utility of the final alternative constant. The intuition for these bounds is that the

[29] One could alternately define the compensating variation as the net increase in income in the case with injury that would make maximum utility before the quality improvement equal to maximum utility after the quality improvement. This is also called willingness-to-accept (WTA). When consumers are rational, WTP and WTA will differ only by a small income effect. In practice, when WTP and WTA are measured, WTA is usually much larger than WTP. This "endowment" effect may be due to consumer mistrust of hypothetical trade offers, or may arise from psychological asymmetries in the treatment of gains and losses (see Thaler, 1990). The effect may also arise from strategic misrepresentation, since the consumer may see an opportunity to gain substantial rents by overstating WTA. For these reasons, valuation exercises concentrate almost entirely on WTP.

ability of the consumer to substitute away from i under restored conditions may increase final utility, requiring additional income reduction to equate final utility to the initial utility level. Similarly, the ability of the consumer to substitute away from k under injury conditions may increase initial utility, requiring less income reduction to equate final utility to the initial utility level. If no alternative becomes less desirable as a result of the quality improvement, then $C_{ii} \geq 0$. If alternative k is unaffected by the quality improvement, then $C_{kk} = 0$. Together, these conditions imply that a consumer whose final choice k is not affected by the quality improvement must have $C(y,\boldsymbol{t}, \boldsymbol{x}', \boldsymbol{x}'',\varepsilon) = 0$.

The distribution of the unobserved ε's in the target population will induce a distribution of income compensation levels $C(y,\boldsymbol{t}, \boldsymbol{x}', \boldsymbol{x}'',\varepsilon)$. The mean of this distribution will be mean WTP for the quality improvement, $EC(y,\boldsymbol{t}, \boldsymbol{x}', \boldsymbol{x}'',\varepsilon)$. The distribution of ε's also induces the choice probabilities $P_j(y,\boldsymbol{t}, \boldsymbol{x})$. Let $P_j' = P_j(y,\boldsymbol{t}, \boldsymbol{x}')$ denote the injury case probability of alternative j, and $P_j'' = P_j(y,\boldsymbol{t}, \boldsymbol{x}'')$ denote the restored case probability of this alternative when compensation is actually paid.[30]

Evaluation of mean WTP $EC(y,\boldsymbol{t}, \boldsymbol{x}', \boldsymbol{x}'',\varepsilon)$ is often difficult, requiring multidimension numerical integration or simulation. However, bounds for mean WTP may suffice for many policy purposes. In most RUM applications, the assumption is made that the ε's enter the indirect utility function additively, so that $U(y\text{-}t_j, \boldsymbol{x}_j, \varepsilon_j) = V(y\text{-}t_j, \boldsymbol{x}_j) + \varepsilon_j$ and there is no interaction between income and the unobservable ε's. In this case, the income reduction C_{jj} for a fixed alternative j satisfies $V(y\text{-}t_j, \boldsymbol{x}_j') = V(y\text{-}t_j\text{-}C_{jj}, \boldsymbol{x}_j'')$ and does not depend on ε. Then, one has the bounds

$$\sum_{j=1}^{J} P_j' C_{jj} \leq EC(y,\boldsymbol{t},\boldsymbol{x}',\boldsymbol{x}'',\varepsilon) \leq \sum_{j=1}^{J} P_j'' C_{jj} \qquad (11)$$

If the choice probabilities depend on income, then the restored choice probabilities P_j'' depend on the compensating income reduction, and will themselves have to be bounded to calculate the right-hand-side of (11). These bounds provide a convenient check on exact computations of population mean WTP. In applications where the quality improvement is small or induces

[30] When income effects are present, P_j'' will differ by a small income effect from the restored case probabilities that prevail when compensation is *not* paid. When the probability of travel to each injured site has a positive income elasticity, the choice probability without compensation will be larger than the choice probability with compensation, and the right-hand bound in equation (11) holds for the uncompensated choice probability.

little substitution among alternatives, an average of the bounds will often provide an adequate approximation to *EC* without further calculation.

The bound (11) is all that is needed to estimate mean WTP for small changes. An infinitesimal change from $\boldsymbol{x}'$ to $\boldsymbol{x}'' = \boldsymbol{x}' + \Delta\boldsymbol{x}$ results in an infinitesimal change $P_j'' = P_j' + \Delta P_j$ in the choice probabilities. From the formula for the compensating income reduction for a fixed alternative *j*,

$$C_{jj} = \frac{\partial V(y-t_j,\boldsymbol{x}'_j)/\partial \boldsymbol{x}_j}{\partial V(y-t_j,\boldsymbol{x}'_j)/\partial y}\cdot \Delta\boldsymbol{x}_j \tag{12}$$

Substituting this expression into the bounds above yields

$$EC(y,\boldsymbol{t},\boldsymbol{x}',\boldsymbol{x}'+\Delta\boldsymbol{x},\varepsilon) = \sum_{j=1}^{J} P_j' \cdot \frac{\partial V(y-t_j,\boldsymbol{x}'_j)/\partial \boldsymbol{x}_j}{\partial V(y-t_j,\boldsymbol{x}'_j)/\partial y}\cdot \Delta\boldsymbol{x}_j \tag{13}$$

For small quality improvements, this formula will approximate closely the target population mean WTP, no matter how income influences choice. Also, note that mean WTP is linear in the fraction of the population visiting the injured sites; thus, mean WTP for small changes at unpopular sites must be very small.

Common Linear Income Effects

Consider the linear indirect utility function, $U(y\text{-}t_j,\boldsymbol{x}_j,\ \varepsilon_j) = \alpha(y\text{-}t_j) + \gamma\boldsymbol{x}_j + \varepsilon_j$, with an income coefficient that is common across alternatives, and let

$$U^*(y,\boldsymbol{t},\boldsymbol{x}) = E\ \mathrm{Max}_j\ U(y\text{-}t_j,\boldsymbol{x}_j,\varepsilon_j) \tag{14}$$

denote expected maximum utility. McFadden (1981, 1997) establishes for this case that U^* is a representative or social utility function for this problem, with Roy's identity giving the choice probabilities,

$$P_i(y,\boldsymbol{t},\boldsymbol{x}) = -\frac{\partial U^*(y,\boldsymbol{t},\boldsymbol{x})/\partial t_i}{\partial U^*(y,\boldsymbol{t},\boldsymbol{x})/\partial y} \tag{15}$$

and mean WTP satisfying

$$U^*(y,\boldsymbol{t},\boldsymbol{x}') = U^*(y - EC(y,\boldsymbol{t},\boldsymbol{x}',\boldsymbol{x}'',\varepsilon),\boldsymbol{t},\boldsymbol{x}'') \tag{16}$$

or

$$EC(y,\boldsymbol{t},\boldsymbol{x}',\boldsymbol{x}'',\varepsilon) = \frac{1}{\alpha}\cdot\left\{\mathrm{EMax}_i U(y-t_i,\boldsymbol{x}_i'',\varepsilon_i) - \mathrm{EMax}_i U(y-t_i,\boldsymbol{x}_i',\varepsilon_i)\right\} \tag{17}$$

Thus, in the common linear income case, mean WTP can be calculated as the income adjustment needed to equate expected utility before and after the quality improvement. However, even this calculation can be difficult if expected utility does not have an analytic form.

In the common linear income case, the bound (11) reduces to

$$\sum_{i=1}^{J} P_i' \cdot \gamma(\mathbf{x}_i'' - \mathbf{x}_i') / \alpha \le EC(y, \mathbf{t}, \mathbf{x}', \mathbf{x}'', \varepsilon) \le \sum_{i=1}^{J} P_i'' \cdot \gamma(\mathbf{x}_i'' - \mathbf{x}_i') / \alpha \tag{18}$$

and the approximation (13) for small changes reduces to

$$EC(y, \mathbf{t}, \mathbf{x}', \mathbf{x}' + \Delta\mathbf{x}, \varepsilon) \approx \sum_{j=1}^{J} P_j' \cdot \frac{\gamma}{\alpha} \cdot \Delta\mathbf{x}_j \tag{19}$$

These approximations will often provide a useful indication of the magnitude of mean WTP without full calculation of (17).

Nonlinear Income Effects

Next consider more general nonlinear dependence of indirect utility on income, $U(y\text{-}t_j, \mathbf{x}_j, \varepsilon_j) = V(y\text{-}t_j, \mathbf{x}_j') + \varepsilon_j = f_j(y\text{-}t_j) + \gamma \mathbf{x}_j' + \varepsilon_j$, where the functions $f_i(y\text{-}t_i)$ are not all linear, or vary by alternative. The income compensation for a consumer who chooses i in the injury case and k in the restored case satisfies $C_{ik} = y - t_k - f_k^{-1}(f_i(y\text{-}t_i) - \gamma(\mathbf{x}_k'' - \mathbf{x}_i') + \varepsilon_i - \varepsilon_k)$. Unlike the linear income case, C_{ik} cannot be written as a difference of utilities normalized by the marginal utility of income, so the formula for expected WTP cannot be expressed in terms of a difference in expected utilities in the injury and restored cases. In particular, because expectations are not preserved by nonlinear transformations, one cannot calculate $EC(y,\mathbf{t},\mathbf{x}',\mathbf{x}'',\varepsilon)$ by solving for the income reduction that equates *expected* utility before and after the quality improvement. One can still define expected utility, but it will not be a social utility function that generates the market demand functions via Roy's identity, and the value C of an income reduction that keeps expected utility constant will not coincide with the WTP $EC(y,\mathbf{t},\mathbf{x}',\mathbf{x}'',\varepsilon)$ derived by averaging the equivalent variation income reductions for each individual.

Calculation of $EC(y,\mathbf{t},\mathbf{x}',\mathbf{x}'',\varepsilon)$ will in general require explicit consideration of the distribution of income compensations induced by the nonlinear transformation, with simultaneous or iterative solution of the random utility model for $C(y,\mathbf{t},\mathbf{x}',\mathbf{x}'',\varepsilon)$. This calculation is in most cases analytically intractable; McFadden (1997) provides a simulation method that is asymptotically unbiased and can be made as accurate as is needed, and applies it to generalized extreme value (GEV) random utility models.

The general bounds (11) on mean WTP can be specialized to the case of indirect utility functions that are linear and additive in the disturbance and a (nonlinear) function of income. The income compensation C_{jj} that equates the utility of alternative j before and after the quality improvement must then satisfy $\gamma(\mathbf{x}_j'' - \mathbf{x}_j') = f_j(y-t_j) - f_j(y-t_j-C_{jj})$. Using the theorem of the mean, $\alpha_j C_{jj} = \gamma(\mathbf{x}_j'' - \mathbf{x}_j')$, where α_j is the marginal utility of income $df_j(y)/dy$ evaluated at some intermediate point between $y-t_j$ and $y-t_j-C_{jj}$. Substituting the expression for C_{jj} in (11), one obtains

$$\sum_{j=1}^{J} P_j' \cdot \gamma(\mathbf{x}_j'' - \mathbf{x}_j') / \alpha_j \leq EC(y,\mathbf{t},p,\mathbf{x}',\mathbf{x}'',\varepsilon) \leq \sum_{j=1}^{J} P_j' \cdot \gamma(\mathbf{x}_j'' - \mathbf{x}_j') / \alpha_j \tag{20}$$

Then, substituting upper and lower bounds, respectively, on α_j, one can use these inequalities to bound mean WTP when the dependence of the indirect utility function on income is nonlinear or varies by alternative. Again, for small quality improvements, the bounds are narrow, and establish that WTP in the target population is approximately equal to the sum over alternatives of the choice probabilities times the quality improvement for an alternative, divided by the marginal utility of income for this alternative.

WTP in Generalized Extreme Value Models

The family of generalized extreme value models, which includes multinomial logit (MNL), yields explicit forms for choice probabilities, and when combined with a common linear income effect, an explicit "log sum" form for mean WTP. Suppose indirect utility can be written as $U(y-t_i,\mathbf{x}_i, \varepsilon_i) = V(y-t_i,\mathbf{x}_i) + \varepsilon_i$, and assume that the ε's have a joint cumulative distribution function (CDF) that can be written as

$$F(\varepsilon_1,\ldots,\varepsilon_J) = \exp(-H(e^{-\varepsilon_1},\ldots,e^{-\varepsilon_J})) \tag{21}$$

where $H(w_1,\ldots,w_J)$ is a non-negative linear homogeneous function of $w \geq 0$, satisfying the properties that if any argument goes to $+\infty$, then H goes to $+\infty$; and the mixed partial derivatives of H exist, are continuous, and alternate in sign, with non-negative odd mixed derivatives. A function H with these properties will be termed a GEV generating function. McFadden (1978, Theorem 1, 1997, Theorem 1), verifies that when the function H is a GEV generating function, (21) is a CDF whose one-dimensional marginals are extreme value distributions. Further, the random utility model $u_i = v_i + \varepsilon_i$ with ε distributed $F(\varepsilon)$ satisfies

$$E\,\text{Max}_i\, u_i = \log H(e^{v_1},...,e^{v_J}) + 0.57721 \tag{22}$$

and the choice probabilities satisfy

$$P_i = e^{v_i} \cdot H_i(e^{v_1},...,e^{v_J}) \,/\, H(e^{v_1},...,e^{v_J}) \tag{23}$$

The linear function $H = w_1 + ... + w_J$ is a GEV generating function which yields the MNL model. The following result (McFadden, 1997, Lemma 2) can be applied recursively, starting from linear functions, to establish families of GEV generating functions that include nested multinomial logit models:

> If sets A, B satisfy $A \cup B = \{1,...,J\}$, $H^A(w_A)$ and $H^B(w_B)$ are GEV generating functions in w_A and w_B, respectively, and if $s \geq 1$, then $H(w) = H^A(w_A^s)^{1/s} + H^B(w_B)$ is a GEV generating function in $(w_1,...,w_J)$.

Consider mean WTP when the unobserved ε's have a GEV distribution and the indirect utility function has a common linear income effect, so that $U(y\text{-}t_i, \mathbf{x}_i, \varepsilon_i) = v_i + \varepsilon_i$, *with* $v_i = \alpha(y\text{-}t_i) + \gamma \mathbf{x}_i$. Let $v_i' = \alpha(y\text{-}t_i) + \gamma \mathbf{x}_i'$ and $v_i'' = \alpha(y\text{-}t_i) + \gamma \mathbf{x}_i''$. Combine equation (17) for mean WTP in this case with equation (22) for expected maximum utility in GEV models to obtain the log-sum formula

$$EC(y,\mathbf{t},\mathbf{x}',\mathbf{x}'',\varepsilon) = \frac{1}{\alpha} \cdot \left\{ \log H(e^{v_1''},...,e^{v_J''}) - \log H(e^{v_1'},...,e^{v_J'}) \right\} \tag{24}$$

This is the formula first developed by Ben Akiva (1972), McFadden (1973), and Domencich and McFadden (1975) for the MNL model, and by McFadden (1978, 1981) for the nested logit model. This formula is valid only when the indirect utility function has a common linear income effect and choice probabilities have the GEV form.

RUM-CONSISTENT CHOICE

Consumer Rationality

The program outlined in the second and third sections for translating choice into an acceptable measure of social value will fail if choices are not consistent with a RUM that accurately reflects preferences. This could happen because (i) consumers do not have the stable, rational preferences required to make benefit-cost analysis work; (ii) consumer information and market structure allow incentives and perceptions to confound the link from utility to choice;

or (iii) true preferences are veiled by cognitive illusions or rule-driven behavior. These potential problems arise in both RP and SP analysis. Regarding the first possibility, cost-benefit analysis can tolerate a degree of irrationality, in the sense that if there is an acceptable way to aggregate values of consumers with different tastes into social value, then these same principles can be applied to intra-individual taste fluctuations. We have already seen this principle applied in deriving mean WTP from a RUM, where the randomness in utility was interpreted as interpersonal taste heterogeneity, but could equally well have been interpreted to include intra-personal randomness in tastes. However, beyond this, there is no accepted economic theory of social choice based on individual tastes. I will assume for the remainder of this chapter that consumers have core preferences that are consistent with social valuation, and examine issues (ii) and (iii) in determining true preferences from RP or SP data.

Information and Incentives

Consumers on the demand side of a market can be described in terms of the information they have about the commodity, from previous experience and from the market itself, and their perceptions of uncertain product features. They can also be described in terms of their familiarity with market protocol, including trading rules, expected payoffs to strategic behaviour, and the consequences of alternative decisions. The strongest case for classical rational behaviour can be made when the consumer has complete information about the products in a market, has extensive experience and familiarity with market operations, and knows that his or her behaviour cannot influence the price or quality of the product purchased. For example, one can argue that travel-mode choice by commuters and site choices by avid anglers accurately reflect preferences for the alternatives. The more complex and heterogeneous the commodity, and the less experienced the consumer, the more difficult it is to convey and absorb the information required to understand what its consumption entails. Consumers deal with complex commodities by drawing heavily on past experience, search that provides information on market conditions (and on preferences), and perhaps a less careful analysis of the consequences of choices. When forced to make quick decisions involving unfamiliar complex commodities, consumers may fall back on more generalized experiences or ideologies, using simple screening rules. For example, some experiments in marketing suggest that when consumers must choose from a small set of multi-attribute commodities, they weigh the trade-offs among the attributes, but when faced with large sets of commodities, they first screen on one or two attributes to reduce the choice set before beginning trade-off analysis. More generally, individuals are perhaps more likely to adopt "black and white" ideologies to guide unfamiliar choices, and to weigh intermediate alternatives when making familiar ones.

An automobile is an example of a complex commodity where the information-gathering process influences choice. Potential buyers actively search for information on alternatives, through advertisements and motoring magazines, discussions with friends, and test drives. The depth and duration of search for information, and the order in which it is assembled, are all under the consumer's control. The weight given to different aspects of cars may depend on which are emphasized in various information sources, and on the sequence in which alternatives are screened out. The search environment involves social interaction, and in many channels is visual and aural rather than written. One might say that the consumer goes through a process of constructing preferences for automobiles. While a classical economist might bridle at the terminology, the presence of learning, and the dependence on history that learning implies, is neither inconsistent with rationality nor with benefit-cost analysis. However, when learning is important, preferences may appear to be context-dependent and unstable. Consequently, RP data may be accurate for simple commodities, but it may be difficult to find real markets for complex, infrequently purchased commodities in which learning has proceeded to a point where consumers' choices reflect core preferences. In principle, SP methods could do more than merely mimic a real market for a complex commodity; they could go beyond it in providing information and decision-making experience to the consumer. However, it is difficult to duplicate in the laboratory the time frame and social setting of complex consumer decisions, or the interaction between the degree to which the consumer actively initiates and controls search and the saliency of search results. Most SP experiments in practice provide abbreviated information to a passive consumer, little or no opportunity for questions or social interchange, and require quick response. I would not be surprised if consumers forced to purchase automobiles under the time and information conditions typical of these experiments behaved differently than consumers in the real auto market. Then, I believe we should not be surprised if stated preferences for innovative and complex goods appear unrealistic when held up to the yardstick of real market behavior.

An additional problem in linking preferences to behavior is strategic misrepresentation. Consumers may deliberately misstate preferences when they believe that they can influence product quality or price by their own behavior. For example, in a market such as the used car market where bid and ask prices are quoted and a transaction price is negotiated, the consumer has an incentive to understate value as a rent-seeking strategy. The issue of strategic misrepresentation is important in SP experiments. Usually, consumers in these experiments have no economic incentives for misrepresentation, as it is clear that the market questions are hypothetical and will not entail purchase and consumption of the chosen commodity. In these circumstances, the consumer may find it easier to be truthful about preferences than to fabricate an alternative. On the other hand, there are no economic penalties for misrepresentation, so that factors that are insignificant when money changes hands may become important. For example, consumer preferences for less convenient environmentally safe commodities may be more positive in a hypothetical setting than in a real market, as the SP experiment makes it less costly to convince the observer (and oneself) that one is socially responsible.

Cognitive Illusions and Rule-Driven Behaviour

Experiments in cognitive psychology that address the effect of context on response suggest that preferences are constructed when the consumer faces a choice, and that this construction is highly sensitive to context. The cognitive illusions uncovered by these experiments appear to be the strongest when the consumer is presented with unfamiliar tasks, and seeks cues from context to shape an appropriate response. Table 15.1 lists some common cognitive illusions; many share the feature that attention is selective, and focuses on the most readily available information. Many economists argue that self-interest, habit, and discipline imposed by the market will limit the degree to which important market choices are clouded by cognitive errors. Clearly there are arenas, such as financial markets where consumers engage in frequent transactions, where this argument is quite persuasive. However, there is a broad sweep of market activity, where transactions are infrequent and inconsistencies in preferences hard to detect, and deviations from rationality are unlikely to be penalized by the market. Decisions in a hypothetical market regarding unfamiliar complex commodities will be particularly vulnerable to contamination by cognitive biases.

An example of a cognitive error that complicates RP and SP analysis is the phenomenon of mental accounting (Thaler, 1990), where consumers treat dollars differently depending on the category and direction of flow. I noted in the second section that consumers weigh out-of-pocket travel costs more heavily than indirect variable travel costs such as depreciation, and described in principle how acceptable calculation of WTP could proceed by identifying the relevant mental account. A problem in SP experiments is that presentation of cost information on complex commodities may lead to a different mental accounting than the analogous real market. Taking automobile choice as an example, consumers presented in an SP experiment with information on initial cost and operating cost may consider these equally salient and combine them in a single mental account. In real markets, there is a temporal and saliency mismatch between initial costs and operating costs, and consumers are likely to account for these components separately. This psychological phenomenon may explain the empirical finding that consumers appear to use irrationally high discount rates in trading off initial and operating costs of durables.

Cognitive illusions are likely to be particularly important in decisions that rely on recall of past experiences. When the consumer draws on memory and on information gathered from social interaction, availability, regression, effects, and representativeness errors are likely to lead consumers to place too much weight on selected experiences, with inadequate allowance for statistical noise, and too little weight on background information. Thus, a few experiences with automobile brands may unjustifiably be taken as representative and permanent, and may then color perceptions. Core preferences are veiled, and measured preferences appear context-dependent and unstable.

Table 15.1.
Cognitive Illusions

Effect	Description
Anchoring	Responses are influenced by cues contained in the question
Availability heuristic	Responses rely too heavily on readily retrieved information, and too little on background information
Certainty effect	The likelihoods of low probability events are misjudged, and treated either as too likely or as zero
Context effects	Previous questions and interviewer interaction color perception
Framing	Question format changes saliency of different aspects of the cognitive task
Focality effect	Quantitative information is stored and/or reported categorically
Intertemporal choice	Temporally inconsistent time discounting
Primacy/Recency	Initial and recently experienced events are the most salient
Projection	Responses are consonant with the self-image the subject wishes to project
Regression effect	Causality and permanence are attached to past fluctuations, and regression to the mean is underestimated
Representativeness Heuristic	High conditional probabilities induce overestimates of unconditional probabilities
Rule-Driven behavior	Motivation and self-control induce strategic responses
Saliency effect	The most salient aspects of the question are overemphasized
Status Quo bias	Current status and history are privileged
Superstition effect	Elaborate causal structures are attached to coincidences

Most market commodities are private goods, and do not carry any ethical baggage. Then, the decision to purchase at a price arguably establishes a value at least equal to this price. If, however, the consumer feels that she has an ethical entitlement or property right to the commodity, or that its purchase promotes or reduces "fairness", or induces externalities, then there is potentially a more complex relationship between purchase decisions and the values that should enter a social benefit calculation. An example of a private commodity with an ethical aspect is use of a seat belt in an automobile. Consumers may feel that freedom of choice on seat belt use is an entitlement, or may be concerned about external effects or social opprobrium. Or, the commodity may become a token for a larger ethical issue, such as the principle of self-responsibility for one's safety. The presence of ethical aspects to a commodity is also likely to make its valuation more dependent on context, since ethical issues tend to be cast in relation to past positions and entitlements.

The use of principles or rules to control behavior can be explained by the conditioning of human behavior (see Ainslie, 1982; Prelec, 1991). Children acquire self-control, learn to delay gratification, until in normal adults some degree of abstinence and control becomes an end in itself. Processes of precommittment, particularly adoption of rules or principles, become an important part of behavior. Conduct in an abstract choice situation is likely to be determined by deciding which principles apply. There is no obvious reason for choices in different situations to be mutually consistent. Money is a prime example of a good where individuals develop what appear to be elaborate rules for how it is accounted for, how it can be used acceptably and unacceptably, and the process as well as the outcome of exchanges of money. "Neither a borrower nor a lender be" is a principle for conduct, not an instruction for rational behavior. Then, mental accounting may be governed not only by accessibility and saliency, but also by ethical rules. The consumer is most likely to rely on rules when there is a mismatch between cost and benefits in terms of time, saliency, or scale. Principles such as "I always use a seat belt" preempt a utilitarian calculation of the expected cost of one missed occasion to fasten your seat belt.

An important observation about rule-driven decision making is that it does not deny the possibility of an underlying utilitarianism. It may be possible to rationalize a system of rules as a consequence of overarching preference maximization, in which the "sensible self" establishes self-control that prevents the "indulgent self" from excesses that would have undesirable consequences latter. For example, a fully rational individual will when entering a car compare the cost of the five seconds of time required to fasten her seat belt with the probability of avoiding various injuries, given the distance to be traveled and driving conditions. A "sensible self" might decide that this calculus is too difficult, or too prone to misjudgment, and conclude that the rule "always fasten your seat belt" is strategically optimal. Even if there is a full utilitarian explanation for seat belt behavior, it may be one in which long-term expectations and rule setting are paramount. Consequently, it might be quite misleading to infer WTP to avoid injury from behavior on a single trip. There is no reason for strategically optimal behavior to be tactically optimal when the purpose of strategy is to avoid strategically dangerous tactics.

Design of Stated Preference Experiments

The major issue in design of SP experiments is how to provide a context that elicits core preferences, and avoids distortions caused by cognitive illusions or strategic behavior. It is plausible that the more realistic a hypothetical market setting, the more likely that stated choice behavior will look like real choice behavior. Therefore, it is useful to point out the differences between real and hypothetical markets that are potential sources of difficulty in producing accurate forecasts from SP data. I conclude this discussion with brief reviews of several studies that have been more or less successful in using SP methods to measure values, and note the features that appear to be critical to success.

The most widespread practical development of SP methods has been in marketing, where a family of techniques called conjoint analysis (CA) have evolved for eliciting information on perceptions, attitudes, and preferences. The essential feature of conjoint analysis is that consumers are asked to make a sequence of choices from (or to rate or rank) sets of hypothetical alternatives. The sets and attribute levels of alternatives are typically specified using a statistical design that allows a value to be assigned to each feature of an alternative. CA procedures usually collect additional responses designed to measure perceptions of commodities, and general attitudes ("values") that influence preferences. The foundation for scaling these features is weak, but they have been used with some success to individualize RUM and to detect clusters of consumer types. Traditionally, marketing applications of CA data treat preferences as purely behavioral, and do not attempt to impose on them any assumptions of economic rationality. In recent years, it has become more common to process CA data using models that assume the existence of economic preferences. However, the emphasis of these marketing methods is on the psychometric problem of scaling responses to achieve predictive accuracy, not on the objective of recovering preferences. McFadden (1986) surveys this literature and discusses its connection to classical consumer theory; other surveys are Green and Srinivasan (1991), Akaah and Korgaonkar (1983), and Leigh, Mackay, and Summers (1984). There is also a substantial literature in resource economics on CV methods, which can be viewed as a special case of CA. However, the multiple questions and experimental designs used in CA to isolate aspects of complex commodities and reduce variance have so far not been much used in environmental economics.

Another strand of development of SP methods comes from experimental economics (Kagel & Roth, 1995). This line of research has investigated a variety of market games, using players in a laboratory setting. Many of these games impose on the players the objective function of maximizing dollar payoff. The focus is then on the formation of expectations and strategies, and on the validity of alternative game solution concepts, but the experimental results also provide information on strategic misrepresentation, learning, and consistency of behavior. Some of the experiments use more complex games that are intended to mimic real markets, such as the market for natural gas, with dollar payoffs that are token shares of actual market scale. Experience with the effects of complexity, information, and learning in these experimental settings illuminates some of the phenomena observed in CA. A particular issue that could be clarified using experimental games is the occurrence of gross misrepresentations of preferences when dollar payoffs are absent and perceptions of fairness or entitlements arise. The techniques of subject instruction and repeated play used in these experiments are also potentially useful in studying how learning leads to convergence to core preferences.

It is useful to review a few more or less successful constructed market studies, as judged by consistency with market behavior or by forecasting accuracy, and identify the features that seem to be associated with success. One can then ask whether these operating conditions are met in potential transportation applications.

1. *Demand for Extended Area Telephone Service.* Kridel (1988) reports on an experiment in which residents of Red Bluff, Texas were asked if they would subscribe to extended area telephone service, giving them the ability to call Dallas without incremental charges. The offering was made at a series of hypothetical monthly charges. This was a service that had been intensely discussed in a local election and petition campaign, and rulings of the Texas Public Service Commission made it clear that the service would be offered in the near future. The SP data were used to estimate a utility model, which was then used to forecast the penetration of the new service and the additional calling volume it would generate. The model was used to establish the price of the new service in its subsequent offering, and the population was followed for one year to test the accuracy of the model. In this case, the model was accurate to within a few percent for both the penetration rate and the calling volume.
2. *Demand for Goose-Hunting Licenses.* Bishop and Heberlein (1979) conducted an experiment in which recipients of free early-season duck hunting licenses in Wisconsin were offered cash to sell their licenses, or alternately asked to state a hypothetical selling or buying price. The study also compared these values with those obtained by the TCM. They found substantial discrepancies between mean willingness to buy and to sell in the hypothetical market, and between the hypothetical and cash markets, with the hypothetical mean selling price about 40

percent higher than the actual mean selling price, and the hypothetical mean buying price about one-third the actual mean selling price. The authors conclude that the reduced realism of the hypothetical offering leads to a significant bias. A companion paper by Bishop, Heberlein, and Kealy (1983) provides further evidence and reaches similar conclusions.

3. *Demand for High-Speed Trains in Holland.* Morikawa, Ben-Akiva, and McFadden (1998) analyzed data in which the actual inter-city mode choices of Dutch travelers were observed, and in addition CA data were collected on the demand for high-speed rail, with a variety of experimental treatments. In these experiments, the high-speed rail alternative was not familiar to most respondents, and the experimental description was quite stylized. The authors found that choice models fitted to the CA data were not consistent with observed choice behavior, but that after permitting the parameters for the hypothetical data to be shifted in location and scale, the real and stated preferences could be reconciled. The pattern of findings suggests that inexperienced consumers confronted with incomplete information on a commodity may make a biased imputation of unobserved attributes, and may make mistakes in weighing these attributes in comparison with observed attributes. In this application, real market data can be used to recalibrate the SP so that they are consistent with RP, and add to predictive power. However, without the recalibration, the SP would be substantially biased.
4. *Demand for Telephone Services.* Hausman and Ruud (1987) analyzed ranking data from a CA experiment, and found significant differences between the attribute tradeoffs in first ranks and subsequent ranks. They found that rescaling model parameters from subsequent rank data eliminated most, but not all, discrepancies in preferences. Thus, apparently innocuous changes in the cognitive task induced behaviors that are inconsistent with maximization of stable preferences. The study suggested that SP results may be sensitive to framing of the task and elicitation format.
5. *Demand for Wilderness Areas.* Diamond and Hausman (1994) and McFadden (1994) analyzed a series of experiments that examined WTP to protect wilderness areas from logging. The studies found statistically and economically significant elicitation effects, in which WTP from referendum questions is much higher than WTP from open-ended questions. They also found strong embedding effects, in which WTP failed to be sensitive to the scale of the resource being preserved. These studies concluded that current CV methods applied to problems of existence value are troubled by psychometric distortions that make them unreliable, and that there is some evidence of fundamental failure of the assumption of classical stable preferences. The primary problem is that one is dealing with an unfamiliar and extremely complex commodity for which ethical

issues and rule-driven behavior are probably sufficient to obscure core preferences even if they exist.

The pattern that runs through these studies is that apparently minor deviations from standard market operating conditions, where the consumer is deciding on an exchange that will actually be executed, lead to substantial biases in behavior. In some cases, these biases are systematic, and can be corrected when there are market data available to be used for recalibration. An overall assessment is that constructed market studies provide information on preferences but that this information must be filtered to extract the preferences from surrounding noise. It is rare that the SP measured in a constructed market study will coincide with true market preferences without recalibration. This suggests that for a constructed market study to have a reasonable probability of success, two conditions must be met: (1) The consumer must be fully informed on the attributes of the commodity, and either experienced in making decisions about it or trained in a manner that provides a context that resembles historical experience; and (2) the exchange must parallel a real exchange sufficiently closely so that the consumer states accurately his or her intentions in the real exchange. Otherwise, there must be real market data against which SP can be recalibrated.

Some transportation applications can meet these conditions, but some of the areas where SP methods would be most valuable, in dealing with new commodities and services with no close real market parallels, are also likely to be the most problematic. In a typical CA experiment, the commodities are presented in configurations that may be unfamiliar or implausible to many consumers. The consumer is a passive recipient of a linear, often brief, written or verbal description of the commodities, with limited opportunity for questions or discussion in a social setting. Response is required in an abbreviated time frame, possibly on unfamiliar scales of the importance of various attributes. As in the real purchase of an automobile, learning and construction of preferences is likely to occur, but the time frame, poverty of information, and lack of economic incentives may be more likely to lead to protocols for this construction that are inappropriate (see Schkade and Payne, 1993, 1994; Whittington et al., 1992). One promising approach to making information-gathering in SP experiments more realistic is information acceleration, a multimedia interactive technique that gives the consumer active control of the search process, while permitting the analyst to monitor the search protocol (see Hauser et al., 1993; Urban et al., 1993, 1994, 1996).

CONCLUSIONS

The first part of this paper lays out the steps necessary to go from RUM-consistent core preferences to mean WTP, an acceptable measure of the benefits of a transportation improvement. The second part examines the question of when RP or SP experiments are likely to reveal core preferences, and the real or hypothetical market conditions and SP experimental designs which appear to be necessary to obtain reliable data on preferences.

ACKNOWLEDGEMENT

This research was supported by the E. Morris Cox Endowment at the University of California, and by National Science Foundation Grant SBR 9601149.

REFERENCES

Akaah, I. and P. Korgaonkar (1983). An empirical comparison of the predictive validity of self-explicated, Huber-Hybrid, traditional conjoint, and hybrid conjoint models. *Journal of Marketing Research,* **20**, 187-197.

Ainslie, G. (1982). Beyond microeconomics: Conflict among interests in a multiple self as a determinant of value. In: *The Multiple Self* (J. Elster, ed.), pp. 133-175. Cambridge University Press, Cambridge, UK.

Ben-Akiva, M. (1972). *The structure of travel demand models.* Doctoral Thesis, Massachusetts Institute of Technology MIT, Cambridge, MA.

Bishop, R. and T. Heberlein (1979). Measuring the values of extra-market goods: Are indirect measures biased? *American Journal of Agricultural Economics*, **61**, 926- 30.

Bishop, R., T. Heberlein and M. Kealy (1983). Contingent valuation of environmental assets: Comparisons with a simulated market. *Natural Resources Journal*, **23**, 619-33.

Chipman, J. and J. Moore (1980). Compensating variation, consumer's surplus, and welfare. *American Economic Review*, **70**, 933-949.

Chipman, J. and J. Moore (1990). Welfare change, consumer's surplus analysism and the gorman polar form. In: *Preferences, Uuncertainty and Optimality: Essays in Honor of Leonid Hurwicz* (J. Chipman, D. McFadden, and M. K. Richter, eds.), pp. 68-120. Westview Press, Oxford.

Deaton, A. and J. Muellbauer (1980). *Economics and Consumer Behavior.* Cambridge University Press, New York.

Diamond, P. and D. McFadden (1974). Some uses of the expenditure function in public finance. *Journal of Public Economics,* **3**, 3-21.

Diamond, P. and J. Hausman (1994). Contingent valuation: Is some number better than no number? *Journal of Economic Perspectives,* **8**, 45-64.

Domencich, T. and D. McFadden (1975). *Urban Travel Demand: A Behavioral Analysis.* North-Holland, Amsterdam.

Green, P. and V. Srinivasan (1990). Conjoint analysis in marketing: New developments with implications for research and practice. *Journal of Marketing,* **54**, 3-19.

Hausman, J. and P. Ruud (1987). Specifying and testing econometric models for rank-ordered data. *Journal of Econometrics,* **34**, 83-104.

Hauser, J., G. Urban and B. Weinberg (1993). How consumers allocate their time when searching for information. *Journal of Marketing Research,* **30**, 452-66.

Hensher, D. (1985). An econometric model of vehicle use in the household sector. *Transportation Research B,* **19**, 303-314.

Hicks, J. (1939). *Value and Capital.* Oxford University Press, London.

Kagel, J. H. and A. E. Roth (eds.) (1995). *The Handbook of Experimental Economics.* Princeton University Press, Princeton, NJ.

Keeler, T. and K. Small (1977). Optimal peak-loading pricing, investment, and service levels on urban expressways. *Journal of Political Economy*, **85**, 1-25.

Kridel, D. (1988). A consumer surplus approach to predicting extended area service (EAS) development and stimulation rates. *Information Economics and Policy,* **3**(4), 379- 390.

Leigh, T., D. Makay and J. Summers (1984). Reliability and validity of conjoint analysis and self-explicated weights: A comparison. *Journal of Marketing Research*, **21**, 456-62.

Manering, F. (1983). An econometric analysis of vehicle use in multivehicle households. *Transportation Research A,* **17**, 183-189.

Manering, F. and C. Winston (1985). *A dynamic empirical analysis of household vehicle ownership and utilization.* Working Paper, Department of Transportation, Massachusetts Institute of Technology MIT, Cambridge, MA.

McFadden, D. (1973). Conditional logit analysis of qualitative choice behaviour. In: *Frontiers in Econometrics* (P. Zarembka, ed.), pp. 105-142. Academic Press, New York

McFadden, D. (1974). The measurement of urban travel demand. *Journal of Public Economics,* **3**, 303-328.

McFadden, D. (1978). Qualitative methods for analyzing travel behaviour of individuals. In: *Behavioural Travel Modelling* (D. Hensher and P. Stopher, eds.), pp. 279-318. Croom Helm, London.

McFadden, D. (1981). Econometric models of probabilistic choice. In: *Structural Analysis of Discrete Data with Econometric Applications* (C. Manski and D. McFadden, eds.), pp. 198-272. MIT Press, Cambridge, MA.

McFadden, D. (1986). The choice theory approach to market research. *Marketing Science,* **5**, 275-297.

McFadden, D. (1994). Contingent valuation and social choice. *American Journal of Agricultural Economics*, **76**, 689-708.

McFadden, D. (1997). Measuring willingness-to-pay in discrete choice models. In: *Essays in Honor of John Chipman* (J. Moore and R. Hartman, eds). In press.

McFadden, D. and K. Train (1976). The goods/leisure tradeoff and disaggregate work trip mode choice models. *Transportation Research,* **12**, 349-353.

Morikawa, T., M. Ben-Akiva, and D. McFadden (1998). Incorporating psychometric data in econometric travel demand models. *Journal of Econometrics.* In press.

Prelec, D. (1991). Values and principles: Some limitations on traditional economic analysis. In: *Perspectives on Socioeconomics* (A. Etzioni and P. Lawrence, eds.), pp. 131-145, M. E. Sharpe, London.

Schkade, D. and J. Payne (1993). Where do the numbers come from? How people respond to contingent valuation questions. In: *Contingent Valuation: A Critical Assessment* (J. Hausman, ed.), pp. 271-304. North Holland, Amsterdam.

Schkade, D. and J. Payne (1994). How people respond to contingent valuation questions: A verbal protocol analysis of willingness to pay for an environmental regulation. *Journal of Environmental Economics and Management*, **26**, 88-109.

Thaler, R. (1990). Savings, fungability, and mental accounts. *Journal of Economic Perspectives,* **4**, 193-205.

Train, K. (1986). *Qualitative Choice Analysis: Econometrics and an Application to Automobile Demand.* MIT Press, Cambridge, MA.

Urban, G., W. Quals and J. Bohiman (1994). *Information acceleration of high tech industrial products: A market feasibility test.* Working Paper, Sloan School, Massachusetts Institute of Technology MIT, Cambridge, MA.

Urban, G., B. Weinberg and J. Hauser (1996). Premarket forecasting of really-new products. *Journal of Marketing,* **60**, 47-60.

Urban, G., J. Hauser and J. Roberts (1993). *Design and Marketing of New Products.* Prentice-Hall, Englewood Cliffs, NJ.

Varian, H. (1984). *Microeconomic Analysis* (2nd ed.). Norton, New York.

Whittington, D., V. K. Smith, A. Okorafor, J. L. Liu and A. McPhail (1992). Giving respondents time to think in contingent valuation studies: A developing country application. *Journal of Environmental Economics and Management,* **22**, 205-225.

16

A MODEL OF HOUSEHOLD CHOICE OF ACTIVITY PARTICIPATION AND MOBILITY

Thomas F. Golob, University of California, Irvine, CA, U.S.A.

OBJECTIVES AND SCOPE

The travel choice model system introduced here is founded on the principle that personal travel choices involve an integrated set of decisions involving what activity to participate in, where and when to participate in that activity, and how to travel between activity locations. In this model system, we treat mobility decisions, involving household car ownership, car usage, and travel by different modes, as the direct results of demands by household members to participate in activities at different locations. But we also recognize that mobility allows a person to engage in activities at different locations. Thus, the model must also capture feebacks from mobility demand to activity demand.

Despite the indisputable linkages between activity participation demand and travel demand, the vast majority of travel choice or travel behavior models assume that most aspects of activity participation demand are given. Almost all travel choice models are based on trips or trip chains, with the endogenous variables of interest being a selection from the list that includes: destination, mode, route, time of day, and frequency of repetition. If we assume that the reasons for making a trip will remain unchanged, then the most important travel choices are already made, and we are left modeling only the minor details. Granted, these details are of interest to planners, because they impact traffic flows on transportation networks, but it is foolish to assume that the motivations for trips do not change. This is why conventional trip-based models fail to account for increased demand for travel due to enhanced accessibility and improved transportation system levels of service; and this is why trip-based models are inappropriate for evaluating the effects of telecommunications on travel (Kitamura, 1991).

One basic group of activity participation decisions is whether or not to work, whether to work full-time or part-time, and whether or not to conduct some or all of the work at home. If we assume that employment status, work schedules, and work locations are given in our travel choice models, we have abdicated forecasting responsibility for a great portion of travel behavior. This might be acceptable only if all of the following were true: (1) We have reliable exogenous forecasts of detailed employment characteristics, which we do not. (2) Working does not affect other travel, which, of course, it does. (3) There are no feedbacks from mobility to employment, which there are. And (4), working at home, versus away from home, is not important, but it is. Likewise, we should not assume a fixed demand for activities such as shopping and personal business (banking, medical services, etc.), because we know that accessibility and mobility substantially influence patterns of demand for these activities. Moreover, home-based shopping and conduct of personal business are major aspects of the telecommunications revolution.

We must also recognize that an individual's activity participation and mobility choices are not made in isolation. Activities are typically shared among household members, and household members and non-household members alike engage in many activities and travel together. For this reason, we should strive to model the activity and travel choices of household members simultaneously.

The objective here is to develop a model system that can explain some basic causal relationships between household activity participation, travel to out-of-home activity sites, car ownership, and household vehicle utilization. The effects of residential location on activity and mobility demand were investigated by Golob and McNally (1996); it was concluded that, as a first approximation, residential location could be treated as an exogenous variable.

The behavioral units in this case study application of the proposed choice model system are the two married or unmarried male and female adults who can be identified as heads of a household in which there might be any number of other members, such as children or other persons. Thus, the focus in this application is on the activity and travel demand of adult heads of household, and their household car ownership and utilization demand.

The case study documented here is viewed as an initial test of a model system that can be expanded to meet many policy evaluation needs related to travel demand. The exogenous variables in the model can be expanded to encompass essentially any variables that have been used in conventional models. The difference is that the current model system is designed to explain the effects of exogenous variables on both activity and mobility demand. This includes network and other spatial data related to residential locations and transportation levels of service. Activity demand in this model system can be measured in terms of any set of activity categories; the extent of the division is limited only by the sample size and the quality of the data. Mobility demand can be measured in terms of travel demand by mode, car ownership demand, and car usage demand. Modal travel demand, in turn, can be measured in terms of travel times, trip rates, rates of journeys or tours, or travel distances. Again, the travel demand component of the model system is, in principle, limited only by data availability.

Another important feature of this model system is its ability to estimate the influences of any endogenous variable on the other endogenous variables. For example, the model system can be used to investigate the effects of an independent increase or decrease in certain activities, such as in-home working or in-home shopping, on all other activity and mobility demands.

The model system could be used to generate household demand for travel (in terms of trips or trip chains by mode), car ownership, and car usage for spatial aggregations of households. As such, it could feed some combination of destination choice, route choice, or traffic assignment models.

BACKGROUND

Travel behavior models based on activity demand make use of the time-use data collection and analysis techniques developed by geographers and sociologists (e.g., Chapin, 1974; Hägerstrand, 1970; Robinson, 1977). The trick is to apply these techniques in a way that usefully improves traditional travel demand forecasting methods. The principle of basing travel demand models on activity patterns was endorsed in the 1970's and early 1980's, by van der Hoorn (1979), Jones et al. (1983), Pas (1985), and others; and several conferences, workshops, and special sessions were held on the subject, leading to numerous publications (e.g., Carpenter and Jones, 1983). Kitamura (1988, 1996) and Pas (1996) provide comprehensive review of the pioneering efforts in activity analysis for transportation planning.

Overall interest in new activity-based models has waned over the last decade. However, there have been efforts to make incremental improvements to conventional planning models by using principles originally developed in activity analyses. For example, Purvis, Iglesias, and Eisen (1996) used early results on "feedback" relationships between travel and activities (Golob, Kitamura, and Lula, 1994) to modify a conventional travel demand model system to incorporate work trip accessibility in non-work trip generation models for the San Francisco Bay Area. Further examples of advances based on activity analyses are Goulias, Pendyala, and Kitamura (1990) and Pas (1988).

Development of activity-based models has been held up in part by the lack of time-use data. Trip diaries do not provide the necessary data; activity diaries are needed. Progress has also been held up by the lack of familiarity in the travel demand modeling community with methods to handle multivariate problems with endogenous variables that are not multivariate normally distributed (e.g., variables measuring activity durations, which have high concentrations of observations at zero and are otherwise constrained to the positive domain). Travel choice modeling has been ruled by the paradigm of a single-equation discrete choice variable.

Today, there are enhanced opportunities for activity-based models. The conventional four-step or five-step transportation planning process in use today was established in all essential aspects 40 years ago. These travel demand models are inappropriate for forecasting latent demand, or for assessing the role of accessibility on car ownership, mode choice, and residential location (Kitamura, 1991). Planners are now being forced to abandon this planning process because it does not meet the requirements of the relatively recent changes to requirements for federal funding in the United States. Also, plans based on the four-step process have been successfully challenged in court by environmental groups concerned with induced demand caused by improvements in levels of service for car travel.

Activity-based models are a potential way out of this dilemma, and metropolitan planning organizations (MPOs) and the U.S. Federal technical support bureaucracy have expressed recent interest in such models. Several MPOs, such as Portland, Oregon, Dallas-Fort Worth Texas, and Raleigh-Durham, North Carolina, have conducted extensive activity diary surveys. In principle, activity-based models applied to such rich datasets can deal with many of today's critical planning issues, if they can be made practical.

THEORY

The choice model system has four sets of endogenous demand variables:

1. *Activity participation.* Choices concerning whether or not to participate in various activities, at home and away from home, and how long to engage in such activities are considered to be the basic determinants of household mobility. In turn, activity participation will be influenced by all of these components of mobility, because access to activity sites and the time available for out-of-home activities is dependent on the accessibility of activity sites and travel times. Activities must be divided into categories in order to specify an operational choice model. Categories such as "work and work-related", "shopping and personal business" and

"discretionary" have been used in time-use studies (Chapin, 1974; Robinson, 1977).

2. *Travel times*. Travel will be strongly and directly determined by activity participation. But travel time will also be determined by vehicle ownership, because some activity sites are more accessible to households with a higher level of vehicle availability. In turn, travel times affect activity participation due to pressures on time budgets. Travel times can also influence vehicle ownership decisions through the link with activity participation. We can group travel times into times for travel to a particular category of activities (Golob and McNally, 1997), or travel times by category of travel mode, such as "car," "public transport" and "non-motorized modes."
3. *Vehicle ownership*. Decisions concerning household vehicle ownership level will be influenced by all of the above sets of demand variables because households acquire vehicles in order to satisfy demands for travel to activity sites, and demand for activity participation depends in part on vehicle availability.
4. *Vehicle miles of travel (VMT)*. This is a variable that is key in many policy assessments. Consequently, why not measure it directly and include it in the demand model system as an endogenous variable? Here we do.

In addition, the following six sets of direct causal effects are identified:

1. *Travel demand* (causal links from *Activity Participation* to *Travel Times*). In this choice model system, demand for travel is derived from demand for activity participation. Consequently, there will be strong direct effects from demand for an out-of-home activity to the travel times needed to access the sites for that activity. Travel times will be determined by travel-mode choice decisions, by land use patterns, and by network architecture and performance.
2. *Time budget effects* (links from *Travel Times* to *Activity Participation*). The total amount of time available for both out-of-home activities and the travel to and from these activities is relatively fixed, assuming that household members have responsibilities at home or desire to be at home for certain minimum periods. If accessibility improves, say, by infrastructure improvements or by reducing congestion, travel times associated with activities might initially decrease. But the individuals engaged in the activities will have more time available. Some of the saved time is likely to be consumed by increasing the duration of the activity at that site. However, some of the saved time might also be consumed by other out-of-home activities, or by substitution of more preferred destinations for the same activities, either of which would lead to increased travel (an effect typically called "latent demand"). Such "feedback" effects from travel to activity duration were

captured in a more limited way in the budget models of travel originally proposed by Zahavi (1979) and studied by Downes and Emmerson (1985) and Golob et al. (1981).

3. *Mobility demand* (links from *Activity Participation* to *Vehicle Ownership*). Activity participation choices, and the life style choices that underlie activity participation choices, will influence car ownership demand (Burns and Golob, 1976). Households that choose to live in areas well served by transit and pedestrian networks will exhibit less need for vehicles, while households in residential areas that are auto-oriented will generally strive to acquire one vehicle for every driver.
4. *Accessibility* (links from *Vehicle Ownership* to *Activity Participation*). Potentially, access to vehicles can enhance activity participation, if certain activity sites are not well served by alternative modes of transportation, so that travel times to such sites are substantially less by car. Also, some household members might forego certain activities, if household vehicles are being used by other household members and there is no vehicle available when they want to pursue the activity.
5. *Vehicle utilization* (links from *Travel Times* to *Vehicle Miles of Travel*). Household vehicle miles of travel (VMT) will be a function of travel time dependent on how households split their travel times between driving a household vehicle and all other modes. Households in which most travel is by solo driving will exhibit a high multiplier between travel times and VMT, while households with more car passenger, public transport, bicycle, and pedestrian trips will put fewer miles on their vehicle(s) per hour of total travel.
6. *Excess mobility demand* (links from *Vehicle Miles of Travel* to *Vehicle Ownership*). For households in which there are fewer vehicles than drivers, intensive use of the household vehicle(s) will pressure the household to acquire an additional vehicle. This effect will be mitigated to the extent that vehicle ownership is "saturated" with at least one vehicle per driver.

The six hypothesized sets of direct causal effects between the sets of endogenous variables effects listed above and portrayed in Figure 16.1, provide numerous paths by which each set of variables can have a total effect or influence on any of the other sets of variables. For example, there are no hypothesized direct effects between vehicle ownership and travel times, but vehicle ownership can influence travel times via the path combining the accessibility and travel demand links through activity participation, and travel times can influence vehicle ownership via the path combining the time budget and mobility demand effects. The identification of the strengths and directions of total effects is one of the objectives in our modeling.

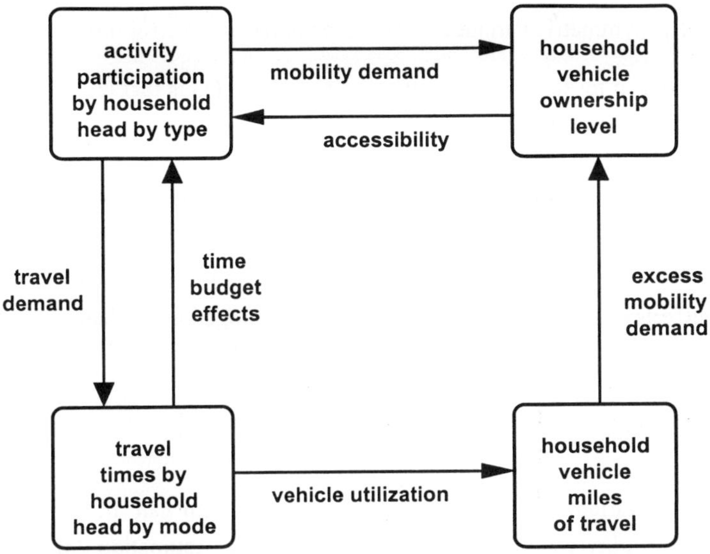

Figure 16.1. Conceptual flow diagram of the model system.

METHODOLOGY

We can test our hypotheses by specifying and estimating a system of activity participation and mobility choice models using the analytical framework of structural equation models with latent choice variables. Other applications of structural equations to data on activity participation and travel are found in Golob et al., (1994, 1996), Golob and McNally (1997), Lu and Pas (1997), and Fujii et al. (1997).

Structural Equation Models

Structural equations model systems are generally defined by the matrix equation system:

$$y = By + \Gamma x + \zeta \qquad (1)$$

where y is a column vector of endogenous variables, x is a column vector of exogenous variables, and ζ is a column vector of unexplained portions of the endogenous variables (error terms). The structural parameters are the elements of the three matrices:

B : matrix of causal links between the endogenous variables

Γ: matrix of regression effects of the exogenous variables,

$\Psi = E(\zeta\zeta')$: symmetric variance-covariance matrix of the error terms.

A necessary condition for identification of equation system (1) is that ($\boldsymbol{I}$ - $\boldsymbol{B}$) is non-singular, where $\boldsymbol{I}$ denotes the identity matrix.

The total effects of the endogenous variables on one other, $\boldsymbol{T}_{yy}$, implied by system (1) are:

$$\boldsymbol{T}_{yy} = (\boldsymbol{I} - \boldsymbol{B})^{-1} - \boldsymbol{I} \tag{2}$$

And the total effects of the exogenous variables on the endogenous variables, $\boldsymbol{T}_{xy}$, are given by

$$\boldsymbol{T}_{xy} = (\boldsymbol{I} - \boldsymbol{B})^{-1} \Gamma \tag{3}$$

which are the parameters of the so-called reduced-form equations.

If the endogenous variables in the structural equation system can all be considered to be continuous and multivariate normally distributed, the estimation of the system is straightforward, using normal-theory maximum likelihood (Bollen, 1989). However, in the present case, the endogenous variables represent imperfectly observed choices, and this requires the use of special estimation methods (Golob and McNally, 1997).

Specification of the Choice Variables

Demand for participation in any activity type can be observed in terms of the duration of time spent in that activity. Likewise, demand for travel on a category of travel-modes can be observed in terms of time spent traveling on those modes. However, durations data must be gathered through the use of survey instruments such as activity or time use diaries (or logs), and these diaries typically cover time periods of from one day to one week, with two-day diaries becoming popular with transportation planners. With such restricted time periods of observation, we can expect to detect many individuals with zero durations for many activity types and travel times. The probability of observing an individual participating in an activity will be directly related to the frequency pattern with which the individual engages in that activity. Extensive examinations of the day-to-day variability in activity and travel patterns are provided by Pas (1986), Pas and Koppelman (1987), and others (reviewed in Kitamura, 1988).

We can get around this problem of limited observation time by treating the activity participation, travel time, and VMT choice variables as being censored. We assume that each of these observed choice variables y_i is associated with a latent (unobserved) normally distributed variable y_i^* with mean μ and variance σ^2. The observed and latent variables are related such that:

$$\begin{aligned} y_i &= y_i^* \quad \text{if } y_i^* > 0 \\ y_i &= 0 \quad \text{otherwise} \end{aligned} \tag{4}$$

For observed durations (or mileage) of zero, all we know is

$$P(y_i = 0) = P(y_i^* \leq 0) \tag{5}$$

The likelihood function for estimation of the unknown parameters is thus

$$L\left(\mu, \sigma^2 \middle| y_i\right) = \prod_{y_i > 0} \frac{1}{\sigma} \phi\left(\frac{y_i - \mu}{\sigma}\right) \prod_{y_i \leq 0} \Phi\left(\frac{-\mu}{\sigma}\right) \tag{6}$$

ϕ and Φ denote the density and distribution functions of the standard normal. Estimation of μ and variance σ^2 for each censored endogenous variable in the choice model system is accomplished using the Tobit model described in, for instance Maddala (1983). The Tobit model was introduced by Tobin (1958), and was further developed by Amemiya (1973) and Fair (1977).

We can treat all endogenous choice variables this way, with the exception of vehicle ownerhip. For this variable, we observe the number of vehicles available to the household, from zero on up, and this distribution can have a long tail, with a few households having large numbers of vehicles. Econometricians have determined that estimations of such observed ordered choice variables can be improved by defining a normally distributed latent variable that is cut by thresholds to generate the observed ordered categories. The latent variable itself is not observed, but the ordinal indicator is related to it in the following way

$$\begin{aligned} y_i &= 1 \quad \text{if} \quad y_i^* \leq \alpha_1 \\ y_i &= 2 \quad \text{if} \quad \alpha_1 < y_i^* \leq \alpha_2 \\ &\vdots \\ y_i &= c \quad \text{if} \quad \alpha_{c-1} < y_i^* \end{aligned} \tag{7}$$

where $\alpha_1 < \alpha_2 < \cdots < \alpha_{c-1}$ are the unknown threshold values. These thresholds are estimated corresponding to the marginal distribution of the population over the categories, using the ordered-response probit regression model developed by Aitchison and Silvey (1957) and Ashford (1959) as an extension of the binomial probit model.

Estimation of the Structural Equations Model with Latent Choice Variables

Structural equations systems are estimated using methods of moments (also called: variance analysis methods). We begin by defining the sample variance-covariance matrix of the combined set of endogenous and exogenous variables, partitioned with the endogenous variables first:

$$\boldsymbol{S} = \begin{bmatrix} \boldsymbol{S}_{yy} & \boldsymbol{S}_{yx} \\ \boldsymbol{S}'_{yx} & \boldsymbol{S}_{xx} \end{bmatrix} \tag{8}$$

where $\boldsymbol{S}_{yy}$ denotes the variance-covariance matrix of the endogenous variables, $\boldsymbol{S}_{xy}$ denotes the covariance matrix between the endogenous and exogenous variables, and $\boldsymbol{S}_{xx}$ denotes the variance-covariance matrix of the exogenous variables, which is taken as given.

It can easily be shown using matrix algebra that the corresponding variance-covariance matrix replicated by model system (1) with parameter vector θ (denoting all parameters in the $\boldsymbol{B}$, Γ, and Ψ matrices) is

$$\Sigma(\theta) = \begin{bmatrix} \Sigma_{yy} & \Sigma_{yx} \\ \Sigma'_{yx} & \Sigma_{xx} \end{bmatrix} \tag{9}$$

where

$$\Sigma_{yy} = (\boldsymbol{I} - \boldsymbol{B})^{-1}\left(\Gamma \boldsymbol{S}_{xx} \Gamma' + \Psi\right)\left((\boldsymbol{I} - \boldsymbol{B})^{-1}\right)' \tag{10}$$

$$\Sigma_{yx} = (\boldsymbol{I} - \boldsymbol{B})^{-1} \Gamma \boldsymbol{S}_{xx} \tag{11}$$

and $\Sigma_{xx} = \boldsymbol{S}_{xx}$ is taken as given for the exogenous variables.

The parameters of the B, Γ, and Ψ matrices are estimated by making $\Sigma(\theta)$ be as close as possible to $\boldsymbol{S}$. With latent choice variables corresponding to censored and ordered categorical (ordinal) observed endogenous variables, this is accomplished using the method known as asymptotically distribution-free weighted least squares, or ADF-WLS. This method produces consistent and asymptotically efficient parameter estimates and unbiased estimates of standard errors and goodness-of-fit (Browne, 1974, 1982, 1984).

The ADF-WLS estimation method proceeds in two distinct steps after estimation of the means and variances of the latent variables corresponding to the observed censored variables, and the thresholds of the latent variable corresponding to the ordinal variable.

The first step is to estimate the correlations between the latent endogenous variables, and the correlations between each of the latent variables and the continuous exogenous variables in the system. For pairs of censored endogenous variables, the problem is to determine the unknown correlation coefficient between the latent variables that maximizes the likelihood of observing the cross-products where below-censoring level observations are assigned normal scores determined by the Tobit model (Raj, 1953). For endogenous variable pairs involving the ordinal endogenous variable and a censored variable, the polyserial correlation coefficient method is used (Olsson et al., 1982).

The final step in the ADF-WLS method is to estimate the parameters of the structural equation model by making the model-implied correlation matrix as close as possible to the sample correlation matrix, where the sample matrix is determined in the previous steps. The fitting function is then:

$$F_{WLS} = [s - \sigma(\theta)]' W^{-1} [s - \sigma(\theta)] \tag{12}$$

where $\boldsymbol{s}$ is a vector of censored correlation coefficients for all pairs of endogenous and exogenous variables, $\sigma(\theta)$ is a vector of model-implicated correlations for the same variable pairs, and $\boldsymbol{W}$ is a positive-definite weight matrix. Minimizing F_{WLS} implies that the parameter estimates are those that minimize the weighted sum of squared deviations of $\boldsymbol{s}$ from $\sigma(\theta)$. This is analogous to weighted least squares regression, but here the observed and predicted values are variances and covariances rather than raw observations.

The best choice of the weight matrix is a consistent estimator of the asymptotic covariance matrix of s:

$$W = ACOV\left(s_{ij}, s_{gh}\right) \tag{13}$$

Under very general conditions

$$W = \frac{1}{N}\left(s_{ijgh} - s_{ij}s_{gh}\right) \tag{14}$$

is a consistent estimator, where s_{ijgh} denotes the fourth-order moments of the variables around their means, and s_{ij} and s_{gh} denote covariances. Browne (1982, 1984) demonstrated that F_{WLS} with such a weight matrix will yield consistent estimators which are asymptotically efficient and corresponding z and chi-square test statistics are asymptotically valid. These properties hold for very general conditions, and consequently such estimators are known as asymptotically distribution free (ADF) estimators (also called: arbitrary distribution function estimators).

ADF-WLS estimators are available in several structural equation model packages. The LISREL/PRE-LIS (Versions 8/2 for Windows) programs (Jöreskog and Sörbom, 1993) are used here.

THE PORTLAND CASE STUDY

The choice model system has been implemented using the Portland, Oregon and Southwest Washington Metropolitan Area as a case study.

Data

The data are from the Portland, Oregon 1994 Activity and Travel Survey conducted in the spring and autumn of 1994. This survey involved a two-day activity diary, which was designed to record all activities involving travel and all in-home activities with a duration of at least 30 minutes for all individuals in the household. Only the spring portion of the sample was used, because the autumn portion is concentrated on weekdays, and we prefer that the model system deals with activities and travel on *all* days of the week. This portion of the survey involved 2,230 households with 5,120 individuals. The importance of a multi-day diary for this type of analysis is made clear by Pas (1986), Pas and Koppelman (1987), and Pas (1995).

The sample used here was composed of households in which there are two persons eighteen years or older living in the same residence, one male and one female, who we can identify as household heads. The household heads need not be married. After processing for missing data, particularly exogenous environmental data, the sample of 1,318 such households (approximately 60% of the 2,230 total household) was reduced to a final sample of 1,047. This is a subset of the same sample of 1,292 households used in the analyses reported by Golob and McNally (1997).

The Endogenous Variables

Specific variables were computed for each of the four sets of demand variables depicted in Figure 16.1. Multiple variables were used in three of the four sets, resulting in a total of fourteen endogenous variables. Each set of variables is described below.

Activity Participation. Total activity duration was computed over each individual's two diary days, divided into three categories of out-of-home activities and a single category of in-home activities. The three categories of out-of-home activities represent a hierarchy that has been used in many time-use studies. At the top of the hierarchy is (1) out-of-home work (subsistence activities). This is followed by (2) out-of-home maintenance, which includes activities that households typically need to perform on a regular basis. Finally, we define (3) out-of-home discretionary, which includes social, recreational, and entertainment activities. This hierarchy was originally proposed in the travel demand context by Reichman (1977). Gunn (1981) found the distinction between mandatory and discretionary activities to be important in explaining travel budgets.

Our only in-home activity is work. Theoretically, the demand model structure can handle multiple categories of in-home activities, but we are constrained by limitations in the data and the desire for a simple initial model. The specific activity categories utilized in collecting the Portland activity diaries are assigned to our four activity categories as follows: (1) Out-of-home work is comprised of the two activities coded as "work" and "work-related" at non-home locations, (2) Out-of-home maintenance is comprised of activities coded as "meals," "shopping (general)," "shopping (major)," "personal services," "medical care," "professional services," "household or personal business," "household maintenance," "household obligations," "pick up or drop off passenger," "school," and "religious" at non-home locations; (3) Out-of-home discretionary is comprised of activities coded as: "visiting," "culture," "civic," "amusements," "hobbies," "exercise or athletics," "rest and relaxation," "spectator athletic events," "incidental trips," and "tag-along trips" at non-home locations; and (4) In-home work is comprise of activities coded as "work" and "work-related" at the home location.

It is possible to use other activity categorizations within the same demand model structure. For example, several studies have focused on shopping activities and activities involving picking up or dropping off passengers (the so-called "serve-passenger" trip purpose), and our maintenance category could be subdivided to preserve categories such as these. However, it is wise to evaluate the performance of the model system at its current level of complexity before expanding the detail of its activity demand component.

These activity participation variables were computed for both the male and female heads of spousal households. Thus we have a total of eight activity duration variables, four for each household head, allowing us to capture interactions between the male and female household heads (Damm, 1978; Hanson and Hanson, 1978; Kostyniuk and Kitamura, 1982; van Wissen, 1989; Golob and McNally, 1997).

The endogenous activity participation variables are listed with descriptive statistics in Table 16.1. Over 62 percent of male heads recorded some out-of-home work activity over the two days; and for those men participating in work activities, the two-day mean activity participation was 13.4 hours. In comparison, almost 48 percent of female heads worked outside the home during the two diary days, and the mean duration of 12.0 hours for those participating in work activities was slightly less than that of the male heads. For both men and women working outside the home, as expected, the median duration was less than the mean, indicating a distribution skewed to the right.

Almost 89 percent of the male heads and over 94 percent of the female heads participated in some non-work out-of-home activities over two days. The mean non-work activity durations for both men and women were approximately 5.5 hours, and the conditional distributions are skewed to the right. Finally, only 11.5 percent of the male heads and 8.4 percent of the female heads recorded any in-home work activities.

Travel Times. Travel demand can be measured in various ways, including numbers of trips, number of trip destinations (often called sojourns), number of home-based trip chains (often called tours), travel time, or travel distance. Each of these measures can be broken down by mode, where modes are defined in terms of combinations of car driver, car passenger, car travel by driving alone (solo driver), car travel by multiple-occupancy car (carpool), bus, rail, and other forms of public transportation, pedestrian, bicycle, etc. We chose to measure travel in terms of total travel time, over two days, for each household head broken down into (1)

Table 16.1

The Endogenous Activity Participation Variables (units of hours per two days)

Activity Duration Variable	% Cases > 0	For Cases > 0		
		Median	Mean	Std. Dev.
Total two-day out-of-home work activity duration - male head	62.6%	15.1	13.4	5.75
Total two-day out-of-home maintenance Activity duration - male head	82.8%	2.0	2.9	2.7
Total two-day out-of-home discretionary activity duration: male head	54.1%	3.0	4.2	3.8
Total two-day in-home work activity duration: male head	12.8%	4.0	6.4	5.9
Total two-day out-of-home work activity duration – female head	47.7%	12.8	12.0	5.8
Total two-day out-of-home maintenance activity duration - female head	89.0%	2.6	3.4	3.0
Total two-day out-of-home discretionary activity duration - female head	57.4%	2.7	3.8	3.2
Total two-day in-home work activity duration – female head	8.9%	3.8	5.6	5.2

Table 16.2
The Endogenous Travel Time Variables (units of hours per two days)

Travel Time Variable	% Cases > 0	For Cases > 0		
		Median	Mean	Std. Dev.
Total two-day travel time by car (driver and passenger) – male head	92.6%	2.0	2.2	1.3
Total two-day travel time by all other non-car travel - male head	27.7%	0.8	1.1	1.1
Total two-day travel time by car (driver and passenger) – female head	93.2%	2.0	2.1	1.2
Total two-day travel time by all other non-car travel - female head	28.6%	0.8	1.1	1.0

total travel times by car, and (2) total travel times by all other modes. The use of travel time is consistent with the measurement of activity demand in terms of activity durations, and all out-of-home time can be accounted for by the addition of the three out-of-home activity durations and two travel times for each individual. It is entirely possible to add trip-based travel measures within the same demand model structure, but this is relegated to future research.

The four travel time variables are listed in Table 16.2. The distributions of the travel time variables are similar for males and females. Thus, even though males and females have substantially different activity patterns, they spend similar times traveling.

Vehicle Ownership. Vehicle ownership for the sample (of 1.047) households breaks down into: 3.3% zero-car, 13.8% one-car, 57.5% two-car, 18.9% three-car, and 6.4% more than three cars.

Vehicle Miles of Travel (Household VMT). VMT was calculated from network distances. More than 96% of the households used at least one of their household vehicles during the two diary days. The median VMT for these households was 53.4 miles, and the mean was 61.6 miles, with a standard deviation of 41.7 miles.

The Exogenous Variables

The exogenous variables used in the choice model system are similar to those used in conventional travel demand models, with the exception that we exclude employment status. We endeavor to forecast work activities endogenously. The thirteen exogenous variables used in the model are listed in Table 16.3. Household characteristics include age of the male head, household membership in terms of the number of children by age category, number of drivers,

Table 16.3
The Exogenous Variables

No.	Variable
1	Number of children under 6 years of age
2	Number of children 6-11 years of age
3	Number of children 12-21 years without driving license
4	Number of drivers in household
5	Household in current home 1 year or less (dummy)
6	Household is renting (dummy)
7	Age of male head
8	Male head less than 31 years of age (dummy)
9	Male head 70 years of age or older (dummy)
10	Household Income less than $20,000 (dummy)
11	Household income $20,000 to $30,000 (dummy)
12	Household Income $60,000 or more (dummy)
13	Total retail employment within 1 mile of residence

housing tenure, and income. It was not possible to include ages of both of the household heads because of strong multicolinearity problems. Detailed personal characteristics, such as education, occupation, and personal income were not included, because it is generally not possible to obtain exogenous forecasts of such variables for planning purposes.

One exogenous accessibility variable (# 13) is also included in the model, as a test of policy sensitivity. This variable was created by Metro, the Portland MPO, using Geographical Information System (GIS) computations. It measures the total retail employment within one mile of each residential location, based on *x-y* coordinates, using the actual street network, and land-use data accurate down to quarter acre parcels. This accessibility variable should explain significant portions of activity demand and the split of mobility demand between car and other modes.

MODEL SPECIFICATION

The structural equations model system with latent demand variables was specified with fortysix direct effects between endogenous variables. Each of these effect represents a free parameter in the beta matrix of equation system (1). The postulated matrix is depicted in Figure 16.2, where $+\beta$ indicates an expected statistically significant positive effects from the variable of the column to the variable of the row, and $-\beta$ indicates an expected negative effect. Table 16.4 provides a key to the numbered endogenous variables. There are postulated effects that correspond to the six types of effects listed above:

1. *Travel demand* (from *Activity Participation* to *Travel Times*). These are operationalized in the effects from variables 1 through 4 to variables 9 and 10 (for the male head) and from variables 5 through 8 to variables 11 and 12 (female head). For each household head, the effects from a given out-of-home activity to the two modes will measure the modal split for that activity. We also hypothesize that in-home work will cause a reduction in car travel, but an increase in non-car travel. In addition, we initialize the model estimation by constraining all eight pairs of these travel demand coefficients to be equal across the two household heads.
2. *Time budget effects* (from *Travel Times* to *Activity Participation*). Based on the findings of Golob and McNally (1997), we expect that the maintenance and discretionary activities of the female household head will be most sensitive to feedbacks from car travel time. We also expect that both male and female heads will exhibit more demand for in-home work as a direct function of their car travel time. These effects are in rows 4 through 8 and columns 9 and 11 in the beta matrix.
3. *Mobility demand* (from *Activity Participation* to *Vehicle Ownership*). We postulate that the major pressures on vehicle ownership demand will come from the out-of-home activities of the male head. These effects are in row 13, columns 1 through 4.
4. *Accessibility* (from *Vehicle Ownership* to *Activity Participation*). We postulate that the vehicle availability will positively influence the maintenance and discretionary activities of female heads. These effects are in column 13.
5. *Vehicle utilization* (from *Travel Times* to *Vehicle Miles of Travel*). These effects are in row 14.
6. *Excess mobility demand* (from *Vehicle Miles of Travel* to *Vehicle Ownership*). This effect is in cell (13,14).

Variable		activity participation demand								travel demand				**vehs**	**VMT**
		male				Female				male		female			
		1	2	3	4	5	6	7	8	9	10	11	12	13	14
Male act.	1	--			$-\beta$										
	2	$-\beta$	--		$-\beta$										
	3	$-\beta$	$-\beta$	--	$-\beta$										
	4				--					$+\beta$					
fem. act.	5					--			$-\beta$						
	6					$-\beta$	--		$-\beta$			$-\beta$		$+\beta$	
	7					$-\beta$	$-\beta$	--	$-\beta$			$-\beta$		$+\beta$	
	8								--			$+\beta$			
Male Travel	9	$+\beta$	$+\beta$	$+\beta$	$-\beta$					--	$-\beta$				
	10	$+\beta$	$+\beta$	$+\beta$	$+\beta$					$-\beta$	--				
fem. Travel	11					$+\beta$	$+\beta$	$+\beta$	$-\beta$			--	$-\beta$		
	12					$+\beta$	$+\beta$	$+\beta$	$+\beta$			$-\beta$	--		
vehs.	13	$+\beta$	$+\beta$	$+\beta$										--	$+\beta$
VMT	14									$+\beta$	$-\beta$	$+\beta$	$-\beta$		--

Figure 16.2. Postulated direct effects between endogenous variables (beta matrix specification; no effects are allowed in the diagonal cells)

Table 16.4
Model Endogenous Variables

No.	Set	Variable
1	Activity Participation	out-of-home work – male
2		out-of-home maintenance – male
3		out-of-home discretionary – male
4		in-home work – male
5		out-of-home work – female
6		out-of-home maintenance – female
7		out-of-home discretionary – female
8		in-home work – female
9	Travel Time	car (driver and passenger) – male
10		Non-car travel – male
11		car (driver and passenger) – female
12		Non-car travel – female
13	Vehicle Ownership	Number of household vehicles
14	Vehicle Miles of Travel	Total VMT on all household vehicles

Finally, the model specification needs to cover interactions among the activity and travel demand variables:

1. *Activity hierarchy* (*Within-activity participation* effects). These effects, based on the findings of Golob and McNally (1997), determine an activity-based time-use hierarchy for both male and female heads. Individuals who spend more time in work activities have less time available for all other out-of-home activities, and individuals who spend more time in out-of-home maintenance activities have less time available for discretionary activities. All within-activity effects were equated for male and female household heads.
2. *Modal substitution* (*Within-travel time* effects). Modal substitution is operationalized by the symmetric effects in cells (10,9) and (9,10) and in cells (12,11) and (11,12). As in the case of within-activity effects, the within-travel effects were specified to be the same for males and females.

The exogenous effects (determined by the free elements in the gamma matrix in system (1)) were initially specified to capture relationships found in empirical studies (e.g., Kostyniuk and Kitamura, 1982; Salomon and Ben-Akiva, 1982). Independent Tobit and ordered-response probit model regressions were also computed for each of the endogenous variables to identify important exogenous effects.

We also specified free disturbance-term correlations between corresponding pairs of variables for male and female heads (the error terms being free off-diagonal elements in the Ψ matrix of equation system (1)). We assume that the unexplained portion of an activity or travel demand of the male head is positively correlated with the unexplained portion of the corresponding activity or travel demand of the female. These error-term correlations, shown to be statistically significant by McNally and Golob (1997), are important for unbiased estimation of causal effects.

It can be shown that such a structural equations model is identified. The model was then estimated using the ADF-WLS method described above, and the model structure was optimized by performing nested hypothesis testing common in simultaneous equation modeling.

RESULTS

Model Fit

The value for the final model was 135.35 with 165 degrees of freedom, corresponding to a probability value of $p = 0.96$. This indicates that the model cannot be rejected at the $p = .05$ level. The overall fit of the model is excellent, as measured by all generally accepted goodness-of-fit criteria for structural equations models (Bollen, 1989).

The endogenous variable causal structure of the final model, differed in some details from the postulated structure of Figure 16.2. Thirty-eight of the postulated forty-six direct causal effects were found to be significant at the $p = .05$ level with the correct sign. One coefficient (shown in parentheses in Figure 16.3) was significant at only the $p = .12$ level, but we judged its inclusion to be theoretically justified. Seven of the postulated effects were rejected at the $p = .05$ level, and these are shown by shaded empty cells in Figure 16.3. On the other hand, four unanticipated effects were found to be important in explaining the causal structure. These are

		Activity participation demand								travel demand				**Vehs**	**VMT**
		male				female				male		female			
Variable		1	2	3	4	5	6	7	8	9	10	11	12	13	14
	1				$-\beta$										
Male	2	$-\beta$			$-\beta$										
act.	3	$-\beta$	$-\beta$		$-\beta$	$-\beta$									
	4					$+\beta$				$+\beta$					
	5								$-\beta$						
fem.	6					$-\beta$			$-\beta$			$-\beta$		$+\beta$	
act.	7					$-\beta$	$-\beta$		$-\beta$						
	8	$+\beta$										$+\beta$			
Male	9	$+\beta$	$+\beta$	$+\beta$	$-\beta$		$+\beta$				$-\beta$				
Travel	10		$+\beta$		$+\beta$										
fem.	11					$+\beta$	$+\beta$	$+\beta$	$-\beta$				$-\beta$		
Travel	12					$+\beta$	$+\beta$	$+\beta$				$-\beta$			
vehs.	13	$+\beta$		$+\beta$											$(+\beta)$
VMT	14				$+\beta$					$+\beta$	$-\beta$	$+\beta$	$-\beta$		

Figure 16.3. Direct effects between endogenous variables (beta matrix coefficients) (boxed cells indicate different structure than hypothesized)

Table 16.5
Estimated Endogenous Variable R^2 values

Endogenous Variable	R^2
Out-of-home work activities – male	0.25
Out-of-home maintenance activities – male	0.13
Out-of-home discretionary activities – male	0.12
In-home work activities – male	0.01
Out-of-home work activities – female	0.19
Out-of-home maintenance activities – female	0.01
Out-of-home discretionary activities – female	0.11
In-home work activities – female	0.02
Car (driver and passenger) travel – male	0.29
Non-car travel – male	0.07
Car (driver and passenger) travel – female	0.30
Non-car travel – female	0.15
Number of household vehicles	0.45
Total VMT on all household vehicles	0.21

shown as shaded cells with positive or negative beta coefficient indicators. We conclude that the postulated endogenous variable structure was essentially upheld.

The freely estimated main-diagonal variances of the error-term correlation matrix Ψ produce R^2 values for the endogenous variables. These R^2 estimates are listed in Table 16.5. The model does a good job of explaining demand for cars, followed by car travel demand for both male and female household heads. Demand for male and female out-of-home work activities is also fairly well explained, as is total vehicle utilization. The least well explained variables are demand for male and female in-home work activities, as well as females' demand for out-of-home maintenance activities and males' demand for non-car travel.

Relationships Between the Endogenous Variables

Travel Demand. The estimated total effects on the travel demand variables from the activity demand variables, calculated according to equation system (2), are listed in Table 16.6. The strongest causes of demand for car travel are demands for out-of-home maintenance activities by both male and female heads. Out-of-home work activities impose less demand on car

Table 16.6
Total Effects of Activity Demand on Demand for Travel by Mode
(showing only effects significant at the p = .05 level)

Causal variable	Affected variable	Total effect	*z*-statistic
out-of-home work – male	car travel - male	0.092	4.19
out-of-home maintenance - male	car travel - male	0.226	7.99
out-of-home discretionary - male	car travel - male	0.353	13.00
In-home work – male	car travel - male	-.241	-4.60
out-of-home work – female	car travel - male	-.086	-5.90
out-of-home maintenance - female	car travel - male	0.100	4.09
out-of-home discretionary - female	car travel - male		
In-home work – female	car travel - male		
out-of-home work – male	non-car travel - male	-.042	-5.69
out-of-home maintenance - male	non-car travel - male	0.130	6.56
out-of-home discretionary - male	non-car travel - male		
In-home work – male	non-car travel - male	0.065	3.27
out-of-home work - female	non-car travel - male	0.005	2.07
out-of-home maintenance - female	non-car travel - male		
out-of-home discretionary - female	non-car travel - male		
In-home work – female	non-car travel - male	-.001	-2.08
out-of-home work - male	car travel – female		
out-of-home maintenance - male	car travel – female	-0.001	-2.08
out-of-home discretionary - male	car travel – female	0.012	3.25
In-home work – male	car travel – female		
out-of-home work - female	car travel – female		
out-of-home maintenance - female	car travel – female	0.438	14.37
out-of-home discretionary - female	car travel – female	0.328	12.31
In-home work – female	car travel – female	-.149	-4.25
out-of-home work - male	non-car travel - female		
out-of-home maintenance - male	non-car travel - female		
out-of-home discretionary - male	non-car travel - female		
In-home work – male	non-car travel - female		
out-of-home work - female	non-car travel - female		
out-of-home maintenance - female	non-car travel - female		
out-of-home discretionary - female	non-car travel - female		
In-home work – female	non-car travel - female		

travel, because less work time implies more time for other activities which require travel. In fact, the total effect of out-of-home work on car travel is insignificant for female heads. In-home work leads to reduced car travel for both males and females, but it leads to increased non-car travel time for males, but not for females. There are also some cross-person effects, in which male travel demand is sensitive to female work and maintenance activities, confirming the descriptive findings of Hanson and Hanson (1978), and female car travel demand is sensitive to male non-work activities.

Time Budget Effects. Significant time budget effects were found for car travel, but not for non-car travel (Table 16.7). These results imply that all male activities are sensitive to the male car travel time, and all female activities, with the exception of discretionary activities, are sensitive to female car travel time. Importantly, demand for in-home work is a positive function of car travel time for both males and females. There are also some effects of female car travel time on male activities. These results indicate that budget models of travel (Zahavi, 1979; Gunn, 1981; Downes and Emmerson, 1985; Golob et al., 1981) are credible if the time-budget concept is expanded to incorporate both activity and travel time (Kraan, 1997).

Table16.7
Total Effects of Car Travel Demand on Activity Participation Demand
(showing only effects significant at the *p* = .05 level)

Causal variable	Affected variable	Total effect	*z*-statistic
car travel – male	out-of-home work – male	-.026	-2.70
car travel – male	out-of-home maintenance – male	-.016	-2.17
car travel – male	out-of-home discretionary – male	-.020	-2.18
car travel – male	in-home work – male	0.172	3.19
car travel – male	out-of-home work – female		
car travel – male	out-of-home maintenance – female		
car travel – male	out-of-home discretionary – female		
car travel – male	in-home work – female		
car travel - female	out-of-home work – male	0.001	2.06
car travel - female	out-of-home maintenance – male		
car travel - female	out-of-home discretionary – male	0.002	2.83
car travel - female	in-home work – male	-.004	-2.30
car travel - female	out-of-home work – female	-.015	-3.24
car travel - female	out-of-home maintenance – female	-.138	-2.95
car travel - female	out-of-home discretionary – female		
car travel - female	in-home work – female	0.131	3.87

Table 16.8
Total Effects of Activity Demand on Car Ownership Demand
(showing only effects significant at the *p* = .05 level)

Causal variable	Affected variable	Total effect	*z*-statistic
out-of-home work – male	Car ownership demand	0.189	5.47
out-of-home maintenance - male	Car ownership demand	-.016	-2.91
out-of-home discretionary - male	Car ownership demand	0.218	5.70
In-home work - male	Car ownership demand	-.058	-5.89
out-of-home work - female	Car ownership demand		
out-of-home maintenance - female	Car ownership demand		
out-of-home discretionary - female	Car ownership demand		
In-home work - female	Car ownership demand		

Mobility Demand. All significant total effects of activity demand on vehicle ownership are from activities of the male head (Table 16.8). Out-of-home work and discretionary activities are positively related to vehicle demand, while in-home work and out-of-home maintenance activities are negatively related to vehicle demand. The causal effect of maintenance activities reflects the fact that maintenance activities negatively affect discretionary activities, and the link from discretionary demand to vehicle demand is strong.

Accessibility. The only important total effect of vehicle ownership on activity demand concerns female out-of-home maintenance activities. Enhanced car availability encourages shopping, personal business, child care, and related activities by the female head.

Vehicle Utilization. There is a slightly stronger total effect on VMT from car travel time of the female head than from car travel time of the male head (Table 16.9). This is probably due to lower mean speeds for trips made by the female head.

Table 16.9
Total Effects of Travel Demand on Car Utilization
(showing only effects significant at the *p* = .05 level)

Causal variable	Affected variable	Total effect	*z*-statistic
car travel - male	total VMT on all household vehicles	0.102	7.86
non-car travel - male	total VMT on all household vehicles	-.125	-14.51
car travel - female	total VMT on all household vehicles	0.137	9.65
non-car travel - female	total VMT on all household vehicles	-.112	-11.98

Table 16.10.
Estimated Error-term Correlations

Between	And	Correlation	*z*-statistic
out-of-home work – male	out-of-home work - female	0.17	7.63
out-of-home maintenance – male	out-of-home maintenance female	0.38	15.20
out-of-home discretionary – male	out-of-home discretionary - female	0.43	18.53
in-home work - male	in-home work – female	0.16	5.65
car travel - male	car travel – female	0.20	8.77
non-car travel - male	non-car travel – female	0.24	8.48

Excess Mobility Demand. There is a positive link from VMT to vehicle ownership, as expected, but the total effect (0.116) is not significant at the $p = .05$ level (z-statistic = 1.08).

Error-term Correlations. The estimated correlations of the unexplained portions of the endogenous variables (off-diagonal terms in the Ψ matrix of equation system (1)), are listed in Table 16.12. The strengths of these correlations underscores the importance of including them in the model specification.

Effects of the Exogenous Variables

Every one of the thirteen exogenous variables is present in the fourteen reduced-form equations (one for each of the endogenous variables) because of the extensive endogenous links between the endogenous variables. Most of the effects are consistent with earlier findings (e.g., Golob and McNally, 1997; Pas, 1984, 1988; Lu and Pas, 1997), but there appears to be some new insights. The strongest effects for each exogenous variable are:

1. Number of children under 6 years of age negatively influences female out-of-home work demand, female demand for non-car travel, and male demand for in-home work. It positively influences male and female demand for discretionary activities and VMT.
2. Number of children 6-11 years of age also negatively influences female out-of-home work demand and male in-home work demand, and it positively influences female demand for both maintenance and discretionary activities and VMT, but it has no significant total effect on male demand for discretionary activities.
3. Number of children 12-21 years without driving license has a positive influence on male demand for out-of-home work, and it has a negative influence on male

demand for in-home work and female demand for out-of-home discretionary activities.

4. Number of drivers in household strongly explains car travel times, car ownership demand, and VMT.
5. Households in current home 1 year or less exhibit higher levels of discretionary activities, lower levels of in-home work for both males and females, lower female car travel, and lower VMT.
6. Households renting exhibit lower car demand for travel, ownership, and VMT, but higher demand for out-of-home discretionary and maintenance activities, for both males and females, and higher demand for in-home female work activities.
7. Age of the male head is related to all activity and mobility demand *except* demand for maintenance activities (male and females) and demand for male non-car travel.
8. Household in which the male head less than 31 years of age exhibit more demand for more out-of-home non-work activities (male and female) and more intensive use of vehicles.
9. Households in which the male head 70 years of age or older, as expected, have less work activities, car ownership, and usage. However, there is positive effects on demand for other activities, and there are no significant affects on travel demand, with the exception of lower car travel demand for female heads.
10. Households with income less than $20,000 exhibit lower levels of demand for work activities, car travel, ownership, and usage, but they maintain high levels of demand for non-work out-of-home activities.
11. Household with income $20,000 to $30,000 are similar to households with lower incomes, but they travel more.
12. Household with income $60,000 or more are opposite to low income households in most aspects, with the exception that high income households also exhibit strong demand for female out-of-home discretionary activities, and lower demand for male non-car travel.
13. Total retail employment within 1 mile of residence was extremely effectual as an exogenous variable. Twelve of the fourteen total effects of this variable, which are listed in Table 16.11, were significant at the $p = .05$ level. Controlling for all other exogenous effects, both men and women in households with higher levels on this accessibility index demonstrated less car travel, and more non-car travel. Accessibility was associated with substantially less VMT, but the affect on car ownership was insignificant, indicating that households with better neighborhood accessibility own essentially the same numbers of vehicles, but they use them less. This measure of neighborhood accessibility also predicted a

higher level of demand for maintenance activities among male heads, and lower levels of in-home work for both males and females.

The performance of these exogenous variables indicates that the demand for *both* activities and mobility is highly sensitive to household demographic and economic factors, as well as to environmental factors such as accessibility. This bodes well for the use of such models in forecasting and policy evaluation.

Table 16.11
Estimated Total Effects of Retail Accessibility

Endogenous Variable	Coefficient	*z*-statistic
out-of-home work activities – male	0.0034	2.23
out-of-home maintenance activities – male	0.1293	5.91
out-of-home discretionary activities – male	-0.0094	-2.31
in-home work activities – male	-0.0228	-2.48
out-of-home work activities – female	0.0445	1.97
out-of-home maintenance activities – female	-0.0054	-0.61
out-of-home discretionary activities – female	-0.0139	-1.93
in-home work activities – female	-0.0112	-2.77
car (driver and passenger) travel time- male	-0.1484	-6.75
non-car travel time- male	0.1677	6.33
car (driver and passenger) travel time – female	-0.0829	-3.78
non-car travel time – female	0.2109	8.83
Number of household vehicles	-0.0249	-1.12
total VMT on all household vehicles	-0.2040	-23.90

Summary and Directions for Further Research

We were successfully able to develop and estimate a joint model of demand for activity participation and mobility. The observed endogenous variables are out-of-home and in-home activity durations and travel times by mode for each of two household heads, household vehicle ownership, and household vehicle utilization. The model system of simultaneous equations involves latent demand variables that represent these observed censored and ordinal variables. The distribution-free estimation method provides accurate model evaluation and hypothesis testing capabilities. The goodness-of-fit is excellent, using activity-diary and accessibility data from the Portland Metropolitan Area. This model system can be used to replace or augment conventional travel demand models for certain planning purposes. It can provide forecasts of the effects on activity participation and mobility of factors that are difficult to include in trip-based models.

The demand model system can be expanded to be made more useful in numerous ways. First, we can test the effects of additional exogenous variables on activity and mobility demand, the most interesting candidates for testing being different types of accessibility measures. Second, we can subdivide activities differently, possibly adding non-work in-home activities if reliable time-use data are available. Third, we can modify or expand the set of mobility variables, using trip-based measures, to see how trip and trip chain generation is related to activity participation and other mobility measures. Fourth, we can divide travel demand differently, by separating car travel into solo driving and carpool modes, or by separating non-car travel into public transport and non-motorized modes. Finally, we can divide activity and travel demand into different periods of the week, if we are anxious to know how weekdays differ from weekends. Such a model system can evolve in many ways while preserving its fundamental assumptions and basic causal structure

Exogenous Variables

Adding additional exogenous variables to a structural equation model will generally not be a problem, as long as the sample size is sufficient, and the exogenous variables correlation matrix is positive definite (the variables are not collinear). Thus, the set of potential exogenous variables is limited only by data availability. The most productive new variables might be spatial accessibility indices and levels of service associated with residential location. One such accessibility variable, computed using a GIS, was retail employment within one mile of the residential location. It was found to have substantial explanatory power in the case study application. Many other variables that can capture changes in accessibility for different types of activities and different modes should be useful in evaluating the effects of changes in such things as traffic flows, land-uses, or pedestrian or bicycle facilities, on activity and mobility demand. The next logical step would be to study how people associate changes in activity patterns and mobility with their perceived quality of life.

Activity Demand Variables

The demand model system can accommodate other activity categorizations. Several trip chaining studies have focused on shopping activities and activities involving picking up or dropping off passengers (the so-called "serve-passenger" trip purpose). The maintenance category used in the case study could be subdivided to analyze demand for these more specific activities. However, a large number of activity categories will require a large sample size because (1) minimum sample size is proportional to the number of estimated free parameters, (2) the estimation method relies on asymptotic theory, and (3) we need to observe sufficient demand for each category of activity.

We anticipate that the greatest potential will be in expanding the set of in-home activity categories. This would allow us to analyze the effects on activity and mobility demand of the substitution of in-home for out-of-home activities. We could thus forecast the travel demand consequences of advances in telecommunication that provide opportunities for in-home shopping and banking services, as well as in-home work activities.

Travel Demand Variables

Many measurements of travel demand are possible without changing the basic structure of the model system. We used travel time, by mode, in the case study. Some other obvious candidates are: (1) trip rates (numbers of sojourns, or destinations visited); (2) numbers of home-based trip chains (or, tours); (3) numbers of trip chains by type (simple out-and-back tours, versus multiple destination tours of different kinds); (4) travel distance.

Due to potential collinearity problems and complexity of the endogenous variable causal structure, it might not be possible to accommodate multiple travel demand variables, such as travel times and trip-chain rates, within the same model. Rather, separate models could be estimated for each travel demand measure of interest.

Modal Definitions

Any of these travel demand variables could be divided according to travel mode, where modal definitions can be tailored to policy evaluation needs. The car mode used in the case study could be broken down into (1) solo driving and (2) car travel with two or more occupants. Our non-car mode could be broken down into (1) public transport, (2) pedestrian, and (3) bicycle. Once again, the detail is dictated mainly by the data. Empirically, we have found that censored endogenous variables with less than five percent non-zero observations can cause problems in estimation of this type of structural equation model. Thus, we can avoid estimation problems if activity and mode categories are defined so that we observe at least five percent of our sample engaging in the activity or using the mode within the activity diary period. A longer diary period is also effective in increasing the incidence of relatively rare activity or travel demand, so the sample size requirement can be partially offset by a longer diary period.

Time Periods

We do not think that it is wise to estimate separate models for weekdays and weekend days, because there are complex activity patterns related to day of the week, and separate models will not allow analyses of the substitution of weekday for weekend activities. For example, compressed work weeks potentially alter the activity relationships between weekdays and weekend days. However, some activity diary surveys conducted by MPOs have concentrated on weekdays, so equal representation of all days of the week is not always possible. If we estimate our model system using only data on weekday activities and travel, we will weaken the explanation of the car ownership and usage components, because we are only considering the mobility needs emanating from weekday activities. If we have data for all days of the week and planners are interested in distinguishing activity and travel by weekday versus weekend day, it is possible to divide the activity and travel demand variables by time period, if we have sufficient observations.

ACKNOWLEDGEMENTS

This research was supported in part by a grant from the University of California Transportation Center. The data were generously provided by Portland Metro Regional Services. Keith Lawton, together with Stephen Perone, Jean Sumida, and others at Metro were very supportive of the research effort. The author has benefited greatly from collaboration with his colleague Mike McNally at the University of California, Irvine, and Will Recker of UCI has been a staunch supporter of this flavor of research. The views expressed here are solely those of the author, who is of course responsible for any and all errors.

REFERENCES

Aitchison, J. and S. Silvey (1957). The generalization of probit analysis to the case of multiple responses. *Biometrika,* **44**, 131-140.

Amemiya, T. (1973). Regression analysis when the dependent variable is truncated normal. *Econometrica,* **41**, 997-1016.

Ashford, J. R. (1959). A approach to the analysis of data for semi-quantal responses in biological response. *Biometrics*, **15**, 573-581.

Bollen, K. A. (1989). *Structural Equations with Latent Variables*. Wiley, New York.

Browne, M. W. (1974). Generalized least squares estimators in the analysis of covariance structures. *South African Statistical Journal*, **8**, 1-24. [Reprinted in D. J. Aigner and A. S. Goldberger (eds.) (1977). *Latent Variables in Socio-Economic Models*. North-Holland, Amsterdam].

Browne, M. W. (1982). Covariance structures. In: *Topics in Applied Multivariate Analysis* (D. M. Hawkins, ed.), pp. 72-141. Cambridge University Press, Cambridge.

Browne, M. W. (1984). Asymptotically distribution-free methods in analysis of covariance structures. *British Journal of Mathematical and Statistical Psychology*, **37**, 62-83.

Burns, L. D. and T. F. Golob (1976). The role of accessibility in basic transportation choice behavior. *Transportation*, **5**, 175-198.

Carpenter, S. and P. Jones (eds.) (1983). *Recent Advances in Travel Demand Analysis*. Gower, Aldershot, UK.

Chapin, F. S. (1974). *Human Activity Patterns in the City: Things People Do in Time and Space*. Wiley, New York.

Damm, D. (1978). Analysis of activity schedules along the dimension of gender. In: *Women's Travel Issues: Research Needs and Priorities* (S. Rosenbloom, ed.), pp. 171-198. U.S. Department of Transportation, Office of University Research.

Raj, D. (1953). On estimating the parameters of bivariate normal populations from doubly and singly linear truncated samples. *Sankhya*, **12**, 277-290.

Downes, J. D. and P. Emmerson (1985). Budget models of travel. In: *Proceedings of 13th PTRC Summer Annual Meeting - Seminar L*, pp. 63-80. London.

Fair, R. C. (1977). A note on the computation of the Tobit Estimator. *Econometrica*, **45**, 1723-1727.

Fujii, S., R. Kitamura and T. Monma (1997). A study of commuters' activity patterns for the estimation of induced travel. *Journal of Infrastructure Planning and Management*. In press.

Golob, T. F., M. J. Beckmann and Y. Zahavi (1981). A utility-theory travel demand model incorporating travel budgets. *Transportation Research B*, **15**, 375-390.

Golob, T. F., M. A. Bradley and J. W. Polak (1996). *Travel and activity participation as influenced by car availability and use*. Paper presented at the annual meeting of Transportation Research Board, Washington, DC.

Golob, T. F., R. Kitamura and C. Lula (1994). *Modeling the effects of commuting time on activity duration and non-work travel*. Paper presented at the annual meeting of Transportation Research Board, Washington, DC.

Golob, T. F. and M. G. McNally (1997). A model of household interactions in activity participation and the derived demand for travel. *Transportation Research B*, **31**, 177-194.

Golob, T. F. and M. G. McNally (1996). *A joint model of household activity participation, travel time, car ownership and residential location.* Paper presented at the world congress of the Regional Science Association International, Tokyo.

Golob, T. F. and L. J. van Wissen (1989). A joint household travel distance generation and car ownership model. *Transportation Research B*, **23**, 471-491.

Goulias, K., R. M. Pendyala and R. Kitamura (1990). Practical method for the estimation of trip generation and trip chaining. *Transportation Research Record, No.* **1285**, 47-56.

Gunn, H. F. (1981). *An analysis of travel time budgets into mandatory and discretionary components.* Paper presented at the PTRC 9th summer annual meeting, University of Warwick, UK.

Hägerstrand, T. (1970). What about people in regional science? *Papers of the Regional Science Association,* **24**, 7-21.

Hanson, S. and P. Hanson (1978). The impact of women's employment on household travel patterns. In: *Women's Travel Issues: Research Needs and Priorities* (S. Rosenbloom, ed.), pp. 127-169. U.S. Department of Transportation, Office of University Research Washington, DC.

Jones, P., M. C. Dix, M. I. Clarke, and I. G. Heggie (1983). *Understanding Travel Behavior.* Gower, Aldershot, UK.

Jöreskog, K. G. and D. Sörbom (1993). *LISREL 8: Structural Equation Modeling with the SIMPLIS Command Language.* Scientific Software International, Chicago, IL.

Kitamura, R. (1988). An evaluation of activity-based travel analysis. *Transportation,* **15,** 9-34.

Kitamura, R. (1991). The effects of added transportation capacity on travel: A review of theoretical and empirical results. In: *The Effects of Added Transportation Capacity: Conference Proceedings* (G. A. Shunk, ed.), pp. 79-95. Report DOT-T-94-12, U.S. Department of Transportation, Washington, DC.

Kitamura, R. (1996). *Activity-based travel demand forecasting and policy analysis.* Paper presented at the TMIP conference on activity-based travel forecasting, New Orleans, LA.

Kostyniuk, L. P. and R. Kitamura (1982). Life cycle and household time-space paths: Empirical investigation. *Transportation Research Record, No.* **879**, 28-37.

Kraan, M. (1997). In search for limits to mobility growth with a model for the allocation of time and money. In: *Activity-Based Approaches to Travel Analysis* (D.F. Ettema and H. J. P. Timmermans, eds.), pp. 89-116. Pergamon, Oxford.

Lu, X. and E. I. Pas (1997). *A structural equations model of the relationships among socio-demographics, activity participation and travel behavior.* Paper presented at 76th annual meeting of the Transportation Research Board, Washington, DC.

Maddala, G. S. (1983). *Limited-Dependent and Qualitative Variables in Econometrics.* Cambridge University Press, Cambridge, UK.

Olsson, U., F. Drasgow and N. J. Dorans (1982). The polyserial correlation coefficient. *Psychometrika*, **47**, 337-347.

Pas, E. I. (1984). The effects of selected sociodemographic characteristics on daily travel-activity behavior. *Environment and Planning A*, **16**, 571-581.

Pas, E. I. (1985). State of the art and research opportunities in travel demand: Another perspective. *Transportation Research A*, **19**, 460-464.

Pas, E. I. (1986). Multiday samples, parameter estimation precision, and data collection costs for least squares regression, trip-generation models. *Environment and Planning A*, **18**, 73-87.

Pas, E. I. (1988). Weekly travel-activity behavior. *Transportation*, **12**, 89-109.

Pas, E. I. (1995). Intrapersonal variability in daily urban travel: Some additional evidence. *Transportation*, **22**, 135-150.

Pas, E. I. (1996). *Recent advances in activity-based travel demand modeling*. Paper presented at the TMIP conference on activity-based travel forecasting, New Orleans, LA.

Pas, E. I. and F. S. Koppelman (1987). An examination of the determinants of day-to-day variability in individuals' urban travel behavior. *Transportation*, **14**, 3-20.

Pendyala, R. M., K. G. Goulias and R. Kitamura (1991). Impact of telecommuting on spatial and temporal patterns of household travel. *Transportation*, **18**, 383-409.

Purvis, C. L., M. Iglesias, and V.A. Eisen (1996). *Incorporating work trip accessibility in non-work trip generation models in the San Francisco Bay Area*. Paper presented at the 75th annual meeting of the Transportation Research Board, Washington, DC.

Reichman, S. (1977). Instrumental and life style aspects of urban travel behavior. *Transportation Research Record, No.* **649**, 38-42.

Robinson, J. P. (1977). *How Americans Use Time: A Social-Psychological Analysis of Everyday Behavior*. Praeger, New York.

Salomon, I. and M. Ben-Akiva (1982). Life-style segmentation in travel demand analysis. *Transportation Research Record, No.* **879**, 37-45.

Tobin, J. (1958). Estimation of relationships for limited dependent variables. *Econometrica*, **26**, 24-36.

Townsend, T. A. (1987). The effects of household characteristics on the multi-day time allocations and travel - Activity patterns of households and their members. Doctoral Dissertation, Department of Civil Engineering, Northwestern University, Evanston, IL.

van der Hoorn, T. (1979). Travel behavior and the total activity pattern. *Transportation*, **8**, 309-328.

van Wissen, L. (1989). *A model of household interactions in activity patterns*. Paper presented at the international conference on dynamic travel behavior analysis, Kyoto University, Japan.

Zahavi, Y. (1979). *The UMOT Project*. Report DOT-RSA—DPW-20-79-3. from the U.S. Department of Transportation, Research and Special Programs Administration, and the Federal Republic of Germany Ministry of Transport

17

HOUSEHOLD TRAVEL SURVEYS: NEW PERSPECTIVES AND OLD PROBLEMS

Peter R. Stopher, Louisiana State University, Baton Rouge, LA, USA

INTRODUCTION

Modeling of travel behavior, as well as understanding of the mechanisms underlying behavior, requires data on actual behaviors and on the contexts and constraints under which the travel behavior takes place. This chapter focuses on some of the issues relating to the collection of adequate data for the purposes of testing theories, calibrating models, and performing analyses aimed at increasing understanding of specific travel decisions. It would be of immense value if the chapter was able to cover the range of survey methods used throughout the world. However, such a coverage, to be thorough, is far beyond its scope and might well be too confusing to be helpful in relation to a focus on the specifics of how the data and theories of travel interrelate. Therefore, without further apology, it should be noted that this chapter focuses most heavily on experience and techniques in the United States, while making reference from time to time to important differences known to the author in other countries.

The focus is specifically on household-travel surveys. Many of the other mechanisms, used and available for collecting travel-related data are not covered in any detail. In particular, this is true of various counting programs and surveys that are used to validate and correct data from sources such as household surveys. Neither are some of the specific issues relating to on-board and other intercept surveys examined in any detail.

The primary purpose is to examine in particular the interaction between theories, modeling and the form and nature of the collection of data on travel behavior. The intention is to raise challenges as to the concepts that may be needed to take household travel surveys into the twenty-first century, based not only on the likely changes that will be required to meet new demands of society in measurement of behaviour, but also those that will be required to collect data for emerging models and procedures for analysis. In the next section, the history of household travel surveys is reviewed briefly and the forces behind the evolution to today's surveys are postulated. This includes both societal changes and changes in the nature of the needs of the travel-demand analyst. In the third section, some of the emerging contextual changes for household travel surveys are identified and thoughts are offered on where we may be heading with new designs of surveys to serve both changing societal norms and changing needs of travel demand theories.

HISTORY AND EVOLUTION OF THE HOUSEHOLD TRAVEL SURVEY

Brief Historical Overview

Household travel surveys are a relatively recent phenomenon, with the real beginning having been in the 1940s, no more than fifty years ago. Prior to that time, transportation planning (which itself was in rapid evolution to the regional type of planning performed today) relied principally on some form of roadside survey that collected origin and destination information from travelers on specific roads. Importantly, however, it should be noted that data about trip-making patterns has been a mainstay of transportation planning from the beginning of such planning earlier in the twentieth century. Surveys of trip making have traditionally provided the basis for formulating transportation policy, developing transportation plans, and developing improvements to transportation operations. Surveys have had two primary roles to play. The first role is to provide descriptions of current travel, in order to increase understanding of the demands made on the system, and to identify more clearly where current and near-future problems can be anticipated. The second role is as input to travel forecasting models, where these models are used to identify potential longer-term future problems and also to provide a means to test the efficacy of proposed solutions. A secondary role that such surveys have played has been to provide a basis for research into new methods for modeling travel behaviour and as a means to provide information for the analysis and testing of new theories and revisions of existing theories of travel behavior.

In the 1950s and early 1960s, household travel surveys in the U.S. were largely confined to large cities, while extension to smaller cities began to occur most markedly in the late 1960s and early 1970s. Household travel surveys are generally an expensive undertaking, and may consume a very substantial amount of several years of planning budgets of local planning agencies. As a result, they are conducted infrequently (often less frequently than planners would desire) and involve a range of compromises on information collected, sample size, and other aspects in an effort to keep costs within reason. Unfortunately, it seems that studies have not been undertaken that demonstrate the amount of potential savings that can be accrued in capital transportation investments as a result of using data from surveys. Therefore, the value of data collection, in relation to capital project costs, has not been demonstrated and planning agencies continue to have to fight for funding of those surveys that can be mounted.

An important characteristic of household travel surveys is that they have not generally been looked on as the means by which to develop complete descriptions of the urban area or nation where they have been conducted. Rather, they are frequently collected close enough to such national data-collection efforts as the decennial census, in order that they can take advantage of the supplementary data provided from such sources (Stopher and Metcalf, 1996). This aspect of reliance on certain supplementary data is an important characteristic of household travel surveys, because of its effects on both the data collected and the sample sizes.

Evolution of Household Travel Surveys

Household travel surveys can be described in a number of dimensions. These different dimensions can be used to trace out the effects of evolution of travel surveys, by contrasting the ways used in prior years with those in use today or expected to be in use in the future. The major dimensions along which surveys can be described are: method of collecting the data, type of survey instrument, method of sampling, size of sample, method of retrieving data, and method of coding and entering data for analysis.

Over the past half century, the household-based travel survey has undergone very substantial growth and change along all of these dimensions. With respect to the method of collecting the data, the household travel survey was conducted almost exclusively by face-to-face interviews taking place in the home of the respondent household, using recall of the previous day's travel by household members, and often involving extensive proxy reporting by one family member for most of the rest of the household. Generally, the interview was conducted as a "cold-contact" process, because no prior contact had been made with the household to request their co-operation or arrange the time for the interview. Face-to-face interviews continued into the early 1980s, although the "cold contact" was often replaced by some form of prior contact, particularly by telephone or by a mail-out, and the last few face-to-face surveys involved the use of some form of travel diary sent to the household in advance of the interviewer for completion by household members at an appointed time. The face-to-face interview was initially replaced with mail-out and mail-back surveys, although response rates from such surveys were often disappointing. Most recently, the household travel survey involves telephone contact to recruit the household, followed by mailing some form of activity diary, that may be either a one-day or two-day diary, with a set day in the future on which the activities are to be recorded. This is followed frequently by telephone interview, during which the data from the survey are collected by the telephone interviewer.

Perhaps one of the most significant changes that has occurred in this process has been the change in the relationship between when the survey is conducted and the day or days that are the subject of data collection. In the early in-home interviews, data collection was entirely retrospective. That is, people were asked to recall the travel they had undertaken during the previous day. Many of the earlier mail-out and mail-back surveys continued to collect data retrospectively. With the evolution of diaries that were mailed out to households ahead of the collection of the data, retrospective collection was replaced by prospective collection, that is, by setting a day in advance for which travel would be recorded. Many instrument designs were developed that were based on the assumption that people would complete the survey diaries as they carried out the activities of their day. This remains the basic design, even though there are substantial doubts as to whether people ever do complete the diaries during the day, as opposed to doing a recall at the end of the day. Nevertheless, there have been significant changes in the success of the data collection with the change from retrospective data collection to prospective.

In this evolution, the survey instrument has changed in form quite radically. Initially, forms were designed for the use of trained interviewers and often included listings of numeric codes that corresponded to specific answers, and space to transfer the data from the body of the form into a coding area to be used later by a key-punch operator. Such forms were often unfriendly in appearance, being cluttered, printed in small type, and laid out in a way that only an engineer would understand, because it was always assumed that there would be adequate time to train interviewers in the use of the forms. Subsequently, the forms changed to permit direct data entry without a coding step, which also forced some degree of improvement in appearance, so that both the interviewer and the key-puncher could interpret the form and perform their particular functions efficiently and effectively. As the transition took place from interviewer-administered survey instruments to self-administered, a gradual change took place in the survey form, in an effort to improve response. Early self-administered surveys forms were actually not much different from the unfriendly interviewer-administered forms, and produced poor results as might be expected. In some respects, it took transportation planners an unexpectedly long time to figure out that major changes needed to be made in the design of the forms. In most recent surveys, significant changes can be seen from the early self-administered forms. Efforts to reduce clutter on the form, remove coding and data-entry-related information, and make forms more self-explanatory and interpretable by people who lack an advanced education are all fairly obvious from recent instruments. Nevertheless, there seems to remain considerable scope for further improvement in this area.

Another aspect of the evolution of the household travel survey has to do with both the sampling method and the sample size. Almost exclusively, early household travel surveys used either simple random sampling of households in the metropolitan region, or a form of cluster sampling, as a means to reduce interviewer travel. Such cluster sampling involved random selection of blocks or block faces, followed by interviewing of all households on the selected block or block face. Simple random samples were often compromised by the lack of a complete sampling frame of households in U.S. urban areas, and inadequate field supervision often resulted in non-adherence to the principles of the simple random sampling that was designed to take place. Along with the simple random samples, and possibly as a result of more liberal budgets in the early days of household travel surveys, sample sizes were often large, with recommendations that as many as one to five percent of households in an urban area be surveyed. These sampling rates resulted in surveys of tens of thousands of households in the early years of such surveys. Only in the early 1970s in the U.S. was the notion of small sample surveys introduced, and this notion took some time to penetrate to actual data collection efforts. The first official guidance published by the Federal Highway Administration (U.S. Department of Transportation, 1973) suggested the possibility of collecting small samples, and it was not until the late 1970s that some urban areas decided to be bold enough to attempt such surveys. The shift to smaller sample sizes also brought with it an interest in exploring stratified samples. In addition, as a means to work around problems of incomplete sampling frames, multi-stage sampling appeared as a method, for example sampling counties, census tracts, and blocks, so as to permit enumeration of addresses to be restricted to a small number of blocks within the region, from which the final sample would be drawn.

Current household travel surveys usually use a form of random stratified sampling, with variable sampling fractions, with the strata typically based on household characteristics determined in the initial recruitment call. The stratified sampling is often performed by using simple random sampling and keeping a tally of the numbers of households contacted in each cell of a two-way or three-way stratification scheme. A sample size for each stratum is established at the outset, and sampling proceeds as simple random sampling until cells begin to fill. Then sampling continues as simple random sampling with discarding of households from the sample, once it is ascertained that the household belongs to an already-filled cell. Additional stratification is sometimes done on the basis of major geographic areas, and is accomplished by selection of the initial three digits of the local telephone number and confirmation of the location of the household at the beginning of the recruitment call. Sample sizes are typically much smaller than those of the early years, with samples of 2,000 to 10,000 encompassing the majority of surveys. Indeed, Stopher and Metcalf (1996) documented an average sample size of 2,500 from a sample of surveys conducted in the early 1990s.

Retrieval of the data and coding and data entry have also undergone significant change. Early household interview surveys relied on pencil and paper recording by the in-home interviewer, with many responses written out in long hand, followed by a coding stage, in which the written responses were reduced to alphanumeric codes, and then the codes were entered onto punched cards. As noted earlier, this changed with the changes in design of instrument, so that coding began to be done directly on the forms, and then subsequent changes resulted in virtually self-coding forms. Most of the time, however, these variations in survey forms still involved pencil and paper recording of the data during the interview, with a subsequent coding and keypunching step. Only very recently has this started to be replaced by the use of direct data entry, particularly through the change to telephone retrieval of data. In the direct-entry method, interviewers read questions from a computer monitor screen and enter responses directly at the keyboard, with the data then being stored directly to a computer disk. This process also permits various interactive checks to be made that may prevent erroneous data from being entered.

Reasons Behind the Evolution of the Household Survey

One of the reasons behind the changes observed in household travel surveys is clearly the desire to improve the accuracy of the survey. In early surveys, it was always found necessary to make differential adjustments to survey data, when comparing to screenline and cordon surveys, which typically showed an undercount of non-home-based trips and short trips from the interview surveys. Not surprisingly, it was suspected that the recall process contributed to this, because many of the trips that were undercounted appeared to be relatively minor activities of the day that would be forgotten quickly and not recalled in the retrospective reporting. The shift from recall reporting to use of a diary designed to be completed at a subsequent date is clearly one of the steps that was taken primarily in an attempt to address this problem of underreporting of certain trip types. The hope was that, by reducing proxy reporting (each family member would have a diary in which to record the trips they made) and by shifting from recall to real-time recording of the person's travel, the number of these short trips reported would increase. There is at least some anecdotal information to suggest that this may have happened, but no clear analysis has established that a significant improvement in reporting non-home-based travel actually occurred through this mechanism. Whether such increases in reporting non-home-based trips are a result of survey mechanisms that do a better job of collecting the data, or the result of changing travel patterns, as two-worker families have become more numerous, is beyond the available information to determine.

Second, the nature of the models and other procedures to be developed from the data collected have changed. Many early household travel surveys tended to ignore the survey expert's notion that a survey should be designed parsimoniously and with the eventual uses of the data always clearly in view. Questions were often included either routinely or because someone thought it would be interesting to find out the answers, while yet having only vague notions of how the answers might be used. More recently, much greater attention has been paid to parsimony in design, with attempts to limit questions to only what will be used in either descriptive or modelling work with the data. The demands of the models have also changed. Early household surveys concentrated on origin and destination information, trip purpose, and mode and time of travel. There was relatively little data collected on descriptors of the household and persons within it, beyond age, sex, and relationship, although whether or not the household owned any vehicles and the number of people in the household were also usually collected. Axhausen (1994) has traced something of the evolution of survey questions and shows how questions were added in many surveys in the 1960s, only to be dropped again as the surveys evolved into the 1980s. Many of the questions that were added in the 1960s have again returned to recent surveys. As interest in modelling transit grew in the late 1960s and 1970s, survey forms began to appear with questions relating to costs of both auto and transit, times taken for specific trips, and information was also collected on access to and egress from the transit mode. Initial mode-choice models were developed strictly in terms of time ratios between transit and auto, but the more sophisticated models that began to appear in the 1970s required the addition of these variables relating to costs and access/egress. In many respects, however, outside the mode-choice model, travel-forecasting models have changed very little over the past thirty years and the theories underlying all of the modelling have not changed, so that other aspects of the survey form have also changed very little.

Third, the change from interview surveys to self-administered surveys required changes in the structure and format of the survey form, a fact that was not recognized in the earliest self-administered surveys. The early surveys showed little design change from the interviewer forms, with the results that response rates dropped and completed responses were often filled with errors. These problems arose because many people were unable to understand how to fill out the survey, or were intimidated by the amount of information packed into a rather small space in the survey form, or were confused by the survey and unable to understand how to answer correctly. This resulted in many surveys being discarded, or major inferential work being required to correct the returned surveys. Some survey designers also noted that including space on a survey form for coding, or even coding numbers under question responses, had a negative effect on response rates. Therefore, these were gradually omitted from survey forms, although they persist on a few forms even today.

With an increasing amount of attention being paid to justifying each item of data to be included, and with a slow evolution of some parts of the travel-forecasting model system, changes have been made in the survey instruments to reflect these contextual changes. The general acceptance of disaggregate multinomial logit models for mode-choice modelling had a significant effect on the design of data-collection instruments during the 1980s, in particular, and is reflected in renewed attention to the sequence of use of different travel modes on a single trip, attention to auto occupancy, parking costs, and vehicle availability, and collection of data on the modes of access to and egress from public transportation. The widely spread acceptance of disaggregate models also brought with it an interest in collecting data about subjective evaluations of travel options, in addition to the standard reporting of "objective" data. Thus, the 1970s and 1980s saw a significant interest in collecting data on attitudes, preferences, and opinions about transportation alternatives. However, so little of this information was found to be helpful in travel forecasting, and sufficient doubts raised as to its usefulness even for descriptive purposes, that the late 1980s showed a significant decline again in the collection of such data.

Notwithstanding this decline in the collection of a variety of attitudinal data of varying rigor, the late 1980s and the 1990s have brought a resurgence of interest in collecting what is, now commonly referred to as stated-preference data, or perhaps, more correctly as discussed by Lee-Gosselin (1996), interactive stated-response data. This specific evolutionary change can be explained largely as resulting from two coincident issues. First, in the late 1980s, the U.S. became increasingly interested in the possibility of investing in high-speed rail. Because high-speed rail is not in operation in the U.S., the idea of determining stated responses for the service profile of high-speed rail seemed to be the best method to estimate potential patronage. The general success of this approach gave credibility to stated-response measurement as a means to deal with a new alternative. Current urban policies in the U.S. are focusing increasingly on new options for handling transportation problems, such as transportation demand management strategies, pricing strategies, etc., resulting in increasing interest in using stated response as a means to estimate how the traveling public will respond to such programs.

Another significant issue in the past ten to twenty years has been a decreasing response rate from surveys. There are no hard facts available to explain the reasons for the decline that has been seen, but several surveys bear out that the decline is both substantial and continuing. Stopher (1996) has speculated on the factors contributing to this, many of which are still most prevalent in the U.S., although many may become more prevalent in Europe and elsewhere in the next few years. Stopher (1996) suggested the following:

1. The increasing use of telephone surveys by various marketing agencies that tend to contribute to "burn-out" of the U.S. public with respect to surveys;
2. The increased use of marketing surveys as a "foot-in-the-door" to sell some product (as evidenced by the frequent response that "I'm not buying anything" when a transportation-survey interviewer calls a household);
3. The increasing pace of life in late twentieth-century U.S. that makes people more and more reluctant to spend time in activities not directly connected to their own busy schedules;
4. The increasing concern over personal privacy and the potential for various outside agencies to know personal details that represent an invasion of privacy, resulting in a decreasing willingness to answer any questions relating to demographics, activities, etc.;
5. The perception of vulnerability to crime through the types of information typically requested in a household transportation survey: and
6. The increasing number of immigrant households in which English may not be spoken, or is not spoken well.

Such factors clearly contribute to difficulties in obtaining adequate responses to household travel surveys in the U.S. There has also been a marked decline in the U.S. in the effectiveness of face-to-face interviewing for three principal reasons. First, in most U.S. cities, there are neighborhoods within the city into which it is impossible to send interviewers and be able to guarantee their personal safety. Second, there is the perception by the general public of vulnerability to crime that makes it less and less likely that a household will allow a stranger to enter their home and spend time to interview them. Third, with the rapid increase of two-worker households and the increase in the amount of time spent by household members in out-of-home activities, there is an increasing problem to find a responsible adult at home to interview. As a result of this third problem, the costs of performing face-to-face interviews have sky-rocketed, while the first two problems impact the effectiveness of such interviews and the ability of the interviews to cover a random sample of households.

Household travel surveys continue to evolve, both in response to these historical issues, and also to changes in the direction of travel-demand modelling and the understanding of the profession as to what is needed to provide information for decision making in transportation. In many respects, household-travel surveys in the U.S. are facing a significant dilemma. On the one hand, changes in modelling and understanding of travel behaviour are generating increasing demands for more detailed information; on the other hand, people are generally becoming more concerned about privacy and the potential for personal information to be misused or abused, resulting in increasing reluctance to provide information in a survey.

Some of the details of these dilemmas and some directions being pursued to resolve them are the subject of the next section.

NEW DIRECTIONS IN HOUSEHOLD-TRAVEL SURVEYS

Policy-Driven Data Needs

Non-Motorized Travel. Perhaps the first significant thing that has happened due to recent policy changes in the U.S. (changes that may bring the U.S. in line with most of the rest of the world!) is that there is an increasing interest in planning for non-motorized modes of travel, principally walking and bicycling. In most travel surveys in the U.S., the emphasis has been entirely on motorized vehicles, and all models have been focused on predicting motorized travel only. Neither walk nor bicycle have appeared as possible modes in mode-choice models, and such trips, if measured, have been excluded routinely from the entire modeling process. This situation has changed and there is now a move to incorporate these modes into all travel modeling. To do so, however, raises issues in the design of household travel surveys, including the issue of the potential for people to not think of reporting such travel. There are also other issues of whether or not to include bicycling, walking, or jogging performed for recreation, rather than as a means to travel to a different location.

The decision to include measurement of walking and bicycling raises at least two issues in the design of surveys. First, there is the issue of what to measure and how to ask the questions within the travel survey itself. Second is the issue of how to locate households that use these modes, because the use of them is still relatively uncommon. (It is estimated that no more than about 5 to 8 percent of urban travel takes place solely by walking, and as little as 1 to 2 percent by bicycling.)

Based on some recent experiences with adding such travel to household travel surveys, there are also problems that arise in reporting or coding such travel, particularly relating to the distances involved. This suggests that considerable care may be needed in the design of the survey instrument, to ensure that walk and bicycle are not reported as the only mode for a trip in which one of these two modes is used primarily as access to another mode or egress from it. Second, it is likely to be necessary to introduce other measurements related to walking and bicycling that will allow appropriate values to be developed for the travel times involved.

At this time, there has been relatively little sign of specific changes that address the issues relating to non-motorized travel modes, although recent surveys are routinely adding these to the collection of data. Travel distance or time seems to be the only consistent data collected on these activities, and there are reports of difficulties arising in the modeling of these modes with travel time as the only available attribute. Perhaps this is an area where theory or empirical development is needed on attributes of relevance.

Travel-Demand Management Strategies. A second area in which emerging policy changes have dictated changes to household travel surveys is that of transportation demand management (TDM) strategies. In the U.S., the focus of long-range transportation planning and, therefore, of the models and supporting data collection has been with major investments in transportation infrastructure. Major investments in new infrastructure are increasingly rare, and much more common is a focus on demand and system management. Many TDM strategies involve different options than those that have traditionally featured in household travel surveys, where TDM strategies included such options as the use of carpools and vanpools, use of High-Occupancy Vehicle (HOV) lanes, and various parking-management and parking-pricing options. One of the evolving changes in recent surveys has been a new focus on collecting data about parking location and price. Vehicle occupancy has also reappeared as an issue of considerable interest and importance. There are various problems associated with these variables, however.

Parking prices are troublesome for several reasons. First, for work trips, many employees do not pay the posted prices because they enjoy some level of subsidization from their employer. Also, when posted prices are high and non-subsidized employees have an option to park further away for significantly lower cost, such an option is often taken. The dilemmas that result from this can be described as in the following paragraph.

We believe that travel decisions are made on the basis of what a person actually has to pay. Where subsidies are offered or cheaper parking is available at a distance, it is often difficult to determine how much a person actually paid, or, alternatively, where they parked. Those who choose an alternative mode will likely have a different view of the parking cost that would be incurred than do those who actually park. Therefore, knowing the amount of parking cost that figured in the decision not to drive and park presents considerable difficulty. Furthermore, forecasting to the future the amounts for subsidized parking, or the locations where people will choose to park based on cost, and who will receive a subsidy and who will not, are all difficulties associated with measuring parking costs.

Parking costs also tend to vary quite significantly from parking facility to parking facility. For example, in downtown Los Angeles, within the same block, there is a surface parking lot that charges $4.00 per day, next to a building with a parking garage, where the cost is $16 per day. Both facilities will charge someone who parks for one hour or less about $1.50. Thus, even if one knew the Traffic Analysis Zone (TAZ) in which a person parked or even the distance they walked from the parking location to their final destination, there may be no reliable way to know the amount of parking charge that they faced. The parking-cost problem has not been solved and has largely been handled in recent surveys by attempting to determine how much a person actually paid, whether this represented a subsidized parking cost, and, for those who did not drive and park, asking how much they would have expected to pay if they had driven and parked.

The second element of difficulty is occupancy. Generally, this question needs to be posed only to those who drive or are driven, although there may often be value to knowing how many people traveled together to determine whether this is a factor in the mode chosen (through increased costs for each person, compared to effectively averaging costs across persons). The principal difficulty here is with the extent to which people do not read the question relating to whether or not to include themselves in the count. In the author's experience, phrasing the question to say "Including yourself, how many people were traveling in the car" has produced the answer of "None" sufficiently often as to invalidate the data for a significant subset of the sample; the converse of asking "Not counting yourself, how many people..." has equally often provided counts of 2 or 3, when subsequent questioning or matching of responses or calculation of average auto occupancies indicates that this number often *included* the respondent.

New Paradigms for Travel Forecasting

Activities Versus Trips. Perhaps the most obvious reflection of a need to change the paradigm of travel forecasting and the method of collecting travel-related data exists in the problems that those of us operating in the English language have with the word "trip". The problem here relates to the fact that the professional transportation planner has created a very specific technical meaning for the word "trip", which is not understood by the non-professional and which involves an awkward definition to the uninitiated. Notwithstanding the fact that most survey respondents will not stop to read and try to understand the definition of a word or phrase, even when a definition is provided and read, it often remains unclear to survey respondents what is meant.

The common meanings of the word trip, apart from those having to do with stumbling or dancing are well captured in current dictionary definitions: "to make a journey" or "to get high on a psychedelic drug." A secondary meaning is also offered of "a voyage, journey; a single round or business errand". None of these cover the transportation professional's definition of a one-way movement from an origin to a destination without intermediate stops to perform another activity. Perhaps this whole problem with the word trip should have alerted the profession to the possibility that the wrong entity is being measured. The author of this paper spent many years developing survey forms that measured trips but avoided ever using that particular four-letter word. While this tended to improve overall response to the questions, it did not resolve the problems related to focusing on the trip as the unit of measurement and modeling.

The U.S. Department of Transportation's Travel-Model Improvement Program (TMIP) (TMIP, 1995; Weiner, 1993) and the general frustration with the inadequacies of the current travel-forecasting models are likely to result in some new paradigms of travel behaviour, and consequent changes in the focus of data collection. As a means to improve the accuracy and completeness of the data collected, and to respond to likely paradigm shifts, there has already been the beginning of a shift to the collection of activity patterns and their consequent travel, as opposed to a focus on collecting trip patterns. Stopher and Metcalf (1996), however, found that this shift is by no means widespread. Whether the activity focus, or some other new paradigm, will emerge as the principal new direction of travel forecasting remains to be seen. However, these shifts may be among the most significant contextual changes to have occurred since the first household travel surveys.

One of the aspects of travel that has brought a sharper focus on this whole issue of the trip as the unit of measure is that of trip tours or trip chains, which have increased as a proportion of total travel in recent years and which pose serious problems both for modelling and measurement in the traditional paradigms and surveys. One of the reasons for the shift to an activity focus in collecting travel patterns is the greater ability of this approach to produce more complete information on trip tours than has been obtained historically from instruments that focus on trips. The trip focus leads respondents to omit intermediate, less-consequential stops in a trip chain, because of the notion that a trip is a significant travel event. Thus, the trip focus produces less complete data and may often lead to not identifying that a trip chain even took place. One of the reasons for this increased focus on trip chains is, of course, the changes in lifestyle that have resulted in more and more multiple-worker households, as well as in an increase of single-adult households with children in which no adults can make trips during the day from home to carry out various chores of the household.

A secondary issue in the trip-activity shift relates to communication changes that are resulting in increased opportunities to carry out various tasks at home, instead of at a specialized location away from home. This relates back in part to some of the TDM issues already mentioned, but also involves another area altogether. With modern communications developments, there is the increasing potential for certain types of workers to work from home (telecommute) at least part of the time. There is also a growing sector of out-of-home businesses. It has been suggested that, in the U.S., businesses done out of the home have increased in the past five years by almost an order of magnitude, and the rate of increase still seems to be increasing. The other side of this revolution is to be seen in the shopping and entertainment areas. Increasing capabilities to shop by catalogue and from the television screen, and abilities to select various forms of home entertainment from cable and similar communications devices are likely to result in further shifts in travel patterns. These shifts can be understood only by increasing detail about the activities undertaken by people, and, in particular, knowing about activities done in the home that could be substitutes for outside activities, or those done outside that could eventually be substituted by in-home activities.

This leads to a new area of challenge for transportation survey designers, relating to issues of privacy, intrusiveness of the survey method, and defining those activities that one would like to measure and those that one does not wish to know about. The household travel survey conducted in Portland, Oregon and several surrounding metropolitan areas may have been the first U.S. household-travel survey to attempt to measure in-home activities. It appears to have succeeded quite well in measuring many of the activities involved in the trade-offs. This was done by restricting the activities to be reported to those that lasted at least 30 minutes. However, it also revealed another dilemma for measurement which is that of multitasking. It seems that there is a substantially higher level of multi-tasking that is performed within the home than outside it. Thus, an individual may not attempt to shop and read at the same time, while recent efforts to collect in-home activities have revealed such things as watching television while preparing a meal, reading while eating, and other forms of multitasking to be very common. Not only does this raise some challenges for recording the data on the survey form, but it leads to issues of modeling that have yet to be resolved.

Time-Use Surveys. A recent development in the evolution of the diary has been seen in the case of the North Central Texas Survey of 1996. In this survey, the activity diary has evolved into a time-use diary (Robinson, 1977). The primary difference between the time-use diary and the activity diary is in two aspects. First, the time-use diary attempts to collect data on all activities in the diary period, in such a way that the ending time of one activity is the starting time of the next activity. Second, the time-use diary explicitly defined travel to be an activity, rather than the customary design of the activity diary, in which activities are defined as things done at a place, and the travel to get to or from the activity is collected separately and differently from the activities.

There are two advantages of this approach. First, it seems to be more user-friendly, judging by the number of times that people try to report travel as an activity in the activity diary. In this way, the time-use diary is actually responding to a design fault in the activity diary, as it is perceived by many respondents. Second, the time-use diary provides a natural mechanism to record much greater detail on activities, particularly those relating to travel, than does the activity diary. A determination of the extent to which this is actually feasible in practice must await at least the results of the North Central Texas Survey which is in progress at the time of writing. However, the time-use design would lead naturally to recording each element of a trip as a separate activity. Thus, a walk to the bus stop is an activity, as is the wait at the bus stop, the ride on the bus, and the walk to the final destination. Further, the specific locations of the bus stops for boarding and alighting are logically obtained in the process. This leads to what Axhausen has defined as collecting "stage" data (Axhausen, 1994). Collecting data on travel with this level of detail has been desired in the past, but has been particularly difficult to do in the context of a trip or activity diary, because of the amount of questioning that seemed to be necessary. However, the burden on the respondent becomes rapidly much larger, with the requirement to fill in details about many more activities within a 24-hour or 48-hour period.

It appears that the time-use diary may involve fewer questions about any given activity than an activity diary, but will lead to recording many more events during the recording period, even if the same restrictions are applied about recording in-home activities, as described in the previous subsection. However, what the length burden of such an instrument would be is still a matter to be resolved. Also, whether this length is acceptable and whether the concepts of the time-use diary represent a simplification of the complexity of the activity diary are also questions for resolution.

Multi-Day Surveys. Traditional household-travel surveys have focused on reporting the travel of sampled households for a single day, with most conventional household travel surveys collecting data for a single weekday, usually in the spring or fall. Another aspect, of the increasing need for data to inform new theories and paradigms of behaviour, is an emerging trend to collecting data on more than one weekday, to collecting data on some weekend days, and to considering data collection in the summer and winter also. This is driven in part by the realization that travel patterns are much more variable than can be captured by restricting all data collection to one day, and that there are dynamics of travel behaviour that have not been caught by one-day surveys. In addition, issues relating primarily to planning for air-quality improvements have shifted the focus towards winter and summer travel, because the winter is the time of most carbon-monoxide violations and summer is the time of most ozone violations. One wave of the Puget Sound Panel Survey used a two-day diary, and the Portland, Oregon 1994-95 Survey also used a two-day diary.

However, while it is known that there is some decrease in response rates with additional days of diary completion, there have been no controlled comparative studies in the U.S. to determine the precise extent of the drop in response (there are even anecdotal reports of increases in response on the second day, or at least increases in the numbers of activities reported), nor the gain in total information that might be offered by a 48-hour instrument as compared to a 24-hour instrument. Indeed, the question of repetitiveness of activities over two consecutive weekdays has not been addressed adequately, so that there is no good information available on whether or not real additional data are obtained from a second day, or how much additional data are obtained.

In the Portland case, another interesting aspect of the two-day diary collection is that households were asked to complete their diaries for two consecutive days, with a fraction of the households completing diaries for the combination of a weekday and a weekend day. Issues of the importance of collecting data for weekend days are also unresolved. From a behavioral viewpoint, it can be argued that households trade off activities and travel between weekdays and weekend days, and that understanding weekday travel patterns is not complete unless there is understanding of the relationships between weekday and weekend activities. It can also be argued that more and more peak congestion is moving to the weekend, and that a majority of air-quality incidents are occurring on weekend days. Both of these phenomena would tend to point to an increasing importance for weekend travel measurement.

Household and Personal Data. Recognition of the trends towards trip chaining, and other behaviors associated with multiple-worker households has resulted in an increasing interest in the notion of microsimulating household lifecycle and lifestyle changes as a part of any new paradigm of travel forecasting. Microsimulation seems to be necessary because there is no existing ability to forecast the incidence of certain household characteristics related to lifecycle, so that microsimulation that is based on simulating the evolution of households provides the most promising avenue for forecasting the future numbers of households in different lifecycle stages in different locations in the urban area. In this discussion, lifecycle is understood to include information relating to the presence and ages of children in the household, the presence of working and nonworking adults, and some indicators of relationships among the adults, at a minimum (Vadarevu and Stopher, 1996).

Further, as understanding increases about the role that personal and household constraints and resources play in the decisions on activities and concomitant travel, the greater becomes the transportation planner's interest in a wide array of characteristics of people. The addition of these two elements of household and personal attributes has the effect of increasing the need to collect more complete and more detailed data on household and personal characteristics. Household size, vehicle ownership, and income are probably far from being sufficient to provide understanding about role allocation, choices of activities, and sequencing of activities. Hence, along with the increase in interest in activities themselves, there is an increase in the amount of descriptive data required on the household and its members.

Panels. Possibly one of the most under-utilized survey devices in household travel surveys is the longitudinal panel. Only one panel of significant duration has been undertaken in the U.S. to date (Murakami and Watterson, 1990), although a second one may be in its first wave in the San Francisco Bay Area. Few such surveys have been undertaken elsewhere in the world. The benefits offered by panels have been discussed in numerous other places and are not elaborated further in this paper (Golob et al., 1997). Briefly, however, with the increasing attention being paid to dynamics, panels should be of much greater interest to transportation modelers than they have proved to be to date. A panel offers the opportunity to track changes that occur in response to evolution in lifecycle, as well as to external influences on travel and activities. A panel also offers the capability to measure with much greater accuracy from a much smaller sample, because one is measuring change without sampling error intruding in the estimation of change for the households in the panel.

At the same time, because of the desirability of keeping households in the panel, and because of the repetitive nature of the panel survey, it is probable that such panels should be paid panels and that they would be selected to represent a cross-section of the population that may eliminate some of the biases perceived to exist in current cross-sectional, telephone-based surveys.

Establishment of panels offer advantages in measuring the dynamics of change, in measuring seasonal variations in travel behaviour, and provide opportunities for more extensive measurement over time, by combining different subsets of stated-response questions at different waves of the panel. The potential to rely heavily on ongoing small, paid panels is a concept whose time may have come. Such a panel can be bench-marked from time to time by conducting a modest cross-sectional survey, so as to determine the extent to which the panel remains representative of the population, and to determine how future panel attrition should be made up. The potential to gain a greater wealth of data from an ongoing panel than from conducting large cross-sectional surveys at long time intervals needs to be explored and defined, so that the potential of this concept may be assessed appropriately. The potential to develop a new paradigm of travel from the understanding offered by a panel seems highly probable, however.

Remote Sensing. One of the technological areas that is advancing rapidly is that of various forms of remote-sensing devices. At least two such devices have potential application for household travel surveys. One is the type of device that can be fitted to the automobile to record various different attributes of automobile operation, such as starts and stops, acceleration, deceleration, time, distance, etc. (Taylor, 1991) Coupled with a diary (activity or time use), the potentials for improved data collection about use of vehicles is enormous. In the

U.S., there has been no report of such a coupling of remote-sensing vehicle devices with multi-day diaries to date, although some work has been done in Canada (Lee, 1985).

The second type of technological advance is global positioning systems (GPS) that could be combined in a variety of ways with collection of diary data. As GPS antennas are increasingly miniaturized, the potentials to use them as an aid to the geographic information collection in travel surveys clearly becomes increasingly attractive. At one extreme, GPS could be connected into the type of automobile-sensing device just discussed, in order to provide continuous position information that would provide vehicle routing, as well as all of the vehicle function, time, and distance data. This would be of immense benefit by automating that part of the data collection, which currently must be done by providing a diary to be completed.

In household travel surveys, one of the biggest problems encountered is that of recording geographic locations, particularly ones that are accurate enough to allow geocoding to small area or point locations. Most people know the addresses of their home and workplace accurately enough to record them in a diary. Few other addresses are known with sufficient completeness to allow accurate geocoding. Among methods used recently in the U.S. to improve this has been the request for a place name of the location (such as a shopping center name, or building name). A recent instance in Canada used an extensive gazetteer to allow interactive prompting for more specific addresses (Ng and Sarjeant, 1993). These methods have proved useful to improve the number of addresses that are geocodable, but there remains a significant problem in completing geocodes, particularly to a point location. GPS appears to offer a potential solution to this problem. If GPS equipment can be sufficiently miniaturized, and if issues of privacy invasion can be resolved, then such equipment could be attached to individuals who are also completing diaries. This would provide fully geocodable data on exactly where people go during a reporting period. An experiment on this type of technology is underway by the Federal Highway Administration of the U.S. Department of Transportation at the time of writing, using personal data devices coupled with a miniaturized GPS receiver.

Expanded Sample Coverage. The final area addressed in this chapter is that of expansion of sample coverage in future household travel surveys. Already, the notion has been discussed of abandoning older concepts that the data collected are to be for weekdays only in the spring or fall, and concentrating on a 24-hour period. In addition to this, the notion that it is acceptable to collect data only from households with telephones must be re-examined. Future data collection needs to target both weekend days and weekdays, for reasons already discussed. Future household travel surveys also need to cover different seasons of the year, including summer (particularly in ozone non-attainment areas) and winter (particularly in carbon-monoxide non-attainment areas), and to be conducted over multiple days for each household included in the sample, also as discussed earlier. What has not been discussed already is the notion that expanded coverage needs to take in non-telephone households, unless it can be established that such households are not different from their telephone-owning counterparts. Initial anecdotal information from the North Central Texas pilot tests indicates that there may be a disproportionate number of transit riders from non-telephone households, so that it seems likely that this group will need to be part of expanded coverage in future surveys.

Another related concept in coverage expansion is that of what has been termed "rare" populations. These are groups of the population that undertake behaviors that are not found commonly in the population at large but that may be quite critical for future forecasts and policy development, or may be essential to the development of new paradigms of travel and activity behavior. Such rare populations in the U.S. include those who ride transit, those who use bicycles for their main mode of transportation, and those who ride in vanpools and formal carpools. It also includes those members of the population who choose to be carless because of accessibility to other options for travel or communications. Currently, it also includes those who telecommute and those who work out of small home offices. Other rare populations will exist in other countries, depending on a variety of cultural and developmental factors. Sampling for such rare populations is likely to be of increasing importance to the transportation planning community, as well as to model theorists and empiricists.

CONCLUSIONS

First, the purposes of data collection should be kept clearly in mind. Primarily, data are collected to provide an understanding of the phenomenon being measured, in this case travel behaviour, and the basis for updating models or constructing new ones. The primary consideration that should guide what data are collected and the quality demanded of those data are the uses to which the data will be put. Uses to which the data are put may change over time, so that when data are collected infrequently, such as every decade or two, the data being collected today may have to be usable in models that are not currently available and be able to answer questions and policy issues that have not yet been thought about. This demands that there be more thought given to what data to collect than to satisfy the immediate policy issues of the day and to operate current models. There must be some level of anticipation of future issues and problems, as well as future developments in the modeling arena.

Second, the quality of the data needs to be considered very carefully. It is all too easy to collect poor quality data without recognizing the poor quality, and to collect data that contain large biases and errors without being aware of it. Many practices in household travel-survey data collection have tended to generate just such problems. Reaching total sample requirements by adding in more and more households from the total population to make up for non-responding households and households that provide only partial data is but one of the ways in which data quality and accuracy can be compromised. Both the principal purposes of data collection noted above are susceptible to the phenomenon of "garbage in - garbage out." Therefore, much attention must be paid to improving the quality of the data collected, so that the purposes to be served by the data are not compromised.

Third, this chapter has outlined a number of areas in household travel-survey design that already pose issues for design and improvement, as well as a number that will pose such issues for the future. Improvements in applied travel choice models can come about only if suitable data are collected to test hypotheses and to develop new empirical model results. Improvements in data collection must also not result in compromises with quality as a result of making the task more onerous for the respondent. Perhaps more than ever before in transportation planning, the profession is poised on a knife edge between demands for more accurate data and more data, and increasing nonresponse and respondent burden from the data-collection task. It will take great skill and possibly the development of new strategies of data collection to meet this challenge successfully.

REFERENCES

Axhausen, K.W. (1994). *Travel diaries: An annotated catalogue*. Working Paper, Centre for Transport Studies, University of London.

Golob, T. F., R. Kitamura and L. Long (eds.) (1997). *Panels for Transportation Planning: Methods and Applications*. Kluwer, Boston, MA.

Lee, M. E. H. (1985). Development of a methodology for estimating the kilometrage of drivers throughout Quebec. In: *Proceedings of the Canadian Multidisciplinary Road Safety Conference IV,* pp: 563-572. Équipe de Sécurité Routière de l'École Polytechnique.

Lee-Gosselin, M. E. H (1996). Interactive stated-response techniques. In: *Household Travel Surveys - New Concepts and Research Needs, Conference Proceedings 10.* Transportation Research Board, Washington DC.

Murakami, E. and T. Watterson (1990). Developing a household travel panel survey for the Puget Sound Region. *Transportation Research Record, No.* **1285**, 40-46.

Ng, J. C. N. and P. M. Sarjeant (1993). Use of direct entry for travel surveys. *Transportation Research Record, No.* **1412**, 71-79.

Robinson, J. P. (1977). *How Americans Use Time: A Social-Psychological Analysis of Everyday Behaviour.* Praeger, New York.

Stopher, P. R. (1996). Household travel surveys: Cutting-edge concepts for the next century. In: *Transportation Research Board Conference Proceedings*, ***10***, 11-23. Washington DC.

Stopher, P. R. and H. M. A. Metcalf (1996). Methods for household travel surveys, NCHRP *Synthesis of Current Practice,* **236**.

Taylor, G. W. R. (1991). *AUTOLOGGER: A long duration vehicle use data collection system.* Paper presented at the sixth international conference on travel behavior, Québec, Canada.

TMIP (1995). *Travel Model Improvement Program Newsletter #2*, 1600 E. Lamar Blvd., Suite 112, Arlington, TX 76011.

U.S. Department of Transportation (1973). *Urban origin-destination surveys,* Technical Report, Federal Highway Administration, U.S. DOT, Washington, DC.

Vadarevu, R. and P. R. Stopher, (1996). Household activities, life cycle and role allocation, *Transportation Research Record, No.* **6**, 77-85.

Weiner, E. (1993). *Upgrading U.S. travel demand forecasting capabilities*. Paper presented at the 4th conference on transportation planning methods applications, Daytona Beach, FL.

18

Multiple Imputation Methodology for Missing Data, Non-Random Response, and Panel Attrition

David Brownstone, University of California, Irvine, CA, USA

Introduction

Modern travel-behavior surveys have become quite complex; they frequently include multiple telephone contacts, travel diaries, and customized stated-preference experiments. The complexity and length of these surveys lead to pervasive problems with missing data and non-random response biases. Panel surveys, which are becoming common in transportation research, also suffer from non-random attrition biases. This chapter shows how Rubin's (1987a) multiple imputation methodology provides a unified approach to alleviating these problems.

Before discussing solutions to problems caused by missing data and selection, it is important to recognize that their presence causes fundamental problems with identifying models and even population estimates. The second section reviews this work and stresses the need to make generally untestable assumptions in order to carry out any inference with missing data.

Once some identifying assumptions are made, the most common method for handling missing data is to remove observations with any missing data. This method, which Rubin calls complete case analysis, is inefficient, but it is easy to implement with standard statistical packages. Complete case analysis will lead to biased estimates if the process generating the missing data is correlated with the endogenous variables in the model. For example, suppose we estimate a travel-mode choice model from a random sample of a metropolitan area. If respondents who never take the bus are more likely to not respond to questions about bus travel times and costs, then complete case analysis on the original random sample will yield inconsistent estimates. In this example, the missing data mechanism is equivalent to choice-based sampling.

One common solution for missing data is to impute missing values. For the mode choice example given above, this is usually done using zonal network data together with the respondent's reported home and work locations. Unfortunately, most commonly used imputation methods do not preserve the variance of the underlying variable and therefore produce inconsistent estimates when variables containing imputations are included in models. Even when this problem is avoided, inference is always conditional on the imputed values and therefore ignores errors in the imputation process. The third section provides general methods for generating correct imputations.

The non-random sample in the above mode choice example was caused by using complete case analysis with a non-random missing data process. Although they only discussed deliberate non-random sampling (e.g., choice-based sampling), Manski and Lerman's (1977) Weighed Exogenous Sample Maximum Likelihood Estimator (WESMLE) provides consistent and asymptotically normal estimates for any sampling procedure. Unfortunately, the WESMLE requires the sampling weights to be known constants. In many cases discussed in the fifth and sixth sections, these weights need to be estimated and it can be quite difficult to modify the WESMLE to properly account for estimation error in the sampling weights. Ignoring this estimation error will lead to downward biased standard errors.

Multiple imputation creates multiple imputed values and weights, and then combines the estimators using each set of values into a final consistent estimator that accounts for the errors in the imputation process. Although multiple imputation estimators are rarely efficient, the method can be applied using standard statistical software packages. Other maximum-likelihood methods require extensive programming and/or computation time. The third section reviews multiple imputation methodology, and the following sections discuss applying multiple imputation to common applied transportation problems.

To simplify the exposition, the second to the fourth sections consider only the case of missing data. Formally, survey nonresponse is just a special case where all of the data for the missing respondents are missing. Similarly, erroneous data (or measurement error) is equivalent to having all observations on the true values missing. The fifth and sixth sections discuss survey nonresponse and panel attrition.

IDENTIFICATION WITH MISSING DATA

When data are missing, the only way to identify population statistics such as means and variances is to make assumptions that determine the distribution of the missing data. Since by definition we do not observe missing data, these identifying assumptions are typically untestable. If one is nonetheless willing to maintain these untestable identifying assumptions, then the third to fifth sections of this chapter describe a general methodology for estimation and inference. This section describes work primarily due to Manski (Manski, 1995, and the included references) which shows that even though population statistics are unidentified, it is sometimes possible to bound them between estimable functions.

To illustrate the construction of these bounds, I will closely follow Horowitz and Manski (1995). Assume that each member of the population is characterized by a value of (y, $\boldsymbol{x}$, and z), where $y \in Y$ is the outcome of interest, $\boldsymbol{x} \in X$ is a covariate vector, and z is a binary variable indicating response to some or all survey questions. Suppose that we are interested in estimating conditional expectations, $E[g(y)|\boldsymbol{x}]$, where $g(\cdot)$ is some specified function. If we observe y only when z=1, then:

$$E[g(y)|\boldsymbol{x}] = E[g(y)|\boldsymbol{x}, z=1]\cdot P(z=1|\boldsymbol{x}) + E[g(y)|\boldsymbol{x}, z=0]\cdot P(z=0|\boldsymbol{x}) \qquad (1)$$

The observed sample identifies the mean $E[g(y)|\boldsymbol{x}, z=1]$ and the response probability $P(z=1|\boldsymbol{x})$, but provides no information about the mean of the unobserved sample $E[g(y)|\boldsymbol{x}, z=0]$. However, this unidentified part must lie between $K_0 \equiv \inf_{y \in Y} g(y)$ and $K_1 \equiv \sup_{y \in Y} g(y)$, which yields the following sharp bounds:

$$E[g(y)|\boldsymbol{x}, z=1]\cdot P(z=1|\boldsymbol{x}) + K_0\cdot P(z=0|\boldsymbol{x}) \leq E[g(y)|\boldsymbol{x}]$$
$$\leq E[g(y)|\boldsymbol{x}, z=1]\cdot P(z=1|\boldsymbol{x}) + K_1\cdot P(z=0|\boldsymbol{x}) \qquad (2)$$

These bounds are equivalent to imputing all the missing data equal to the highest and lowest possible values of y. Note that the length of this bound is $(K_1 - K_0)\cdot P(z=0|\boldsymbol{x})$, so if $g(\cdot)$ is bounded, then decreasing the nonresponse probability will sharpen the bound. One important case where $g(\cdot)$ is bounded is when it is an indicator function. Horowitz and Manski (1995) give an example trying to estimate the employment probability from the 1979 U.S. National Longitudinal Survey of Youth. The nonresponse rate is 18.4% ($P(z=0|\boldsymbol{x})$), and the employment probability in the observed sample is 78% ($E[g(y)|\boldsymbol{x}, z=1]$, where y is 1 if the respondent is employed and zero otherwise). Since the probability of employment in the missing observations is bounded between 0 and 1, equation (2) shows that the desired overall employment probability lies between 63% and 82%. The length of this bound is much larger than the sampling variation of the probability estimators, and increasing the sample size with the same nonresponse probability will not decrease the width of the interval.

Of course, many researchers are willing to make identifying assumptions about the unobserved mean $E[g(y)|\boldsymbol{x}, z=0]$. A popular assumption, due to Heckman (1976), has $g(y|\boldsymbol{x})$ follow a normal distribution. The most common identifying assumption is that the nonresponse process is *ignorable* (see Rubin, 1987a, for a formal definition), which here means that the conditional distribution of $y|\boldsymbol{x}$ is the same for both respondents and nonrespondents. As long as no $\boldsymbol{x}$ values for the unobserved sample are outside the range of $\boldsymbol{x}$ values for the observed sample, the assumption of ignorability allows nonparametric identification of $E[g(y)|\boldsymbol{x}]$. One reason why ignorability is so commonly invoked is that without additional information there is no direct evidence to contradict the ignorability assumption (Rubin,1987a).

Many other identifying assumptions are possible, but they must all yield estimates within the bounds given by equation (2). Horowitz and Manski (1995) show that weighted means using standard poststratification (or ratio) nonresponse weights (which are estimates of the inverse of the probability of response) may fall outside these bounds. Weighted estimators are only guaranteed to yield sensible estimates if the weights are constructed conditioning on $\boldsymbol{x}$. However, imputation estimators which explicitly impute missing values and then average over observed and imputed values will always lie inside the bounds given in equation (2).

From a Bayesian perspective, Horowitz and Manski's bounds are similar to Leamer's (1983) "extreme bounds analysis". Leamer derives his bounds by looking at the envelope of the Bayesian confidence regions over all possible prior distributions. In many cases, both Leamer's and Horowitz and Manski's bounds are infinite, and even when they are bounded they are frequently quite wide.

The analysis reviewed in this section implies that we need to be much more humble about our empirical results when there are missing data. A more constructive implication is that researchers involved in collecting data should concentrate more resources on reducing missing data and non-response. Another useful approach is to carry out separate "validation" studies which use intensive interviewing techniques to survey a sample of non-respondents. These validation surveys provide direct evidence about non-respondents to the main survey, so they can be used to identify and measure the unknown $E[g(y)|\boldsymbol{x}, z=0]$. Brownstone and Valletta (1996) show how multiple imputation techniques can be used to combine information from the validation and main surveys to estimate econometric models.

MULTIPLE IMPUTATION

Rubin's (1987a) multiple imputation methodology first requires a model for producing proper imputed values. These imputed values must be conditioned on all observed data, and different sets of imputed values must be drawn independently so that they reflect all sources of uncertainty in the imputation model. Although Rubin developed the theoretical properties of this methodology for Bayesian models, in Rubin (1987a) and Rubin (1996) show that these results apply asymptotically to classical statistical models.

Suppose we are interested in estimating an unknown parameter vector θ. If no data are missing, then we would use the estimator $\tilde{\theta}$ and its associated covariance estimator $\tilde{\Omega}$. If we have a model for predicting the missing values conditional on all observed data, then we can use this model to make independent simulated draws for the missing data. If m independent sets of missing data are drawn and m corresponding parameter and covariance estimators, $\tilde{\theta}_j$ and $\tilde{\Omega}_j$, are computed, then Rubin's Multiple imputation estimators are given by

$$\hat{\theta} = \sum_{j=1}^{m} \tilde{\theta}_j \,/\, m \tag{3}$$

$$\hat{\Sigma} = \boldsymbol{U} + \left(1 + m^{-1}\right)\boldsymbol{B} \tag{4}$$

where

$$\boldsymbol{B} = \sum_{j=1}^{m} \left(\tilde{\theta}_j - \hat{\theta}\right)\left(\tilde{\theta}_j - \hat{\theta}\right)' \Big/ (m-1) \tag{5}$$

$$\boldsymbol{U} = \sum_{j=1}^{m} \tilde{\Omega}_j \Big/ m \tag{6}$$

Note that $\boldsymbol{B}$ is an estimate of the covariance among the m parameter estimates for each independent simulated draw for the missing data, and $\boldsymbol{U}$ is an estimate of the covariance of the estimated parameters given a particular draw. $\boldsymbol{B}$ can also be interpreted as a measure of the covariance caused by the nonresponse process.

Rubin (1987a) shows that for a fixed number of draws, $m \geq 2$, $\hat{\theta}$ is a consistent estimator for θ and $\hat{\Sigma}$ is a consistent estimator of the covariance of $\hat{\theta}$. Of course $\boldsymbol{B}$ will be better estimated if the number of draws is large, and the factor $(1 + m^{-1})$ in equation (4) compensates for the effects of small m. Rubin (1987a) shows that as m gets large, then the Wald test statistic for the null hypothesis that $\theta = \theta^0$,

$$\left(\hat{\theta}-\theta^{0}\right)'\hat{\Sigma}^{-1}\left(\hat{\theta}-\theta^{0}\right) \tag{7}$$

is asymptotically distributed according to an F distribution with K (the number of elements in θ) and ν degrees of freedom. The value of ν is given by:

$$\nu = (m-1)(1+r_m^{-1})^2$$
$$r_m = (1+m^{-1})\ \mathrm{Trace}(\boldsymbol{B}\boldsymbol{U}^{-1})/K \tag{8}$$

This suggests increasing m until ν is large enough (e.g., 100) so that the standard asymptotic Chi-squared distribution of Wald test statistics applies. Meng and Rubin (1992) show how to perform likelihood ratio tests with multiply-imputed data. Their procedures are useful in high-dimensional problems where it may be impractical to compute and store the complete covariance matrices required for the Wald test statistic (equation 7).

The key to successful implementation of multiple imputation is to use a proper imputation procedure. The full definition of a proper imputation procedure is given in Rubin (1987a). Loosely speaking, if the estimates computed with the true values of the missing data ($\ddot{\theta}$ and $\ddot{\Omega}$) are treated as fixed, then $\hat{\theta}$ and $\boldsymbol{U}$ must be approximately unbiased estimators of $\ddot{\theta}$ and $\ddot{\Omega}$. In addition $\boldsymbol{B}$ must be an approximately unbiased estimator of the variation in $\hat{\theta}$ caused by the non-response mechanism. The safest way to generate proper imputation procedures is to explicitly draw from the (Bayesian) posterior predictive distribution of the missing values under a specific model. Other proper multiple imputation procedures also require no explicit Bayesian calculations, and some of these methods are described in the following sections. Any proper imputation procedure must condition on all observed data, and different sets of imputed values must be drawn independently so that they reflect all sources of uncertainty in the response process.

Linear Model

This subsection summarizes results from Brownstone (1991) on using multiple imputation techniques with the linear regression model. Suppose we are interested in estimating a K-vector, θ, in the standard linear model:

$$\boldsymbol{y} = \boldsymbol{X}\theta + \varepsilon, \tag{9}$$

where, conditional on $\boldsymbol{X}$, the components of ε are independent and identically distributed random variables with mean 0 and variance σ^2. In the absence of missing data, θ would be estimated by the ordinary least squares estimator, $\hat{\theta}$, and inference would be based on:

$$\left(\hat{\theta}-\theta\right) \approx N\left(0, s^2 (X'X)^{-1}\right), \tag{10}$$

where $s^2 = y'\left(I - X(X'X)^{-1}X'\right)y \big/ (N-K)$ and I is the $K \times K$ identity matrix.

Suppose further that the first N_0 observations contain missing data in exogenous (*X*) and/or endogenous (*y*) variables, but that there are no missing data in the remaining $N_1 = N - N_0$ observations.

A proper multiple imputation procedure for the linear model above must produce imputed values for the missing observations, y_0^* and X_0^*, which at least match the first two asymptotic moments of the unobserved variables, y_0 and X_0. For any such set of proper imputations, the completed least-squares estimator is given by:

$$\tilde{\theta} = \left[\left(X_0^{*'} X_0^*\right) + \left(X_1' X_1\right)\right]^{-1} \cdot \left[X_0^{*'} y_0^* + X_1' y_1\right] \tag{11}$$

and the associated covariance estimator:

$$\tilde{\Omega} = s^{*2}\left[\left(X_0^{*'} X_0^*\right) + \left(X_1' X_1\right)\right]^{-1} \tag{12}$$

where s^{*2} is defined as in equation (10) with imputed values, y_0^* and X_0^*, replacing the unobserved variables. Since the first two asymptotic moments match, it follows that the probability limits of $\tilde{\theta}$ and $\tilde{\Omega}$ must equal the probability limits of the least squares estimator with no missing data. Note that $\tilde{\Omega}$ is a downward biased estimate of the variability of $\tilde{\theta}$ since it does not account for the variation induced by the non-response process. Drawing repeated independent sets of proper imputations allows consistent estimation of this variance component (***B*** in equation (5)).

As discussed in the second section, before we can construct a proper imputation procedure we need to make some assumptions about the process generating the missing data. To simplify the notation, we will assume that only the exogenous variables, X_0, are missing. If we further assume that the missing data process is ignorable, then the most straightforward way to generate proper imputations for the linear model is:

$$X_0^* = E(X_0 \mid y_0) + \eta_0^* \tag{13}$$

where η_0^* are independent draws from the distribution of the residuals, $X_0 - E(X_0 \mid y_0)$. The ignorability assumption implies that $E(X_0 \mid y_0)$ is identical to $E(X_1 \mid y_1)$, and can therefore be consistently estimated using parametric or nonparametric regression techniques (see Manski, 1991) on the observed data. Ignorability also implies that the distribution of the residuals can also be estimated from the distribution of residuals on the observed data.

If we further assume that the conditional distribution of $(\boldsymbol{X}|\boldsymbol{y})$ satisfies the standard linear model assumptions, then the imputation equation (13) can be implemented by first regressing the observed $\boldsymbol{X}_1$ on $\boldsymbol{y}_1$ to get the estimated normal sampling distributions of the slope parameter(s) and the residual variance. To draw one set of imputed values, first draw one set of slope and residual variance parameters and then draw the imputed residual vector, η_0^*, from independent normal distributions with mean zero and variance equal to the imputed residual variance parameter. The imputed values are then computed by adding this imputed residual to the predicted value from the regression using the imputed slope parameters. Additional sets of imputed values are drawn the same way beginning with independent draws of the slope and residual variance parameters. This imputation method, which Schenker and Welsh (1988) call the "normal imputation" procedure, is equivalent to drawing from the Bayesian predictive posterior distribution when $\boldsymbol{X}$ and $\boldsymbol{y}$ follow a joint normal distribution with standard uninformative priors.

The assumption of joint normality in the previous paragraph is convenient but not necessary for producing proper imputations in this model. Nonparametric regression estimators can be used to estimate nonlinear conditional mean functions, $E(\boldsymbol{X}_1|\boldsymbol{y}_1)$, and bootstrapping methods can be used to sample from residual distributions without assuming a normal distribution. However, the assumption of ignorability is crucial for the validity of these methods.

The simple model analyzed above is not very interesting from a practical perspective since the resulting multiple imputation estimator will have approximately the same distribution as the least-squares estimator calculated from the observed data, $\boldsymbol{X}_1$ and $\boldsymbol{y}_1$. However, if there are additional fully observed variables, then these can be added to the conditioning set in equation (13) to yield improved estimators. The next section describes some circumstances where these additional variables may be readily available to the data collectors. Conditioning on additional variables, even if they are not directly related to the model in equation (9), can also make the crucial ignorability assumption more palatable.

Public Use Databases

According to its developer (Rubin, 1996, p. 473), "multiple imputation was designed to handle the problem of missing data in public-use databases where the database constructor and the ultimate user are distinct entities." The database constructors certainly have more information about the sampling design and surveying process than the ultimate users, and may also have access to confidential information (such as exact addresses) which cannot be released in a public-use file. The users of these data are assumed to have access to standard statistical packages which typically estimate a wide variety of models assuming random sampling and no missing data.

Multiple imputation methodology allows the database constructors to present their superior knowledge about the nonresponse process so that the end users can easily incorporate this knowledge in their analysis. The database constructors produces 5-10 multiple imputed values for key variables (such as income) with substantial missing data. The end users only need to repeat a standard complete-data analysis for each set of imputed values and then combine the results according to equations (3)-(6). These additional calculations can, and have, been incorporated as macros in standard statistical packages. The only other practical problem may be the extra storage required for the multiply imputed values, but this seems to be an increasingly small price to pay for statistically valid inference. Finally, Rubin (1996) points out that there are no general alternatives to multiple imputation for public-use files.

In spite of multiple imputation's obvious advantages for public-use files, there has been considerable controversy about the use of the methodology. Fay (1991, 1992) provides several examples where the multiple imputation variance estimator ($\hat{\Sigma}$ in equation (4)), is an inconsistent estimator of the sampling variance of the multiple imputation estimator ($\hat{\theta}$), even when the imputation model is correctly specified. Fay's examples do not violate Rubin's (1987a) results because the end-users' model conditions on an irrelevant variable which is not included in the imputers' model. Meng (1994) shows that this situation, which he calls "uncongeniality," typically leads to conservative inference (actual coverage probabilities higher than nominal coverage probabilities) from the standard multiple imputation formulas. In spite of this conservative bias, Meng shows that the multiple imputation intervals are still sharper than the "standard" inference from only non-missing observations. A simple partial solution to these problems is to make sure that the imputation models and assumptions used by database constructors are fully documented.

Fay also criticizes practitioners of multiple imputation for not properly accounting for the highly stratified multistage sampling techniques used in many large-scale surveys. This is not a criticism of multiple imputation methodology, which assumes that the complete data covariance estimator ($\tilde{\Omega}$) is consistent. Rubin (1996) points out that multiple imputation can easily be combined with modern jackknife and bootstrap techniques for estimating variances in complex samples. Fay's criticism unfortunately applies to practically all end-users of complex public-use databases, and the biases in estimated standard errors and inferences caused by ignoring complex sampling schemes may be larger than those caused by ignoring missing data.

The case for using multiple imputation in public-use databases is especially strong when the database creator has access to validation or non-response studies. These studies allow identification and imputation with nonignorable nonresponse. In these cases inference from non-missing observations will yield inconsistent point estimates in addition to inconsistent variance estimates. Unless these validation studies are also publicly available, the imputations will be conditioned on more information than can be made available to end users. Providing multiply imputed public-use databases is currently the only general method for giving the end-user the access to the database constructor's crucial confidential information about the nonresponse processes.

There are many opportunities for constructors of public-use transportation databases to obtain information about nonresponse (or erroneous response) mechanisms. Most large transportation surveys ask questions about household's vehicle holdings and vehicle miles traveled. Household reports of vehicle holdings can be checked against vehicle registration files. Household reports of vehicle miles traveled are known to be highly inaccurate (Lave, 1996), but in many states annual vehicle miles traveled can be checked against vehicle inspection records. Although there are substantial legal problems in many countries, household and personal income reports can sometimes be checked against tax records. Collecting this information can be difficult and time-consuming, but the examples described in the next section show large gains in statistical accuracy with relatively small validation study samples.

Modern activity analysis also depends on large-scale travel diary surveys. These surveys are difficult and expensive to collect, and even the best self-reported diaries contain substantial measurement error. New technology using on-vehicle computers equipped with global positioning system satellite receivers can almost totally automate collection of automobile trip details. While this technology is too expensive to deploy for a large-scale activity survey, recent work with electric vehicle trials show that it is feasible to use these new accurate data collection techniques for a small validation sample (see Golob and Gould, 1998).

MISSING OR ERRONEOUS DATA

This section describes some studies where the multiple imputation methodology described in the previous section is used to compensate for missing and/or erroneous data. The studies are chosen to represent the wide range of applications that are possible, but this is not intended to be a review of all known applications of multiple imputation. Rubin (1996) and Meng (1994) provide more complete references to applications of multiple imputation. Heitjan and Little (1991) also give an interesting example applying multiple imputation to fill in missing blood alcohol content measures in the U.S. Fatal Accident Reporting System. Glynn et al. (1993) give another application to missing data on alcohol drinking behavior in a study on the effects of drinking on retirement. They use a follow-up validation study to estimate a non-ignorable response model. Brownstone and Valletta (1996) use similar methodology in their study described in the next sub-section.

Linear Models

Wages (or labor earnings) are a key variable in labor economics, and they are also important in many transportation models. Wages and income are typically measured by directly asking survey respondents, but there is considerable evidence that these reports contain large errors. In particular, both Bound et al. (1990, 1994) and Bound and Krueger (1991) find that measurement error in earnings is negatively correlated with true earnings and positively autocorrelated over time. These findings are based on two validation surveys: the Panel Study of Income Dynamics Validation Survey (PSIDVS) and the 1978 Current Population Survey-Social Security Earnings Records Exact Match File.

Brownstone and Valletta (1996) use the PSIDVS together with the main Panel Study of Income Dynamics (PSID) to examine the effects of wage measurement error on standard linear models used to explain and predict wages. They use the PSIDVS to estimate an imputation model which is then used to multiply impute "true" wages in the main PSID.

Accounting for measurement error in a PSID sample of approximately 2,000 household heads for the survey years 1983 and 1987 increases the estimated return to general labor market experience, reduces the negative effect of blue-collar status, reduces the return to union status, and may affect the returns to tenure and other variables, in both a cross-section and longitudinal setting.

Since the PSID is a large panel study that has been running continuously since 1968, it was infeasible to carry out the validation study on a subsample of the PSID respondents. Therefore, the PSIDVS consists of approximately 400 employees surveyed from a large Detroit area manufacturing firm. An initial set of 534 interviews was attempted in 1983, of which 418 were completed. Reinterviews were successfully conducted with 341 individuals in 1987, of whom 275 were respondents in both 1983 and 1987. An additional sample of 151 hourly workers was interviewed in 1987. The resulting data set matches standard PSID survey responses with company personnel records on a variety of employment variables, including earnings, fringe benefits, hours, unemployment spells, and employment tenure. The company records are highly accurate and were treated as error-free.

The basic modeling assumption is that the true value of the logarithm of annual labor earnings, y^*, and the reported value, y, follow a bivariate normal distribution conditional on exogenous variables X and Z. Because y^* is unobserved in the main sample but is observed in the validation sample, while y is available in both the main and validation samples, it is convenient to write the model as:

$$\boldsymbol{y}^* = \boldsymbol{X\beta} + \varepsilon,\ \varepsilon \sim N(\boldsymbol{0},\ \sigma_\varepsilon^2 \boldsymbol{I}) \tag{14a}$$

$$\boldsymbol{y}^* = \boldsymbol{y}\gamma_0 + \boldsymbol{Z}\gamma_1 + \eta,\ \eta \sim N(\boldsymbol{0},\ \sigma_\eta^2 \boldsymbol{I}), \tag{14b}$$

where $\boldsymbol{X}$ has N rows and K columns and $\boldsymbol{Z}$ also has N rows. The first equation (14a) represents the conditional distribution of $\boldsymbol{y}^*$ given $\boldsymbol{X}$, which is the relationship we want to estimate. The second equation represents the conditional distribution of $\boldsymbol{y}^*$ given $\boldsymbol{y}$ and $\boldsymbol{Z}$, and this is used to impute values for true earnings ($\boldsymbol{y}^*$) in the main sample.

Rubin (1987a) shows that the validation study data must be pooled with the main study data to estimate equation (14a), since otherwise the completed data estimators are not conditioned on all observed data. Similarly, $\boldsymbol{Z}$ must contain all variables in $\boldsymbol{X}$ for equation (14b) to generate proper multiple imputations. To help satisfy these constraints, Brownstone and Valletta used a subsample of the main PSID designed to more closely match the PSIDVS sample, and they also included additional variables in $\boldsymbol{X}$ and $\boldsymbol{Z}$ to facilitate pooling. The PSID subsample includes 2504 male and female household heads for 1983. Ideally, the validation sample is a probability sample from the main study, in which case pooling is not an issue.

The values of γ and σ_η^2 in equation (14b) can be estimated by $\hat{\gamma}$ and s_η^2, the least squares estimates from regressing $\boldsymbol{y}^*$ on $\boldsymbol{y}$ and $\boldsymbol{Z}$ using the validation study. Therefore, $\hat{\gamma}$ follows a $N(\gamma,\ \sigma_\eta^2 \Delta^{-1})$ distribution (where $\Delta = [\boldsymbol{y}\ \boldsymbol{Z}]'[\boldsymbol{y}\ \boldsymbol{Z}]$) and $d s_\eta^2/\sigma_\eta^2$ follows an independent Chi-squared distribution with degrees of freedom d equal to the number of observations in the validation data minus the number of columns in $[\boldsymbol{y}\ \boldsymbol{Z}]$. Brownstone and Valletta use the following steps to create one set of valid imputations for the model in equations (14) according to equation (13):

1) set $\sigma_\eta^{2\,*} = s_\eta^2 \chi^* / d$, where χ^* is drawn from a χ_d^2 distribution,

2) γ^* from a $N(\hat{\gamma}, \sigma_\eta^{2*}\Delta^{-1})$ distribution,

3) draw η^* independently from a $N(0, \sigma_\eta^{2*})$ distribution, and (15)

4) set $\hat{y}^* = y\gamma^*_0 + Z\gamma^*_1 + \eta^*$.

Table 18.1 reproduces a subset of Brownstone and Valletta's results. The first column gives the uncorrected results from regressing the reported values on the X variables. In the corrected regressions reported in the second column, the log of reported earnings for each observation in the main PSID sample is replaced by multiple imputed values of the log of true earnings, which are obtained through repeated application of algorithm (15) above. The results in Table 18.1 show substantial changes from correcting for measurement error. Error correction increases the coefficient on potential experience and the absolute value of the coefficient on its square by about 33%. The negative effect of blue-collar status on earnings is almost halved due to error correction. The return to union status in 1983 is substantially reduced and becomes insignificant, the coefficients on tenure and tenure squared are reduced by 20% and 33%, and the negative effects of being black are substantially reduced. Finally, although the magnitude and statistical significance for most coefficients is reduced by error correction, this shrinkage is not uniform: the coefficients on blue-collar and black status remain relatively constant across the corrected and uncorrected specifications.

Table 18.1
Partial Earnings Models Using Pooled 1983 PSID and PSIDVS Sample
(coefficients; standard errors in parentheses)

	Uncorrected		Corrected	
Schooling (in years, topcoded at 17)	.0646*	(.0067)	.0584*	(.0108)
Potential Experience (=age-schooling- 6)	.0172*	(.0047)	.0230*	(.0061)
(Potential Experience)^2/100	-.0320*	(.0092)	-.0413*	(.0121)
Company tenure	.0379*	(.0043)	.0300*	(.0071)
(Company tenure)^2/100	-.0666*	(.0129)	-.0452*	(.0167)
Black	-.114*	(.052)	-.0707	(.0496)
Blue-collar	-.166*	(.032)	-.0975*	(.0366)
Union coverage	.184*	(.029)	.122	(.090)
Number of Observations	2848		2848	

* indicates significance at 5% level, two-tailed test

Brownstone and Valletta also estimate a longitudinal fixed-effects wage equation using the 1983 and 1987 matched observations in the PSIDVS to correct for measurement error. As in the 1983 cross-section, error correction increases the size and significance of the coefficient on potential experience but decreases the negative effects of blue-collar and union status, including their changes. The measurement error process is highly autocorrelated, so the differencing required to remove the fixed effects does not significantly increase the relative measurement error variance from the cross-section case.

In addition to Brownstone and Valletta's multiple imputation methodology, there are a number of alternative approaches for estimating the model in equations (14). The simplest of these is complete case analysis, which is to estimate the earnings equation using the validation sample only. The simplicity gains of this alternative approach must be weighed against two main benefits of the multiple imputation approach: (1) combining information from both samples increases estimation efficiency; (2) the ability to account for possible differences across the two samples in the parameters of the earnings equation. In Brownstone and Valletta's longitudinal equation, the standard errors of the multiple imputation parameter estimates are approximately 25%-50% lower than estimators just using the validation sample. Since validation samples are typically expensive to collect, their sample sizes are typically an order of magnitude smaller than the surveys they are validating.

Another approach which yields asymptotically efficient estimates is due to Lee and Sepanski (1995). They propose (but do not implement) an instrumental variables method of moments estimators to combine validated and unvalidated data. Their techniques require more programming than multiple imputation and cannot easily be generalized to more complicated models with discrete and limited variables.

Discrete Choice Models

Although there are a number of feasible methods for estimating the linear models discussed in the previous subsection, these more traditional methods become very difficult to implement with discrete choice models. Multiple imputation, on the other hand, can easily be used in these cases, as shown by Clogg et al. (1991). They consider the problem of changing industry and occupation codes between the 1970 and 1980 U.S. Census. These coding changes were so dramatic that it is impossible even to compare major occupation groups across the coding schemes. As a result, it would be very difficult to use Census data to track changes in occupations across the 1970s. Of course, similar problems occur for transportation researchers when the census tract (or traffic analysis zone) boundaries are changed for each decennial census.

To alleviate problems caused by the changed coding system, the U.S. Census bureau randomly sampled about 125,000 individuals from the 1970 Census with known 1970 codes and recoded them using the new 1980 coding system. Clogg et al. (1991) used this double-coded sample to create five multiply imputed 1980 occupation codes for each respondent in the 1970 Census Public Use Sample. Researchers studying occupational change could then estimate five models using these multiply imputed 1980 codes and then use the multiple imputation combining rules in equations (3)-(8) to carry out their analyses. If the five multiply imputed codes are all different for most observations, then it will obviously be very difficult to estimate any reasonable change models. Conversely, if the five multiply imputed codes are identical, then the coding system change will not affect the analyses.

The hard part is modeling the variability in the double-coded sample so that proper multiple imputations can be produced. Clogg et al. (1991) use separate Bayesian multinomial logit models for each of the approximately 500 three-digit 1970 occupational codes. The dependent variables in these models are the 1980 codes assigned to each respondent in the double-coded sample with the same 1970 occupational code. The independent variables include sex, race, age, education, employment status, and region. Due to the small sample sizes and skewness of the dependent variables for many 1970 occupational codes, maximum likelihood estimation of these models was infeasible. Instead Clogg et al. (1991) use an empirical Bayes procedure that shrinks the coefficients of the covariates towards a model with just alternative-specific constants.

Since they use a full Bayesian model, Clogg et al. (1991) sample from the posterior predictive distribution to draw the multiple imputations for the 800,000 Public Use Sample respondents. Recent advances in Bayesian computations have made direct sampling from these posterior distributions feasible for very complex models. Clogg et al. (1991) use Rubin's sampling-importance resampling algorithm (Rubin, 1987b). Clogg et al. (1991) report extensive Monte Carlo and other evaluations of this methodology which show that the imputation method and the resulting multiple imputation inferences are valid for this application.

The application highlights many of the strengths of the multiple imputation methodology. Even if end-users had access to the double-coded sample, it is very unlikely that they would have the statistical and computational resources to carry out the modeling and imputations performed by Clogg et al. (1991). The multiple imputation combining equations (3 - 6) only require the end-users to perform some simple matrix computations to take advantage of all of this modeling and data collection effort. Also, the presence of five different 1980 occupation codes makes it very difficult for even the most casual end-users to ignore the errors in the recoding process. The current practice of filling in missing data in public use files by a single imputation makes it easy for end-users to ignore any potential problems with the missing data process. In addition, standard end-use analysis of singly-imputed data always underestimates parameter standard errors.

Brownstone and Golob (1992) used multiple imputation to compensate for missing data in a key variable explaining commute mode choice in Southern California. The main commute modes in this region are drive alone (about 80%) and carpool. The number of employees at the respondents work site is an important mode-choice predictor. It is easier to find suitable carpool partners from a larger set of employees, so larger worksites should have more carpoolers. Unfortunately, Brownstone and Golob's unusual sampling design caused employer size information to be missing for 30% of the sample, and the remainder of the sample were employed in a small geographic area near the city of Irvine. This design makes complete case analysis using only the respondents with employer-size information very inefficient.

Since Brownstone and Golob's mode choice model used only a dummy variable for whether the employer had more than 200 employees at the respondent's worksite, they used a binomial probit model calibrated on the subsample where employer size is observed. The dependent variable for this imputation model is equal to one if the respondent's employer size is greater than 200, and zero otherwise. Important explanatory variables include respondent age, carpool status, household size and income, commuting distance, and the presence of various employer-provided carpooling incentives. These covariates improved the prediction success rate of the model from 56% to 78%. Brownstone and Golob's procedure assumes that the non-response process is ignorable, which is justified since the non-response is due to sample design features.

Drawing proper multiple imputations from this probit model requires accounting for two sources of errors: the error in the parameter estimates and the error in predicting the actual value of the dummy variable from the estimated probit probability. Brownstone and Golob created their imputations by first sampling from the asymptotic multivariate normal approximation to the sampling distribution of the maximum likelihood probit estimators. Conditional on a single draw from this distribution, they computed the probit probabilities for each respondent with missing data. The imputed values were then drawn from a binomial distribution with probability of success given by the probit probabilities. The next set of multiple imputations then began with a new independent draw from the multivariate normal sampling distribution.

The specification of Brownstone and Golob's imputation equation might be criticized by those unfamiliar with multiple imputation methodology since the dependent variable, mode choice, is being used to help impute an "exogenous" variable in the main model. In fact, the imputation model must condition on all observed data, including the "endogenous" ones, for the multiple imputation method to yield consistent estimates. Inference for the main model is conditioned on the true values of the missing data, so the completed data estimators need to be consistent for this conditional model. The "simultaneous equations bias" occurs when inference is not conditioned on the true values of the missing data.

Panel Data

Panel studies, where a sample is followed over time with repeated interviews, are becoming increasingly important in transportation demand analysis (Golob et al., 1997). Even if the missing data process is completely random, the compounding of a 20% nonresponse rate (which is common for income measures) over only three interviews can leave very few respondents with complete data. In addition, small amounts of completely random measurement error can cause serious problems when using common panel data models which are based on differences between panel waves.

These problems are balanced by the fact that since transportation decisions and most human behavior exhibit inertia, observed data for the panel respondents provide excellent information about the missing data. For example, attempts to impute missing income data in cross-section studies using either standard methods or multiple imputation usually do not produce substantial efficiency gains over complete case analysis. The reason for this is that models for predicting income in cross-sections typically have very poor fit, with R^2 measures below .2. However, if previous or future income values are available from other waves of a panel study, then the fit of the imputation models greatly improves. In a study using the Swedish HUS panel, Brownstone's (1990) models for imputing missing income values had R^2s around .7. The key to this improved fit is that previous and/or future income values for the same household are typically available in panel studies. This suggests that multiple imputation of missing values may yield substantial efficiency gains in panel studies.

The same inertia governing observed human behavior might also characterize measurement error processes in panel data. Brownstone and Valletta (1996) found that measurement error in annual earnings reports were highly autocorrelated. They found that this autocorrelation was high enough so that fixed-effects models were no more affected by measurement error than cross-section models. Of course, the methods used by Brownstone and Valletta crucially depend on the existence of a validation study, and these studies are currently very rare.

Multiple imputation techniques are well-suited to panel data since they require very little additional computation and programming beyond that required for a complete data analysis. Transportation models using panel data, such as dynamic models of mode choice and dynamic vehicle type choice models, are based on complex dynamic discrete econometric models. Even without considering missing data and attrition problems, these models are very difficult to estimate - frequently requiring simulation estimators (McFadden and Ruud, 1994). Although joint full-information maximum likelihood estimation of linear models with missing data is feasible (see Fuller, 1987), these methods are not computationally feasible with many panel data applications.

PANEL ATTRITION AND CHOICE-BASED SAMPLING

Panel studies are often plagued by the attrition of survey respondents. Attrition can bias the sample and limit the usefulness of the panel for long-term dynamic analysis. If the attrition process is correlated with the endogenous variables in the model (non-ignorable attrition), then standard estimation techniques ignoring attrition will yield inconsistent inferences and estimators. Even if the attrition process is independent of the endogenous variables (ignorable attrition), uncorrected attrition may bias forecasts and policy simulations based on the remaining sample. Both of these problems occur in transportation panels. If survey questions are concentrated on the adoption of a new mode or technology, then users of this new mode or technology may be less likely to attrite and the attrition process will be correlated with the endogenous choice variable. Since the main purpose of many analyses of transportation panels is to produce forecasts of the effects of proposed policy changes, it is important to account for the effects of attrition on model forecasts and policy simulations.

This section, (see also Brownstone and Chu, 1997), describes a multiple imputation methodology for obtaining consistent estimates and forecasts from panel models where non-ignorable attrition is present. If the non-attrition probabilities are known, then their inverses can be used as weights in Manski and Lerman's (1977) Weighted Exogenous Maximum Likelihood Estimator (WESMLE). These weights make the weighted sample look like the initial panel wave.

Manski and Lerman (1977) show that a simple modification of the standard Maximum Likelihood estimator for discrete-choice models yields consistent and asymptotically normal parameter estimates in the presence of choice-based sampling when the proportion of the population choosing each discrete alternative is known. If $L_i(\theta, \boldsymbol{x}_i)$ is the log likelihood function for the ith observation, then Manski and Lerman's WESMLE maximizes:

$$\sum_i \omega_i L_i(\theta, \boldsymbol{x}_i) \qquad (16)$$

where θ is a vector of parameters to be estimated, x_i is the vector of observed characteristics for the ith observation, and the sampling weight, ω_i, is the inverse of the probability that the ith observation (individual) would be chosen from a completely random sample of the population. Of course, if the sampling scheme were completely random, then all of the sampling weights would be equal and the WESMLE would simply be the usual maximum likelihood estimator. Manski and Lerman (1977) show that the WESMLE is consistent and asymptotically normal, but not fully efficient (see Imbens, 1992, for fully efficient alternative estimators). Manski and Lerman's proof actually shows that the WESMLE's properties hold for any regular maximum likelihood estimator as long as the sampling weights are known with certainty.

A major advantage of the WESMLE is that it can be computed very easily by modifying existing maximum likelihood programs. The WESMLE for both the linear regression model and the multinomial logit model can be computed by appropriately weighting the variables and applying standard maximum likelihood programs. Unfortunately, this procedure yields downward biased standard error estimates, but the consistent estimates given by Manski and Lerman are easy to compute (see DuMouchel and Duncan, 1983, for a similar analysis of the linear model). This downward bias can be substantial in common applications. Brownstone and Valletta (1996) found a 30% downward bias in their weighted regressions from using the incorrect weighted regression covariance estimator.

A panel survey can always be viewed as the result of the original sampling process and the attrition process. Although in a well-designed panel study the properties of the sampling process are known with certainty, the properties of the attrition process are typically unknown. If they *were* known, then the sampling weights could be easily computed as the inverse of the product of the sampling and attrition probabilities and the WESMLE could be applied to get consistent parameter estimates. Fortunately, there is at least one wave of information about panel attriters, and with some modeling assumptions this information can be used to estimate a model of the attrition process. Unfortunately, the resulting predicted attrition probabilities cannot be used to generate weights for the WESMLE, since this would violate the assumption that the weights are known with certainty.

Suppose we have a procedure for making independent simulated draws from the sampling distribution of the attrition probabilities (which are given from our estimated attrition model). Conditional on this set of simulated attrition probabilities, we can compute a vector of sampling weights (as the inverse of the product of the attrition probabilities and the sampling probabilities for the first wave of the panel). This weight vector can in turn be used to get a consistent (conditional on that particular set of weights) estimate of θ and its covariance using the WESMLE. After drawing a number of independent attrition probabilities, equations (3)-(6) can be used to combine the resulting WESMLE estimators for final inference. This procedure appears to have been first proposed in Brownstone (1991) but it is a simple modification of Rubin (1986).

If the attrition model is correctly specified, then the resulting multiple imputation estimators, $\hat{\theta}$ and $\hat{\Sigma}$, are consistent whether the attrition process is ignorable or not. The standard unweighted maximum likelihood estimators, $\overline{\theta}$ and $\overline{\Sigma}$, which ignore the sampling and attrition weights, are efficient if both the sampling and attrition processes are ignorable, but inconsistent otherwise. Therefore the statistic:

$$T = \left(\hat{\theta} - \overline{\theta}\right)' \left(\hat{\Sigma} - \overline{\Sigma}\right)^{-1} \left(\hat{\theta} - \overline{\theta}\right) \tag{17}$$

is a valid Hausman (1978) test statistic for the null hypothesis that both the sampling and attrition processes are ignorable. Under the null hypothesis, T has a chi-squared distribution with degrees of freedom equal to the rank of $\left(\hat{\Sigma} - \overline{\Sigma}\right)$.

Relative to joint maximum likelihood estimation of the attrition and choice model, the methodology described above is inefficient. However, this methodology is much easier to calculate than joint maximum likelihood, which is frequently intractable in complex models. Simple Hausman (1978) tests can be applied to test for the non-ignorability of the attrition (or missing data) process. Since the WESMLE was originally designed to provide consistent estimates with choice/response-based sampling designs, the methodology proposed here can be trivially modified to yield consistent estimates and forecasts for choice-based panels with non-ignorable attrition.

Brownstone and Chu (1997) applied the above multiple imputation methodology using a dynamic commute mode choice model calibrated from the University of California Transportation Center's Southern California Transportation Panel. The study region and survey methodology are more fully described in Uhlaner and Kim (1993). The panel was selected from respondents to a mail survey, and was initiated in February 1990. The first wave of data were drawn from the original sample and from a refreshment sample introduced three months later. The overall response rate for the first-wave mail survey was approximately 50%. The total sample size for the first wave was 2,189 commuters (approximately 1,850 had complete data). Almost all respondents were employed full-time. The fifth wave of the panel was collected beginning in July 1991. The attrition rate (from Wave 1) was 40%, leaving 1,107 respondents whose data were suitable for dynamic analysis.

Table 18.2
Binomial Logit Attrition Model

Dependent Variable	Count	Percent
In Both Waves	1107	59.97
Attrited (dropped out before Wave 5)	739	40.03
Independent Variables[a]	Estimated Coefficient	*t*-statistic
Annual household income<=$75,000	-0.20233	-1.81388
High school graduate	-0.90640	-2.08486
Some college, but no degree	-1.03234	-2.48198
College degree, including graduate	-0.96502	-2.30214
Older than 24 and younger than 35	-0.40301	-1.95426
Older than 34 and younger than 45	-0.31492	-1.52823
Older than 44 and younger than 55	-0.46445	-2.08844
Older than 54 and younger than 65	-0.47694	-1.80652
Production/manufacturing	0.85561	3.86404
Sales	0.61101	2.83996
Other occupation	0.60538	2.30767
Survey received at work site	-0.25986	-2.37420
Always lived in Southern Ca.	0.29458	2.77893
Considered moving next year	0.29522	2.52530
Non-white	0.47706	3.42080
Arrived at work between 7:00 and 9:00	-0.13672	-1.17095
Years lived at present address (years)	-0.01998	-2.20902
Reserved parking for rideshare	0.30232	2.54872
Household owned vehicles<=3	0.33727	2.00087
Home owner	-0.17925	-1.50927
Always rideshare in last two weeks	-0.69192	-1.71807
Always rideshare and household income<=$75,000	0.57605	1.44266
Always rideshare and moving next year	0.75718	1.80477
Always rideshare and having kids under 16	0.43372	1.20571
Sometime rideshare in last two weeks	1.12516	2.52191
Sometime rideshare and college degree	-0.73210	-2.57659
Sometime rideshare and age>24and<35	-0.74236	-2.53687
Sometime rideshare and household vehicles<=3	-0.73703	-1.74108
Sometime rideshare and having kids under 16	0.49162	1.83682
Constant	0.65782	1.35897
Auxiliary statistics	At Convergence	Initial
Log likelihood	-1164	-1279.5
Number of observations	1846	
Percent correctly predicted	64.6	

[a]All variables defined as dummies except for years lived at present address.

Since Brownstone and Golob (1992) found that the Wave 1 nonresponse process is ignorable, Brownstone and Chu just modeled the attrition process. Table 18.2 reproduces Brownstone and Chu's binomial logit attrition model results for attrition between Waves 1 and 5 of the panel. The model is specified so that positive coefficients favor attrition. Since at least some of the coefficients on the mode choice variables and their interactions are significantly different from zero, the attrition process is not ignorable. The many interactions between mode choice and the demographic variables show the complexity of the process. These results imply that white, middle-aged homeowners with an annual household income of less than $75,000, more education, and more than three vehicles are less likely to attrite from the panel. Those respondents who receive the survey at their work sites (and presumably fill it out during their normal working hours) are also less likely to attrite.

The significant coefficients on the mode choice variables suggest that non-ignorable attrition is a problem for this application. However, the Hausman test given in equation (16) does not reject the null hypothesis that the attrition process is ignorable for Brownstone and Chu's dynamic mode choice model. Their model is quite large, so it is easier to examine the effects of correcting for non-ignorable attrition in a policy experiment simulated from their model. Table 18.3 shows the results of giving all commuters in the sample an employer-paid guaranteed ride home in emergencies. As expected, there is an increase in the number of commuters remaining or switching to ridesharing and a corresponding decrease in drive-alone commuting. However, these results are not significantly affected by correcting for non-ignorable attrition.

Table 18.3
Estimated Effects from Giving Everyone Access to a Guaranteed Ride Home

	Ignoring Attrition		Multiply Imputed WESMLE	
	% Change	Std. Error	% Change	Std. Error
RS[a] → RS	21.43712	8.08602	18.21683	8.79526
DA[b] → RS	68.47014	12.81228	70.49838	14.19468
RS → DA	-37.61644	6.00235	-44.80742	6.53906
DA → DA	-16.15972	5.80547	-14.19958	6.56034

[a] "RS" means 'Rideshare at least once in last 2 weeks'.
[b] "DA" means 'Always drive alone'.

Non-ignorable attrition did not turn out to cause serious biases in Brownstone and Chu's application, but there is no reason to believe that this will be true in other applications. The multiply imputed WESMLE estimator described in this section provides a simple way of testing and correcting for biases caused by non-ignorable attrition.

NON-RANDOM SURVEY RESPONSE AND SYNTHETIC SAMPLING

Transportation researchers use surveys for two purposes: calibrating behavioral models and to provide a representative sample for microsimulation forecasts. Modern transportation surveys place heavy burdens on respondents. They frequently require multiple telephone contacts as well as detailed travel and activity diaries. Unless the survey response mechanism is completely random, the resulting sample will not be representative. If the response mechanism is independent of the endogenous variables, then the response process is ignorable and behavioral models can be efficiently estimated by standard unweighted maximum likelihood techniques. However, even if the response process is ignorable for estimation purposes, the sample will probably still need to be "adjusted" to make it representative for forecasting purposes.

This section briefly reviews current methods for reweighting survey samples for forecasting and estimation. All of these methods ignore any estimation error in creating these weights, and I will indicate how this can be remedied using multiple imputation techniques similar to those discussed in last section. A more important problem is that it is impossible to make any progress on these problems without external data and/or untestable assumptions (see second section). Fortunately, in many cases external data from censuses and administrative records such as vehicle registration files and tax records are useful for the methods described here. No method can test or correct for non-ignorable response caused by variables that are not measured in external data.

There are frequently good external data for commute mode choice available from transit meter counts and highway traffic counters. These data have been used by transportation demand analysts to implement choice-based sampling where the survey sample is stratified by the mode choice variable. Choice-based sampling is clearly a non-ignorable sampling scheme for calibrating mode-choice models. However, choice-based samples are much cheaper to collect when some modes have very small shares. Manski and Lerman (1977) show that the WESMLE yields consistent estimators when the sample strata are weighted to match the known mode shares.

When external data are available, some procedure is needed for estimating the sampling weight for each sample respondent. This sampling weight model can then be used to multiply impute sampling weights, and the WESMLE can be used as in last section to get consistent inferences. The Hasuman test in equation (17) can also be used to test the null hypothesis that the response process is ignorable. Of course, there is no point implementing any of this unless the external data contain information on endogenous variables for the models under consideration.

There are a variety of statistical matching (or poststratification) methods available for estimating sampling weights to match an external reference sample. The simplest is to discretize all variables common to both the survey and external datasets and use the discrete values to define "bins." Respondents assigned to the same bin are assumed to be identical, and the population in the bin is estimated by the sum of the weights of all of the external sample respondents assigned to the bin. The estimated weight for the survey sample is estimated by the ratio of this estimated population and the number of survey respondents assigned to the particular bin. Brownstone and Golob (1992) used this method to match their commute-choice sample to the U.S. Census Bureau's Current Population Survey (CPS). Although the CPS contains excellent demographic and labor market data, it has no information on vehicle holdings or travel time. Therefore, Brownstone and Golob were only able to test for non-ignorability to the extent that the response process was correlated with demographic and labor data.

As long as the response process only depends on exogenous variables which are included in the behavioral model, standard unweighted maximum likelihood techniques will yield efficient estimators and hypothesis tests. Of course, any stratification variables included in the sample design should normally be included in the model to control for design effects. Even if the sample does not need to be weighted for estimation purposes, consistent forecasts require weights unless the response and sampling processes are totally random (independent of all endogenous and exogenous variables). Reweighting is also needed when forecasts are needed for populations (or geographic regions) which are different from the sampled populations. For example, behavioral models may be estimated based on a regional or national sample, but separate forecasts are required for smaller urban regions.

As long as the survey sample has at least one respondent in each bin, the simple matching procedure used in Brownstone and Golob (1992) will produce weights which make the weighted survey sample exactly match the joint distribution of the discrete match variables from the external sample. The main source of error in this procedure is the estimation error in the target joint distribution. Using a standard multinomial model for the joint distribution, it is straightforward to multiply impute estimates, and the imputed distributions can be used to generate multiply imputed weight vectors for the survey sample. In many cases the external sample is very large, in which case the estimation error in the weights is likely to be small enough to ignore.

The simple bin matching procedure has a number of drawbacks in some applications. If there are many matching variables and/or many discrete values, then the size of the joint distribution can become very large and some of the estimated cells (or bins) will be empty. These empty cells cause no problems if the true population count for the cell is zero, but this is rarely the case in practice. Rubin (1986) proposes a predictive mean matching algorithm which is also used by Heitjan and Little (1991). This uses multivariate regression models to predict match variables for both the main and external sample. These predicted variables are then combined to a univariate distance measure, and matches are randomly chosen from the closest respondents using this metric. Additional matching and weighting methods are discussed in Little (1988) and Imbens and Hellerstein (1995).

If the target population is a small geographic area such as a traffic analysis zone or census tract, then data privacy restrictions preclude direct access to census observations. This problem arises in generating synthetic samples for microsimulation models of detailed travel behavior (Kitamura, Chapter 12 in this volume). In this situation the external data typically consist of univariate (occasionally bivariate) marginal distributions of census variables for the small region as well as a public use sample for a larger region with a population of at least a few million and containing the smaller region. Beckman, Baggerly, and McKay (1996) give a procedure to use iterative proportional fitting to estimate the joint distributions for the small regions given this external data. If their method is applied to all subregions comprising the larger region, then the estimated joint distributions for the subregions will aggregate up to exactly match the joint distribution for the larger region.

Little and Wu (1991) give the sampling distribution of the iterative proportional fitting estimators used by Beckman et al. (1996), and Gange (1995) shows how to draw values from these sampling distributions. These results can be used to implement a multiple imputation method to account for the estimation errors in these matching processes by drawing multiple estimated joint distributions. Little (1988) shows how other matching algorithms can be modified to yield approximately proper multiple weights.

While the methods discussed in this section allow researchers to make the most use from expensive transportation survey data, they are no substitute for obtaining a well-designed survey sample to begin with. For example, if the sample only contains a small number (relative to the population proportion) of respondents corresponding to a particular minority group, then the weighting schemes discussed here will yield very high weight values for these minority respondents. This means that weighted estimates and forecasts will depend very heavily on these few respondents, especially if the results are stratified by minority group status. Researchers should therefore always examine the weights resulting from these reweighting procedures. Having a few sample members in a particular "post-strata" is only slightly better than having no observations. In either case, the only practical solution is to further aggregate the match variables.

CONCLUSION

This chapter has shown how multiple imputation methods can be used to help alleviate problems caused by survey non-response and missing data. Multiple imputation is like an adjustable wrench - it is rarely the ideal tool for any particular job but works well for a wide variety of problems. The examples given in this chapter show that multiple imputation can be successfully implemented for real applied problems using existing software packages. Furthermore, Brownstone and Valletta's (1996) application shows that using this methodology can make a substantial difference in the qualitative conclusions.

Manski's (1995) work reviewed in the second section shows that missing data causes serious problems with identification and inference from even simple models. The best way to circumvent these problems is to put more resources into reducing response biases during survey administration. The next best solution is to collect external validation data which allow identification of the non-response process. If these validation data become more widely available, then the multiple imputation methods presented in this chapter provide an easy and consistent way for transportation researchers to incorporate this information into their modeling and forecasting efforts.

ACKNOWLEDGEMENTS

This revision benefited from comments from the editors, Chandra Bhat, Ami Glazer, and especially Kenneth Small. The research described in this chapter was partially supported by the University of California Transportation Center.

REFERENCES

Beckman, R. J., A. K. Baggerly and M. D. McKay (1996). Creating synthetic baseline populations. *Transportation Research A*, **30**, 415-429.

Bound, J., C. Brown, G. J. Duncan and W. L. Rodgers (1990). Measurement error in cross-sectional and longitudinal labor market surveys: Validation study evidence. In: *Panel Data and Labor Market Studies* (J. Hartog, G. Ridder, and J. Theeuwes, eds.), pp. 1-19. North Holland, Amsterdam.

Bound, J., C. Brown, G. J. Duncan, and W. L. Rodgers (1994). Evidence on the validity of cross-sectional and longitudinal labor market data. *Journal of Labor Economics*, **12**, 345-368.

Bound, J. and A. B. Krueger (1991). The extent of measurement error in longitudinal earnings data: Do two wrongs make a right? *Journal of Labor Economics,* **9**, 1-24.

Brownstone, D. (1990). *Multiple imputations for panel studies*. Unpublished manuscript, Department of Economics, University of California, Irvine, CA.

Brownstone, D. (1991). *Multiple imputations for linear regression models*. Technical Report MBS 91-37, Research Unit in Mathematical Behavioral Sciences, University of California, Irvine, CA.

Brownstone, D. and X. Chu (1997). Multiply imputed sampling weights for consistent inference with panel attrition. In: *Panels for Transportation Planning: Methods and Applications* (T. Golob, R. Kitamura and L. Long, eds.), pp. 259-276. Kluwer, Boston, MA.

Brownstone, D. and T. F. Golob (1992). The effectiveness of ridesharing incentives: Discrete-choice models of commuting in Southern California. *Regional Science and Urban Economics*, **22**, 5-24.

Brownstone, D. and R. G. Valletta (1996). Modeling earnings measurement error: Amultiple imputation approach. *Review of Economics and Statistics*, **78**, 705-717.

Clogg, C. C., D. B. Rubin, N. Schenker, B. Schultz and L. Weidman, (1991). Multiple imputation of industry and occupation codes in census public-use samples using Bayesian logistic regression. *Journal of the American Statistical Association,* **86**, 68-78.

DuMouchel, W. H. and G. J. Duncan (1983). Using sample survey weights in multiple regression analysis of stratified samples. *Journal of the American Statistical Association,* **78**, 535-543.

Fay, R. E. (1991). A design-based perspective on missing data variance. In: *Proceedings of the 1991 Annual Research Conference.* pp. 429-440. U.S. Bureau of the Census, Washington DC.

Fay, R. E. (1992). When are inferences from multiple imputation valid? In: *Proceedings of the Survey Research Methods Section,* pp. 227-232. American Statistical Association, Alexandria, VA.

Fuller, W. A. (1987): *Measurement Error Models*. Wiley, New York.

Gange, S. J. (1995). Generating multivariate categorical variates using the iterative proportional fitting algorithm. *The American Statistician*, **49**, 134-38.

Glynn, R. J., N. M. Laird and D. B. Rubin (1993). Multiple imputation in mixture models for nonignorable nonresponse with follow-ups. *Journal of the American Statistical Association*, **88**, 984-993.

Golob, T., R. Kitamura and L. Long (eds.) (1997). *Panels for Transportation Planning: Methods and Applications.* Kluwer, Boston, MA.

Golob T. F. and J. Gould (1998). Projecting use of electric vehicles from household vehicle trials: Trial and error? *Transportation Research* **B**. In press.

Hausman, J. A. (1978). Specification tests in econometrics. *Econometrica,* **46**, 1251-1271.

Heckman, J. J. (1976). The common structure of statistical models of truncation, sample selection, and limited dependent variables, and a simple estimator for such models. *Annals of Economics and Social Measurement*, **5**, 475-592.

Heitjan, D. F. and R. J. A. Little (1991). Multiple imputation for the fatal accident reporting system. *Applied Statistics*, **40**, 13-29.

Horowitz, J. L. and C. F. Manski (1995). *Censoring of outcomes and regressors due to survey nonresponse: identification and estimation using weights and imputations.* Unpublished manuscript, Department of Economics, University of Iowa, Iowa City, IA.

Imbens, G. (1992). An efficient method of moments estimator for discrete choice models with choice-based sampling. *Econometrica*, **60**, 1187-1214.

Imbens, G. and J. K. Hellerstein (1995). *Imposing moment restrictions by weighting.* Unpublished manuscript, Department of Economics, Harvard University, Cambridge, MA.

Lave, C. (1996). Are Americans really driving so much more? *Access*, **8**, 14-18.

Leamer, E. E. (1983). Lets take the con out of econometrics. *American Economic Review,* **73**, 31-43.

Lee, L. and J. H. Sepanski (1995). Estimation of linear and nonlinear errors-in-variables models using validation data. *Journal of the American Statistical Association,* **90**, 130-140.

Little, R. J. A. (1988). Missing data adjustments in large surveys (with discussion). *Journal of Business and Economic Statistics*, **6**, 287-297.

Little, R. J. A. and M.-M. Wu (1991). Models for contingency tables with known margins when target and sampled populations differ. *Journal of the American Statistical Association*, **86**, 87-95.

Manski, C. F. (1991). Regression. *Journal of Economic Literature*, **29**, 34-50.

Manski, C. F. (1995). *Identification Problems in the Social Sciences.* Harvard University Press, Cambridge, MA.

Manski, C. F. and S. Lerman (1977). The estimation of choice probabilities from choice-based samples. *Econometrica*, **45**, 1977-1988.

McFadden, D. and P. Ruud (1994). Estimation by simulation. *Review of Economics and Statistics*, **76**, 591-608.

Meng, X. (1994). Multiple imputation with uncongenial sources of input. *Statistical Science,* **9**, 583-574.

Meng, X. and D. B. Rubin (1992). Performing likelihood-ratio tests with multiply-imputed data sets. *Biometrika,* **79**, 103-111.

Rubin, D. B. (1986). Statistical matching using file concatenation with adjusted weights and multiple imputations. *Journal of Business and Economic Statistics*, **4**, 87-94.

Rubin, D. B. (1987a). *Multiple Imputation for Nonresponse in Surveys*. Wiley, New York.

Rubin, D. B. (1987b). Discussion of Tanner and Wong. *Journal of the American Statistical Association,* **82**, 543-546.

Rubin, D. B. (1996). Multiple imputation after 18+ years. *Journal of the American Statistical Association,* **91,** 473-489.

Schenker, N. and A. H. Welsh (1988). Asymptotic results for multiple imputation. *Annals of Statistics*, **16**, 1550-1566.

Uhlaner, C. J. and S. Kim (1993). *Designing and implementing a panel study of commuter behavior: Lessons for future research.* Working Paper 93-2, Institute of Transportation Studies, University of California, Irvine, CA.

19

FORECASTING DEMAND FOR NEW TRAVEL ALTERNATIVES

Andrew Daly, Hague Consulting Group, The Netherlands
Charlene Rohr, Hague Consulting Group, England

INTRODUCTION

Depending on the viewpoint of the agency conducting the work, the evaluation of a proposed transportation infrastructure investment may require assessment of three aspects of its likely impact: *accessibility* or the benefit enjoyed by users of the infrastructure in the conditions of travel or transport, for instance through a saving in time that will be given by the infrastructure, above any charge that is made for its use; *financial performance* or the net income or deficit accruing to the constructors and operators of the infrastructure and *external impacts* or the environmental and other impacts imposed on third parties including noise and other emissions to which may be added a consideration of the safety impact of the infrastructure; further consideration may also be given to the operation of other transport systems, for instance, those competing with or complementing the proposed infrastructure.

In some cases, evaluations may be made on a restricted set of these criteria. However, the assessment of *any* of these aspects requires a forecast to be made of the traffic flows that will use the infrastructure and the changes that can be expected in traffic flows on other transport systems. The procedure used for forecasting these flows, however simple or complicated it may be, can be called a demand model.

The favoured approach in the transport sector is to base demand models on RP data, observations of behaviour that has actually taken place. In the context of major infrastructure, the use of reliable and convincing forecasting procedures is particularly advisable because of the size of the investments involved and the need to convince a substantial group of decision-makers, including, for example, investors. In this context the nature of RP data gives some guarantee of the reliability of the forecasts.

However, in some cases it is not possible to base models on RP data, perhaps because the proposed infrastructure will offer alternatives to travellers that do not currently exist. In these contexts it is necessary to base forecasts on alternative data collection procedures. The transfer of information from other areas can also be considered, but it is often difficult to find similar circumstances for major infrastructures in other parts of the world and travellers' attitudes are of course different in differing contexts.

An approach exploiting local data that has been applied in many recent studies of this type is the use of SP data in which respondents are asked to indicate what their preferences would be if they were faced with hypothetical choice situations. SP data has advantages in that the choice situations presented to respondents can be carefully controlled in a number of ways to maximise the amount of information collected from each respondent. However, it also has disadvantages relative to RP data, particularly that it is not certain whether the results obtained can be applied in a way that can inspire confidence.

For these reasons there has been significant recent interest in the use of RP and SP data together, attempting to use the better aspects of each data type.

COMBINING RP AND SP DATA

The special issue of *Transportation* published in May 1994 gives an overview of the state of practice and the current problems at that time. In particular, the papers in that issue by Hensher, by Swait et al. and by Morikawa discuss issues in the combined use of SP and RP data. Hensher gives a historical overview of the development of SP methodology and "anticipates a burgeoning literature of RP-SP applications." The paper by Swait et al. also gives an extensive summary of the literature in SP analyses, dating from the early 1970's and from the early 1980's focusing on experimental choice rather than other methods of eliciting preferences from respondents. However, it has been only in the 1990's that formal joint analyses of SP and RP data began to be made, following the work of Ben-Akiva and Morikawa (1990).

The advantages of the joint analysis approach, referred to by Hensher (1994) and conveniently summarised by Morikawa (1994), are essentially that the two data types are complementary, that is, the strengths of the one cover the weaknesses of the other. In particular, the credibility and realism of the RP data combines well with the efficiency and flexibility of the SP data. Because of this advantageous combination, a number of applications studies using formal joint analysis have been made, in particular by Ben-Akiva, Morikawa, and their associates, by Hague Consulting Group and by other researchers. The key aspect of these approaches is the *simultaneous* estimation of the key parameters (e.g., the coefficients of the policy variables) from *both* RP and SP data.

A key group of workers in the history of SP analyses have been Louviere and his associates. Their work has been seminal to many of the important advances in this area of consumer choice analysis. The approach proposed in the 1994 paper (Swait et al.) for the analysis of SP and RP data on freight choices is therefore of considerable interest as it differs from the other approaches to combined RP/SP analyses. Swait et al. propose an approach using both SP and RP data and using a two-stage estimation: (1) the main coefficients are estimated from the SP data only, that is, obtaining both the relative and the absolute values of these coefficients from SP data only; and (2) the alternative-specific constants are estimated from the RP data which is simultaneously scaled to match the SP estimation scale; that is, the current market shares only are estimated on the RP data.

The justification of this approach given by Swait et al. is that it exploits the strengths of each of the data types. In particular, they note that in joint estimation with RP and SP data together the alternative-specific constants cannot be matched across the data sets by the same scaling used for the remainder of the coefficients.

In the studies conducted by the present authors, described in this paper, we have accepted some but not all of the arguments put by Swait et al.; of course, like them, we attempt to use the distinct strengths of RP and SP data. However, we have not been willing to accept that the main coefficients of the model should be based only on SP data. The reasons for this are the following. Firstly, in the corridors we have studied there have been large RP surveys - with tens of thousands of observations - already in existence prior to the start of the SP work; the balance of importance between RP and SP clearly shifts when data of this type is available. In the work described in Swait et al. the RP data is collected only from SP-survey respondents and is therefore of relatively limited size. While it has been possible in our work to collect RP responses during the SP surveys, this has not added significantly to the large RP data base available from other sources. It is therefore understandable that we should give more relative weight to RP than do Swait et al. Second, in principle, both the RP and the SP data contain information relating to choice. It makes sense to use all of this data to estimate the forecasting model. Moreover, it is efficient from a statistical point of view to make the estimation considering all of the data simultaneously, rather than using a sequential estimation that does not fully optimise the model. Third, Swait et al. are correct that it is not in general possible to estimate the alternative-specific constants consistently between the RP and SP data; the overall scaling on its own is not adequate to account for the differences that are found in practice. However, these differences can be overcome simply by estimating separate constants for the two data types, rather than rejecting the whole concept of joint estimation. Fourth, in long-distance or international corridors such as those we have studied the journeys made are complex, with many characteristics of the alternatives significant in influencing travellers' choices. In SP surveys of moderate size and complexity (and requiring a moderate amount of the respondents' time) it is not possible to design adequate variance in all of these variables. Thus only a subset of the relevant parameters can be estimated from the SP data and the others must be estimated from RP data only. Fifth, the RP data we have used is typically drawn from surveys that are modally based, inevitably involving an element of choice-based sampling. The surveys are also sometimes out of date. Thus, there is in any case a need to adjust the alternative-specific constants to account for the choice-based sampling and shifts in the market since the surveys were taken. Further adjustments are also required when the model is subsequently re-adjusted (after estimation) to take account of newer data. Finally, the fundamental reason for requiring the use of RP data is that only from this source can the real-world variances in measurement and perception be determined, for instance relative to the data describing those journeys generated by transportation networks which are often used for forecasting changes in journey times and costs. For this reason we require that the ultimate scale of the model be determined by RP data; this is the key difference in our approach from that of Swait et al.

For these reasons we have insisted on a joint estimation using both SP and RP data simultaneously with a scaling factor, also estimated simultaneously, to account for the different error variance in the RP and SP data. This approach is now fairly standard (see references listed by Hensher, 1994). However, because of the contexts and data structures specific to our work, we have found it necessary to extend the standard method in several ways.

The approach we propose incorporates a base estimation step, in which both RP and SP data are used, with a scaling factor applied to the SP data and with alternative-specific constants estimated separately for the two data types (as far as this is possible with the data available). Subsequently, using the RP data alone, alternative-specific constants are estimated to take account of the true market shares of the alternatives in the current market place. Finally, because the main interest of our work has been in new alternatives, which are obviously not present in the RP data, it is necessary to adjust the alternative-specific constants for those alternatives, using the SP data, in the light of the new values of the constants for the existing alternatives. These steps are explained in more detail in the following section of the chapter.

In this work we have used exclusively models of the generalised logit form, introduced into joint RP-SP modelling by Bradley and Daly (1991). The earlier approach of Ben-Akiva and Morikawa (1990), using models of the probit form, is not at present suitable for large-scale modelling under the tight time and cost constraints of practical studies, although advances in computer hardware and software may make probit models more practical in the future.

Experience of estimation of models based on mixed RP and SP data is now fairly extensive and it is reasonable to claim that a reliable standard procedure can be followed. Problems of efficiency remain in a number of areas. While it is possible to take account of the 'repeated measures' problem in the analysis of SP data by the use of advanced analysis procedures, methods for use in practical studies are not yet available. Some progress is possible by the use of 'jack-knife' and similar techniques (see Cirillo et al., 1996) but the consequences of this work for combined RP/SP analysis have not yet been worked out.[31] It can be difficult in practice to specify the most productive set of alternatives and characteristics to present in SP games. This is essentially a balance between experimental design, local realism, and foreseeing the policy requirements accurately. While some assistance can be gained from simulation and piloting, problems remain for work being done in unfamiliar contexts. Software development continues, but is not yet advanced adequately to support all aspects of survey design, survey conduct, data processing, model estimation, and model application. A number of packages are available for each stage of the process.

A further difficulty is that several important studies are made in commercial contexts where publication of the results is prohibited or at best delayed for several years. This feature of the work means that it is difficult to take advantage of the work done by others. Much of our own work is done under such constraints and it is the objective of the current chapter to expose the general methods that we have employed, if not the confidential results from the specific studies.

In contrast to the relatively advanced state of model estimation based on joint RP and SP data sets, model application is less well developed. This is the main focus of the chapter.

MODEL APPLICATION

While considerable progress has been made with the problems of model estimation, as outlined above, less attention has been given to the issues that arise in applying the models.

[31] A further possibility would be to exploit the covariance matrix adjustment suggested by McFadden (Chapter 15 in this volume).

Two particular points of concern may be noted. The appropriate model structure for use in application, and the differences of context between model estimation and application.

Model Structure

An important issue is the way in which the choice situation with new alternatives is represented. Two differing approaches may be considered. (a) The choice among existing alternatives may be modelled without considering the new alternatives; then, diversion to the new alternatives may be modelled for each of the existing alternatives. (b) The new alternatives may simply be entered into the choice structure along with the existing alternatives.

The first approach has a number of advantages. First, it allows the forecasting to relate more closely to an SP exercise in which travellers using a specific existing alternative are interviewed about their willingness to change to a new alternative; in particular, this approach mirrors the common SP presentation in which two alternatives, often the current alternative and the new alternative, are presented to respondents, rather than the full set. Second, it permits a dependence on the chosen alternative to be used to indicate aspects of travellers' preferences. For these reasons, this approach has been used in some important studies, particularly for short-term forecasting.

However, in most of our work we have preferred the second approach, which is to assign a role to the new alternative or alternatives that is not basically different from those assigned to the existing alternatives. The fundamental reason for this is that, in investigating infrastructure investment, the modelling has a focus on the longer term. Over a period of years, the concept of the "current choice" becomes less relevant: habits change; new travellers enter the market and older ones drop out; further new alternatives emerge, other than those directly considered in the SP; and some existing alternatives may drop out under the competition of the new. All of these changes make the use of a diversion model less satisfactory. Further, in long-distance and international markets, many travellers are making their journeys for the first time: The concept of diversion does not apply. For those to whom it does apply, it is possible to take account of inertia, which we believe to be the most important impact of current on future behaviour. When the new alternative is singled out in the way implied by the first approach, the sensitivity of choice between it and the other alternatives cannot be controlled by observations of existing choice. Finally, a model of the first form makes it more difficult to model increases in total travel: which are caused not only by the introduction of new alternatives but also by changes in the existing alternatives, perhaps in competitive reaction.

Unless account is taken of changes in all the alternatives, generation cannot be modelled in a satisfactory way.

For these reasons, we have preferred to work with a model in which all of the alternatives, existing and new, play essentially comparable roles. This approach avoids a number of problems, while the first approach does not offer a corresponding reduction in the problems of implementation.

Application Context

An important issue in setting up a forecasting model is that the context in which forecasts are made differs from the context in which the SP data are collected. The forecasting model must then be set up with these changes in mind.

The most important such changes are the following. (1) In the SP survey it is usual to offer the respondent the restricted choice between his or her currently chosen alternative and a new alternative. However, in forecasting, for the reasons explained above, it is necessary to introduce all of the relevant alternatives. This change is essentially from the binary choice of the SP experiment to the multinomial choice of the forecast situation. (2) Many of the journeys in long-distance and international corridors are complicated, with several changes of mode. In the SP survey it is not always possible to present all of the characteristics of the alternatives explicitly, for example, because of time constraints in the interview situation. However, for forecasting it is necessary to include these characteristics, at least for those that may change. (3) The current alternatives against which the new alternatives are compared in the SP survey have unmeasured characteristics (often, image and quality aspects) that are current at the time of the SP survey. The forecasting model will, however, have as its base situation an RP survey, often of another date, in which the unmeasured characteristics may well have different values.

Because of these differences, it is necessary to adjust the model for the changes between the estimation and forecasting stages. Further adjustment is also necessary because of the choice-based nature of the RP and SP surveys. In the interests of estimation efficiency and simplicity, we prefer to undertake the estimation without weighting or other correction, correcting subsequently to match the base shares of the alternatives.

Thus, for several reasons it is necessary to make corrections to the model between estimation and application. The procedure we have devised for these corrections involves two secondary estimations, subsequent to the main model estimation. This procedure is set out in the following section.

ESTIMATION PROCEDURE

The procedure that has been devised to deal with the need for model adjustments outlined in the previous section involves the following three-stage estimation:

1. The *base estimations* use both SP and RP data to determine values for all of the coefficients of the model. The logit scaling procedure of Bradley and Daly (1991) is used to obtain the appropriate scales between the two data types.[32] Alternative-specific constants for RP and SP data may be distinct (preferable) or equal as required by the data.
2. The *A Runs*, the first adjustment procedure applied to the base model, use the RP data alone to adjust the alternative-specific constants of the existing alternatives to match the base-year observations.
3. The *B Runs*, the second adjustment procedure, use the SP data alone to re-adjust the alternative specific constants of the new alternative or alternatives.

Two further steps may also be necessary to prepare the model for forecasting, depending on the context in which estimation and application are carried out:

4. The model is adjusted of the model to allow for the introduction of multiple alternatives in the forecast situation.
5. Any dependency on the current choice is treated.

Further details of the reasons for these steps and how they are implemented are given below.

Base Estimation

The main objectives of the base estimation procedure are: (a) to estimate the trade-off ratios in the model, for example, the relative values of the coefficients for travel time, cost, etc., based on both RP and SP data; and (b) to estimate the absolute values of the coefficient values from the RP data and a corresponding scale to adjust the SP data.

The basic procedure used is as outlined in Bradley and Daly (1991), where the coefficients are jointly estimated from SP and RP data sources. In this base estimation, variables that occur

---

[32] Further data types, for instance different types of SP surveys, can also be included with further scaling as appropriate.

only in either the RP or SP data are also included in the relevant utility equations. One or more scaling parameters are applied to the SP data to account for the differing variance of the error term in the model which arises from the different contexts of the RP and SP data.

The model estimated is of the hierarchical logit form, with indirect utility functions given for alternative i in data set P (P = RP, SP_k, where k may index several SP data sources) by

$$V_{Pi} = \alpha_{Pi} + \beta X_{Pi} + \gamma Y_{Pi} + \varepsilon_{Pi}$$

where α, β, γ are the unknown parameters to be estimated, X_P are the measures in source P of data items measured in both RP and SP sources, Y_P are the data items measured only in source P, and ε_{Pi} is the error term. The assumption is that ε is distributed independently and identically for each i with mean 0 and variance σ^2_P, with $\sigma^2_{SPk} = \theta^2_k\sigma^2_{RP}$; the parameters θ are estimated simultaneously with the other unknown parameters[33].

The estimation procedure used is explained by Bradley and Daly (1991). It is important to note that generally the alternative-specific constants α should be different between different data sources[34], the point made by Swait et al. (1994) that the scaling cannot be expected to apply to these parameters is thus taken into account in the modelling. In practice, however, for various reasons it is not always possible to identify separate data-specific constants and therefore this requirement may have to be relaxed, recognising that this situation is not ideal.

In our work, the RP and SP data are typically from independent sources which avoids potential correlations between RP and SP responses for the same individual, a problem identified and treated by Morikawa (1994). However, there are almost always several SP responses from each individual in an SP source, and often the various SP sources relate to successive experiments carried out with the same set of individuals[35]. The assumption that the ε variables are distributed independently is therefore incorrect. This problem (known as the repeated measures problem) is a feature of almost all practical work involving SP data. Recent work (Cirillo et al., 1996) suggests that the errors caused by this problem may not be too serious in practice and proposes methods to eliminate any biases. Clearly this undesirable feature of the analysis should be eliminated as soon as reasonably possible.

---

[33] It is possible to introduce further complexity by specifying tree structures within either or both of the RP and SP data sets, relaxing the assumption on the variance of ε. However, this complication is not important for the present discussion.

[34] The assumption of zero mean for ε is therefore quite reasonable.

[35] For example, it is common to carry out both within-mode and between-mode games with each respondent in an SP survey. Such games would then be treated as separate SP sources in the analysis.

The basic model estimation is carried out using disaggregate unweighted data records; this procedure which effectively assigns equal information content to each observation is believed to be the most statistically efficient use of the data, apart from the problem of ε correlation noted. However, since both RP and SP data are frequently collected by procedures that are choice based, it is necessary to adjust the model subsequently. This is the function of one of the secondary estimations, the A Runs, described in the following section.

All the coefficients are estimated relative to an RP basis by the use of the θ scaling parameters. Thus the absolute level of the coefficients is appropriate to their use with data prepared by the mechanisms used in preparing the RP data - usually transportation networks. Since the mechanisms are usually the same as will be used in forecasting, it is reasonable to assume that the errors will be of similar magnitude and that the scale of the RP coefficients is therefore appropriate for use in forecasting.

The procedure of Bradley and Daly (1991) provides an estimation procedure for models of the form given above by the use of a hierarchical logit estimation program. When using this program, it is relatively straightforward to introduce further hierarchical structures within the alternatives in either or both the SP and RP data. Generally, we have tried to use the same tree structure in the SP and RP models, although in some instances this is not possible because different alternatives may be defined; for example, not all RP alternatives may be included in the SP exercise or the SP alternative definitions may be more detailed than is able to be identified in the RP data.

It is clear that no claim is made that the procedure proposed is a proper maximum likelihood estimate of the model. The repeated-measures problem, as well as the treatment of choice-based samples, make any such claim impossible. However, the procedure is believed to make reasonable estimates of the coefficients in a complicated situation. Since the circumstances described are both common and important, further research is highly desirable.

Secondary Estimation Using RP Data (A Runs)

Because the base estimation uses unweighted choice-based data, the alternative-specific constants are not correct in that they do not reproduce the observed alternative shares, even if separate RP and SP constants are estimated. These constants therefore need to be adjusted so that the "base" alternative shares, that is, those of the alternatives present in the base year of the study, are replicated. When the structure of choice among these alternatives is multinomial logit, these adjusted constants are the maximum likelihood estimates of the true constant values, conditional on the estimated values of the other coefficients; this procedure resembles the correction procedure for pure choice-based data (see Lerman and Manski, 1979).

An automated procedure has been developed where the *weighted* RP data is used to re-estimate the alternative specific constants of the existing alternatives while constraining the other coefficients to the values estimated in the model calibration process (estimated either from the RP data only or jointly estimated from the SP/RP data). This automated procedure is called the A Run.

A similar procedure can be used to update the alternative-specific constants for subsequent years when new count information is available, by making the appropriate adjustments to the record weights. Level-of-service changes and changes to other variables, for instance income, may also be required. New modes which may have appeared between the study base year and the forecast years can be introduced into the model structure in a similar manner, by introducing a utility equation for the new alternative with the appropriate level of service and/or socio-economic variables. The relevant coefficients are constrained as in the above procedure and new constants are estimated such that the observed modal shares are reproduced.

In some circumstances it may be preferable not to use an automated calibration procedure but to employ an iterative process of continual adjustment until the model predicts the observed alternative shares. In the simple case this is just a more time-consuming version of the automated procedure. However, in more complex cases it may be easier to use this method. For example, in some cases the numbers choosing the alternatives may not be given directly, or may not be given for each market segment, or further details may be given. To use this data in an automated procedure may require specialised programming and it is sometimes simpler to use continual adjustment. Of course the computer can be used to semi-automate this process.

Secondary Estimation Using SP Data (B Runs)

Here the issue is one of consistency between the SP choice context which provides the basis for the new alternative-specific constants and the RP context to be used for forecasting. In order to transfer the SP results to the RP forecast environment, a number of adjustments are required; these are discussed in turn below.

New-Alternative Constants. One of the primary assumptions in the use of the simultaneous joint SP/RP estimation approach is that the mean value of the unmeasured component of utility, in both the SP and RP data sets, is 0. The presence of separate SP and RP alternative specific constants ensures that this condition is met (Bradley and Daly, 1991). As noted above, there are instances where this condition has to be relaxed for practical purposes, that is, when separate SP and RP alternative-specific constants cannot be identified; this relaxation, however, is not desirable and should be avoided if possible.

Assuming that separate alternative-specific constants can be identified for the RP and SP models, there are still, however, inconsistencies in the definitions of the constants. In the RP models, the alternative-specific constants have been estimated on the basis that all alternatives are available for an individual (in the absence of further information such as car availability). The SP constants, however, are typically estimated on the basis of binary choices, where respondents have made choices between their observed alternative and the new alternative. Each observed alternative SP constant is estimated only by respondents who are using that mode for their current journey and the new alternative-specific constants are therefore measured relative to these.

Thus, while it is desirable to estimate the trade-off coefficients using a model structure with separate sets of constants for the different data types, when it is required to use the resulting constants in a RP context it is essential to ensure that all travellers' behaviour is consistent with the observed alternative-specific constants. This can only be done by maintaining the same constants in the RP and SP observations. Therefore, in the secondary estimation phase of the SP data (the B Runs), it is necessary to constrain the values of the alternative-specific constants for observed modes to be equal to the RP estimates as obtained in the RP secondary estimation phase described above (A Runs); strictly, these should relate to the same time (e.g., year) as the SP data. The resulting new alternative-specific constants will then be consistent for forecasting.

Variables Present in Only one Data Type. While it is important to include coefficients which can only be estimated from the SP data in the main estimation phase in order to estimate the trade-off parameters, if this information is not available in the forecasting data then these variables must be deleted in the secondary SP estimation so that the SP and forecasting contexts are consistent.

Conversely, there may be cases where a specific attribute may not be included in the SP choice exercise, because of limits of what respondents can be expected to comprehend in the SP exercise, etc., but which are to be included in the forecast model structure. The issue is how these variables affect the values of the new alternative constants and therefore how they should be treated in the secondary estimation.

Simply excluding variables such as these from the secondary estimation phase and then including them in forecasting implies that these variables affect each of the observed alternatives and the new alternatives equally (since it is only the new alternative constants which are being estimated). We would recommend instead that these variables be added to the secondary estimation SP-utilities. Although this information was not included in the SP exercise, it implies that respondents were aware at some level of their effect and considered their effect when making their appropriate choice response. Obviously, neither approach is ideal and optimally one would include all the variables which are to be included in the forecasts in the SP structure (even if these are presented for information only).

Any change of this kind in the utility specification for the alternatives will of course affect the choice probabilities. To get back to the level indicated by the SP survey it is necessary to recalibrate the new-alternative constants. Fortunately, the correction for these changes can be introduced in the B Runs at the same time as the adjustments to take account of the changes in existing-alternative constants are performed.

Multiple Alternatives in Forecasting

The SP choice structure is between the respondent's current alternative and the new alternative. To maximise consistency with the SP survey, it would be natural in forecasting also to model the choice between the currently chosen alternative and the new alternative. The notion of a "current" choice in forecasting is, however, difficult to sustain. We therefore prefer to include the full set of alternatives in the forecast structure, all of which are considered to be available to every traveller, in the absence of further information, such as car availability.

The new-alternative constants, however, have been estimated on the basis of comparisons to one alternative and the inclusion of a set of competing alternatives will in itself increase the utility of the observed choices. To maintain consistency with the choice observed in the SP survey it is necessary to correct the new-alternative utility, in the forecast structure, by a constant (for each individual), which is equal to the increase in measured utility given by the full set of alternatives relative to the single chosen alternative. Fortunately, this constant can be calculated exactly.

The calculation that can be made is just to exploit the logit model formula

$$\log p_c = V_c - \log \Sigma_k \exp V_k$$

where p_c and V_c are respectively the choice probability and measured utility of alternative *c*. This formula states that the difference between the utility of a single alternative and the utility of the full set is just the log of the predicted probability of the alternative. The log probability of the chosen alternative is exactly the correction that is required.

Dependence on Current Choice

In some cases the model of choice in the SP context will include some representation that relates to the current choice, for instance, that travellers may be more willing to switch their route than their mode of travel. Such effects are generally classified as inertia or habit.

When significant inertia effects are found, these variables should be included in both the main and SP-based secondary estimation procedures. The resulting new-alternative constants can then be implemented directly in the forecast models. The issue of how to treat the inertia term itself requires consideration of decay.

In the long run, it can be expected that the experience of travel in the base situation will not affect travellers' choices. The impact of inertia (or other current-choice-dependent variables) on a new mode will be greatest immediately after the new alternative is introduced and can be expected to decay over a finite period to zero. Judgements are required to determine what is the appropriate decay level for the scope of the forecasts.

In passing, it is worth pointing out that there can be a problem of identification in an inertia effect. In some circumstances, this variable can in reality be wholly or partially a representation of a simple permanent preference for the existing alternative by its users. This effect will of course *not* decay over time and account must be taken of this possibility in deciding how to set up the model for forecasting.

MODELLING RESULTS

To give an indication of the success of the procedures described in the previous section, we have drawn together some summary results from a number of major studies conducted by us in recent years.

Base Estimation Results

Three studies of transport infrastructure are presented, in each of which the primary objective was to estimate demand for a new transport alternative. Two of the studies concerned major fixed link structures (models of mode and route choice), the third an urban light railway (only the mode choice model is discussed here).

In each study, an SP survey was conducted for the primary purpose of obtaining trade-off data between respondents' observed travel alternatives and the new alternative. It should be noted that each survey also contained at least one other SP, investigating trade-offs between other variables which were particularly relevant to the specific study (e.g., design options for the infrastructure). In all the studies, the RP and SP data sources were independent, thus avoiding the potential correlation problems identified by Morikawa (1994).

Joint RP/SP models were developed using the procedure outlined by Bradley and Daly (1991), that is, that coefficients which are common to the separate data sets, including structural parameters, are estimated simultaneously using all the data by appropriate scaling of the different data. It should be noted that data-specific coefficients, that is, RP or SP-specific parameters, may also be included in the model specification when required. Of course, the data-specific alternative-specific constants are an example of this separate specification which can also be applied in the same way to other coefficients.

In each of these studies, scale factors were introduced to scale the variance of the SP coefficients relative to the RP value. In principle, data from separate SP experiments should be assigned separate scaling factors; in practice, however, problems of identification of the scaling coefficients were encountered. Unfortunately, tests for non-identification (or worse, near non-identification) in tree logit models have not yet been developed and therefore the estimation process in the practical studies involved much trial and error, a problem made worse by the fact that when a model is over-specified it usually requires a large number of iterations before this fact becomes apparent. It is clear that these problems are less prevalent in situations where there is plenty of data, that is, large sets of both RP and SP data, and where the majority of the coefficients are defined over both data sets.

The solution most often adopted to the problem of non-identification is to reduce the number of parameters. In these models this has largely been a matter of eliminating data-source and alternative-specific constants and if non-identification was still a problem, then by reducing the number of scaling parameters, etiher by aggregating parameters across SP experiments or by constraining the values to 1.0 (i.e., assuming that the SP data has the same scale as the RP data). It is perhaps worth noting that the number of parameters estimated remains relatively large, ranging from 20 in the simplest models to about 240 in the most in the most complex, the number of data records (SP plus RP) ran into tens of thousands. Powerful hardware and software is needed for models of this scale but complicated interactions remain difficult to explain.

Table 19.1 shows some typical SP scale parameters for within-mode and between-mode data in the three studies (*t*-statistics are shown in brackets). In the first study, scale parameters for a number of segments (those with reasonable numbers of SP data records) are presented; in this study there were large RP data sets. For Segment 2, separate within-mode and between-mode scales could not be identified, so they were assumed to be equal. For all the studies, the values indicate that SP-to-RP scaling is necessary; all values are well determined and are very much different from one. The Study 1 values also indicate that the within-mode responses have significantly less unexplained variance than the RP data (scale factor is greater than 1) and the between-mode responses have significantly more unexplained variance when compared to the RP data (scale factor is less than 1). In this particular study these results were considered to be reasonable.

The scale parameter values presented from Study 2 are both values for between mode experiments. The first a between-mode experiment comparing the respondent´s observed mode with another existing mode with a new alternative. Again, the values are well determined and mostly significantly different from 1.0.

Table 19.1
Typical SP Scale Values

Segment	Within-Mode Experiment	Between-Mode Experiment(s)	Constrained Scale
Study 1:			
Segment 1	1.98 (4.3)	0.33 (4.9)	
Segment 2			2.12 (7.9)
Segment 3	4.29 (8.7)	0.74 (9.3)	
Segment 4	4.93 (7.9)	0.75 (8.8)	
Segment 5	3.90 (4.3)	0.41 (2.8)	
Segment 6	15.70 (4.8)	0.50 (5.9)	
Study 2:			
Segment 1		1.08 (7.6) ; 1.26 (8.2)	
Segment 2		1.77 (6.2) ; 1.83 (7.8)	
Segment 3		1.12 (4.8) ; 0.81 (5.8)	
Segment 4		0.76 (7.5) ; 0.62 (8.5)	
Segment 5		0.61 (5.0) ; 0.52 (4.8)	
Study 3:			
Segment 1	8.23 (10.7)	2.31 (6.7)	
Segment 2	15.42 (2.5)	6.53 (2.5)	
Segment 3	5.90 (7.7)	2.95 (6.7)	

In Study 3, both between-mode and within-mode scales are greater than 1, although the within-mode scales remain significantly higher. The poorer estimation in Segment 2 is connected with a reduced number of records here.

It can be concluded that scale parameters can be estimated with a reasonable degree of accuracy, given sufficient data of both RP and SP types, confirming the results of previous research. The scale values depend on the modelling context. It is noticeable that although between mode scales are generally lower than within mode (i.e., between mode choices are less well determined), which could be considered to be generally a reasonable result, there is more consistency within particular studies than within data of a particular type. That is, the scale appears to be determined by the local context and the presentation of the interview, rather than the formal type of choice that is made.

A-Run Results

As might be expected, depending on the sampling employed to create the base data files, the A-Run procedure can significantly change the values of the alternative-specific constants. Typical base estimation and A-Run constants from Study 1 and Study 3, for selected segments, are compared in Table 19.2. Absolute changes of up to 6 utility units (logit scale) are observed, changing as a function of the sampling procedures in the studies.

It is noticeable that the changes in Study 3 are much more consistent across the segments than the changes in Study 1. This is because the same sampling scheme was used for the RP data for all of the segments in that study. In Study 3 there was significant correlation between the membership of the segments and the probability of passing specific sampling points.

Table 19.2

Changes in Alternative-Specific Constants in A Runs

Segment	Alt. 1	Alt. 2	Alt. 3	Alt. 4
Study 1:				
Segment A	+2.2	+0.8	+6.2	+6.2
Segment B	+0.2	+0.2	+2.6	+1.6
Segment C	+0.2	-0.8	-0.8	+1.7
Segment D	+2.0	+1.7	+1.5	+1.2
Study 3:				
Segment 1	-1.2	+0.2	0	-0.7
Segment 2	-1.2	+0.5	+0.1	-0.9
Segment 3	-1.2	+0.4	+0.4	-0.8

As in the case of the SP scale coefficients, these changes in the constants need to be checked, in this case against the sampling procedures that were used to ensure that reasonable changes are being made. However, little else can be concluded from the specific values.

B-Run Results

As a consequence of the various changes to the model specification required for forecasting as explained above, including the changes to the constants for the existing alternatives made by the A Runs, it is necessary to repeat the estimates of the constants of the new alternative constants. These new estimates are the B Runs.

The B-Run procedure can change the values of the constants substantially. For Study 1, Table 19.3 compares the base estimation and B-run constants for the new alternative for a number of segments; the standard errors of the estimates are also presented. The observed changes in the constants vary substantially between the segments, ranging from very minor changes to absolute changes in the order of 5 utility units.

Confidence in the values of the new-mode constants is crucial for forecasting. The table indicates that the B-run values tend to be more accurately estimated than those estimated during the calibration phase, generally because of the constraining of some coefficients which are strongly correlated with these constants. However, one segment (iii) has a value which was not considered acceptably accurate. In such cases, judgement is required as to the most appropriate constant to be used in the forecasts. In this particular study, a judgement was taken and either the B-run value was used directly (where it seemed reasonable) or the values were averaged with similar segments (as was done in the case of segment iii).[36]

Table 19.3
Comparison of Base Estimation and B-Run New-Mode Constants: Study 1

Segment	Base Estimation		B-Run	
	Constant	standard error	Constant	standard error
Segment i	-7.74	3.63	-2.36	0.27
Segment ii	-4.08	1.23	-2.15	0.22
Segment iii	-1.37	0.49	-0.90	0.70
Segment iv	-1.16	0.15	0.40	0.11
Segment v	-1.66	0.25	-0.96	0.066
Segment vi	-3.91	1.18	-1.48	0.15

[36] A weighting procedure was used where the constants were weighted by the inverse of their variance.

Table 19.4

Comparison of Base Estimation and B-Run New Mode Constants: Study 3

Segment	Base Estimation		B-Run	
	Constant	standard error	Constant	standard error
Segment 1	0.12	0.013	0.10	0.095
Segment 2	-0.50	0.23	-0.69	0.14
Segment 3	-0.38	0.16	-0.19	0.048

In Study 2 the model tree structure was set up to explicitly allow the new-alternative constants to be estimated across several segments (segment and data-specific constants were estimated in the main estimation phase). In this study, strong inertia effects were also identified in the estimation and B-run phase, measuring both respondents' general reluctance to change modes as well as their specific reluctance to change from a private vehicle to public transport and vice versa.

B-run results for Study 3 are presented in Table 19.4. In this study, the absolute changes to the new-mode constants are much smaller, reflecting both the small changes made by the A runs and the fact that, because this was an urban rather than a long-distance study, the presentation of the alternatives to the SP respondents had been much more complete.

CONCLUSIONS

The procedure of joint exploitation of RP and SP data, based on the scaling procedure in a tree logit model described by Bradley and Daly (1991), has proved effective in practice and has been applied intensively by the authors and their associates; independent applications have also been made (see Hensher, 1994, for an overview). This procedure permits the estimation of a model based on the combined data sources, scaling the SP data to take account of the differences in the error structure between it and the RP data. The objective is thus to base the forecasting on a data base similar in properties to the RP data used for model estimation.

However, a number of problems have had to be solved in order to put these models into practice. Two broadly different approaches are available: one representing the future decision whether or not to switch to the new alternative as a binary choice between it and one of several existing alternatives, the other representing the future choice as among a series of essentially comparable alternatives, including the new alternative. For longer-term forecasting we have preferred the latter approach as leaving fewer problems.

To deal with the problems of implementation that remain, a sequence of secondary estimations has been devised. This sequence deals with the problems of implementation in a rational way and appears to avoid serious biases. However, it cannot be claimed that maximum likelihood estimates are obtained for all of the model parameters. In any case, given the nature of the base-year data, such estimators would not necessarily be acceptable, because of the wish to reproduce exactly the base situation, a feature not achieved by maximum likelihood estimators unless the model is of the simple logit form.

The advantage offered by this procedure is that it makes possible the application of reasonably simple SP methods in very complicated choice situations. The authors have applied the method in transport corridor infrastructure studies, a group of problems with considerable practical importance, but it is possible that applications can be made in other areas, such as residential choice.

The results of the applications that have been made indicate that the scale of the SP data does indeed need adjustment for forecasting; that is, the scale values found are significantly different from those of the RP data which corresponds with the description of the forecast situation. Further, the corrections made in the A and B Runs are substantial, indicating that the corrections that have been made are essential to avoid serious error.

The summary presentation of results in this chapter indicates that the method works satisfactorily in its own terms. The base estimations yield reasonable scale values - as has been shown by other studies - while the secondary estimations also give logical values.

Further research appears to be needed into the practicalities of presentation to survey respondents. Balancing the needs of presentation of a complicated hypothetical situation, maintaining credibility, while using experimental designs that yield information-rich responses, remains a challenging problem. The authors are currently involved in work to improve the presentation capabilities of interview support software which is hoped will reduce some of these difficulties.

Further, improvements could be made in estimation theory to give a better foundation or an alternative for the A and B Run procedure applied here. In any case, the size of the models is such that efficient software is essential. The authors are currently involved in research to improve the estimation software dealing with data of different types, but once again this will reduce rather than eliminate the problems.

A particular theoretical problem concerns the repeated-measurements problem affecting SP data, a problem addressed in parallel work by Cirillo et al. (1996). The consequences of that work for joint RP/SP analysis have yet to be worked out.

REFERENCES

Ben-Akiva, M. and T. Morikawa (1990). Estimation of switching models from revealed preferences and stated intentions. *Transportation Research A,* **24**, 485-495.

Bradley, M. and A. Daly (1991). *Estimation of logit choice models using mixed stated preference and revealed preference information*. Paper presented at the 6th international conference on travel behavior research, Quebec, Canada.

Cirillo, C., A. Daly and K. Lindveld (1996). *Eliminating bias due to the repeated measurements problem in SP data.* Paper presented at the 24th PTRC European Forum, London.

Hensher, D. (1994). Stated preference analysis of travel choice: the state of practice. *Transportation,* **21**, 107-133.

Lerman, S. and C. Manski (1979). Sample design for discrete choice analysis of travel behaviour: The state of the art. *Transportation Research A,* **13**, 29-44.

Morikawa, T. (1994). Correcting state dependence and serial correlation in the RP/SP combined estimation method. *Transportation,* **21**, 153-165.

Swait, J., J. Louviere and M. Williams (1994). A sequential approach to exploiting the combined strengths of SP and RP data: Application to freight shipper choice. *Transportation,* **21**, 135-152.

20

STATISTICAL THEORY: SUMMARY AND FUTURE RESEARCH POSSIBILITIES

Chandra Bhat, University of Texas at Austin, TX, USA

INTRODUCTION

The preceding chapters in this part cover four important and inter-related subject areas: (a) survey design and data collection techniques; (b) imputation for missing data in surveys; (c) model estimation; and (d) model application. This summary has four remaining sections. Each section discusses one or more chapters within the subject-oriented classification identified above and identifies future research possibilities.

SURVEY DESIGN AND DATA COLLECTION TECHNIQUES

Stopher's chapter provides a history of survey design methods and presents possible future directions for survey methods. It traces the evolution of survey design techniques within the context of the changing needs of the day and the changing perspective of travel behavior. Specifically, as the definition and unit of travel behavior has changed, so has the form and nature of data collection efforts. The chapter serves as an excellent overview of the state-of-the-art and the state-of-the practice in survey methods.

An important issue raised by Stopher relates to the value of survey data collection. Stopher points to the lack of adequate analysis on the potential savings "that can be accrued in capital transportation investments as a result of using data from surveys." While this is a valid point, such an analysis is anything but straightforward because there are several intermediary controlling factors such as sampling design, response bias issues, modeling methods employed, and empirical context that affect the final outcome. For example, the value of data is certainly related to the modeling methodology employed. With the same data source, different models might provide quite different forecasts and, consequently, quite different estimates of the potential savings that can be accrued in capital investment expenditure. Hence, there might be substantial variation in the benefit to cost ratio of survey data collection (*vis-a-vis* capital investments) within even one empirical context. It behooves us, therefore, to develop a detailed evaluation framework that accommodates variations in intermediary controlling factors and different empirical contexts while assessing the value of data.

There are also several other areas in survey design that would benefit from detailed benefit-cost analysis, including multi-day versus single-day data collection, testing of alternative instrument forms, and the evaluation of alternative sampling schemes. It is probably fair to say that though we have qualitative (and some quantitative) understanding of the advantages and disadvantages of each type of design, there is much to be desired by way of systematic evaluations of the benefits and costs associated with each design. For example, there is a general perception that a two-day survey diary is adequate in capturing the variation in individuals' activity-travel behavior over a week. However, the adequacy of a two-day diary in capturing day-to-day variations is surely related to the unit of travel behavior being used (i.e., whether we are focusing on number of shopping trips per day or entire daily activity-travel patterns). Similarly, Stopher discusses the lack of research on the value of additional data versus cost by way of lower responses between 24-hour and 48-hour survey diaries. This again should be tied to the unit of travel behavior of interest. If the unit is the number of work trips, then a 24-hour diary may prove to be more than adequate, while if we are addressing the issue of number of tours and stops in a day, a two-day diary may offer substantial gains in information.

Toward the end of his chapter, Stopher makes a distinction between time-use diaries and activity diaries. The primary distinction is that time-use diaries explicitly define travel to be an activity while activity diaries define travel as a means to get to or from an activity. It appears to me that this is largely a semantic distinction. Stopher points out that travel-related information is collected separately and differently from activities in an activity diary. However, I believe this is inevitable even in a "time-use" diary given the different attributes of an activity and of travel associated with that activity. Attributes of an activity may include activity type, start time, end time, and location. Attributes of travel include travel mode used, travel time, cost of travel, etc. I think the important point is to clearly define what constitutes an activity and to develop a diary format consistent with that definition.

IMPUTATION FOR MISSING DATA IN SURVEYS

Brownstone's chapter presents the potential problems caused by missing data in surveys and proposes a general method to generate correct imputations. The multiple imputation method discussed in the chapter accounts for the errors in the imputation process. The method requires a model for producing imputed values and involves independent draws of imputed values (based on the imputation model) to capture all sources of uncertainty in the imputation model.

Brownstone emphasizes the need to make generally non-testable assumptions about the distribution of missing data. Without such assumptions, the bounds for the expected value of a variable conditional on other variables which have missing data can be very large. The simplest assumption about the distribution of missing data is that it is no way systematically different (in terms of correlation with the endogenous variable of interest in a model) from the distribution of observed data. This assumption, which Brownstone refers to as the *ignorability* assumption, is appealing since it is as good as any other assumption in the absence of any additional information about the distribution of missing data.

Setting aside the issue of developing an imputation model, the multiple imputation approach presents a simple way of accounting for the errors in the imputation model and obtaining appropriate estimates and covariance matrices of the parameters of interest in a model which has the variables with missing data as exogenous or endogenous variables.

An issue that does not come out explicitly from Brownstone's chapter is how many independent draws of the imputed variable(s) are needed to obtain an accurate estimate of the covariance matrix of the parameters of interest. I believe that a systematic simulation evaluation of this subject is warranted.

The multiple imputation approach is applicable to missing data occurring in many different contexts such as survey nonresponse, panel attrition, choice-based sampling, and synthetic sampling. Brownstone's chapter should be a valuable reference because of its nice exposition and comprehensive treatment of the subject. The empirical illustration of the value of a small supplementary validation sample to allow identification of the non-response process is encouraging. Of course, the best way to circumvent the problems caused by missing data is to reduce missing data in the first place by careful survey data collection design. This is an area where, as discussed in the earlier section, more research needs to be done to better understand the relationship between survey collection methods, instrument design, and sampling schemes on the one hand and non-response/attrition on the other.

MODEL ESTIMATION

Golob in his chapter focuses on participation in various types of activities, travel times expended in out-of-home activity participation, and mobility choices characterized by car ownership and vehicle miles of travel. Most activity-based models to date have focused on the detailed scheduling of sequencing of activities at the individual level, without adequately addressing the problem of activity generation (i.e., the generation of the agenda of activities for participation and the aggregate times spent on specific activity types) or the issue of allocation of activities among household members (see, Bhat and Koopelman, 1993, for a detailed discussion). Decisions on activity generation and allocation of activities, the focus of Golob's chapter, may be viewed as precursors to the detailed activity scheduling and sequencing decisions that individuals make.

Golob argues that it is not appropriate to consider work-related activities as being exogenous in travel choice models, nor is it correct to assume a fixed demand for other activities such as shopping and personal business. This is because of the need for accurate employment related forecasts and the need to accommodate the interaction between demand for activities and accessibility/mobility choices. In transportation planning, employment-related forecast are relegated to simple aggregate-level side-calculations rather than being based on causal models that address the behavioral factors underlying employment decisions and income-earning potential. Such aggregate-level forecasts fail to adequately represent the distribution of changes in work and other activity demands across various socio-demographic groups. This is likely to lead to misleading and inaccurate forecasting of travel-related variables. A causal disaggregate model of employment and other activity demand decisions within the framework of mobility/accessibility choices can be used as part of an overall activity-travel analysis process. Golob's chapter is an important contribution to an emerging trend among travel behavior researchers to examine time-use decisions. The terminology used to characterize the direct causal effects among the endogenous variables is insightful and apt. However, the motivation for, and importance of, the chapter within the broader framework of the overall activity-based framework for modeling travel choices could have been better highlighted. As it stands, it is not clear how the model in the chapter would help in travel demand forecasting or transportation policy analysis. It would have been particularly useful to apply the model in addition to the behavioral insights that the model provides. At the least, identifying specific contexts in which the model could be used for policy analysis would have made the importance and value of the model more transparent. Also, a detailed interpretation of the error-term correlations would have been helpful. An empirical extension of Golob's work is to apply the analysis to household types other than those with one male adult and one female adult.

On a general note, activity-based travel research has provided us a better understanding of the process underlying travel decisions and we will certainly continue to improve our understanding of this process in the years to come. But I believe we know reasonably enough about the process now that we should be directing more attention to demonstrating the applicability of our process models to analyzing transportation policy options and to travel forecasting.

MODEL APPLICATION

Daly and Rohr develop procedures to apply a model estimated from both revealed preference (RP) data and stated preference (SP) data. The main reason to collect SP data is that a proposed improvement in transportation infrastructure may entail the provision of a currently non-existent service. In such a case, modeling the choice situation with the new service would require estimation of alternative-specific constants associated with the new alternative. These parameters, of course, are not estimable using current RP data.

There has been substantial research in estimating trade-off (of level-of-service) coefficients from joint RP and SP data. Daly and Rohr indicate that there has been relatively little work, however, in applying the resulting model.

Daly and Rohr's approach represents a practical and convenient procedure to achieve forecasts after "introduction" of a new service. It does not, as the authors acknowledge, accommodate for repeated SP measurement from the same individual. This might have the result of overstating the standard errors of level-of-service and other parameters. The approach also does not address the issue of accommodating unobserved stationary and/or decaying heterogeneity in the utility of alternatives common to the RP and SP contexts.

There are several other possible approaches to estimation and application that have been proposed in the literature. An issue that has, in my opinion, not received adequate (if any) attention is the evaluation of the various approaches relative to each other and relative to actual choices. Until extensive validation studies are available, it is difficult to assess the performance of the RP-SP methods in relative terms and in absolute terms. Small-scale (and relatively inexpensive) studies that generate a scenario with a new alternative and compare the performance of models based on earlier recorded RP and SP data with actual choices made in the scenario are required. An alternative that might become feasible is to use virtual reality and "virtually" place individuals in the new scenario in experiments and observe their choices. These choices can serve as the validation test beds for the RP-SP choice approaches.

McFadden's chapter develops bounds and exact formulas for Willingness to pay (WTP) for improvements in attributes. The WTP computations have widespread applicability in many contexts. McFadden presents some travel contexts in his chapter. Other examples might include willingness to pay tolls for an improved service facility or for using a recreational/scenic roadway.

McFadden's bounds for the WTP computation (for the case when the error terms enter the indirect utility functions additively and there is no interaction between income and the random components) provide useful and easy-to-compute limits for the population mean WTP. The bounds, however, can be quite broad when the quality improvement is large. One useful avenue of research might be to explore possibilities to narrow the band-width by using "before-after" information on similar improvements made earlier elsewhere. Of course, when income enters additively and linearly in the indirect utility function, McFadden shows that one can obtain exact formulas for the WTP by solving for the income reduction that maintains a constant expected utility before and after a quality improvement. The issue of narrowing the band-width of the bounds is, therefore, relevant primarily for the case when income enters non-linearly and as an alternative-specific variable. In this case, McFadden shows that the expected utility method provides an increasingly biased estimate of the actual (but generally non-closed form) mean WTP with large improvements in quality. However, it would be interesting to accommodate the variation in the WTP around its mean (caused by sampling error) and assess the difference between the expected utility method and the actual mean WTP relative to this sampling error band. Such an assessment might be performed using simulation methods.

ACKNOWLEDGEMENTS

This research was supported by National Science Foundation grants DMS 9208758 and DMS 9313013 to the National Institute of Statistical Sciences (NISS).

REFERENCES

Bhat, C. R. and F. S. Koppelman (1993). A conceptual framework of individual activity program generation. *Transportation Research 27A*, **6**, 433-446.

SUBJECT INDEX

D

F

G

N

O

P

R

U

V

W

NAME INDEX

A

B

C

D

E

F

I

J

K

L

N

O

P

R